Lecture Notes in Physics

W0042787

Springer-Verlag Berlin Heidelberg GmbH

The Editorial Policy for Proceedings

The series Lecture Notes in Physics reports new developments in physical research and teaching – quickly, informally, and at a high level. The proceedings to be considered for publication in this series should be limited to only a few areas of research, and these should be closely related to each other. The contributions should be of a high standard and should avoid lengthy redraftings of papers already published or about to be published elsewhere. As a whole, the proceedings should aim for a balanced presentation of the theme of the conference including a description of the techniques used and enough motivation for a broad readership. It should not be assumed that the published proceedings must reflect the conference in its entirety. (A listing or abstracts of papers presented at the meeting but not included in the proceedings could be added as an appendix.)

When applying for publication in the series Lecture Notes in Physics the volume's editor(s) should submit sufficient material to enable the series editors and their referees to make a fairly accurate evaluation (e.g. a complete list of speakers and titles of papers to be presented and abstracts). If, based on this information, the proceedings are (tentatively) accepted, the volume's editor(s), whose name(s) will appear on the title pages, should select the papers suitable for publication and have them refereed (as for a journal) when appropriate. As a rule discussions will not be accepted. The series editors and Springer-Verlag will normally not interfere with the detailed editing except in fairly obvious cases or on technical matters.

Final acceptance is expressed by the series editor in charge, in consultation with Springer-Verlag only after receiving the complete manuscript. It might help to send a copy of the authors' manuscripts in advance to the editor in charge to discuss possible revisions with him. As a general rule, the series editor will confirm his tentative acceptance if the final manuscript corresponds to the original concept discussed, if the quality of the contribution meets the requirements of the series, and if the final size of the manuscript does not greatly exceed the number of pages originally agreed upon. The manuscript should be forwarded to Springer-Verlag shortly after the meeting. In cases of extreme delay (more than six months after the conference) the series editors will check once more the timeliness of the papers. Therefore, the volume's editor(s) should establish strict deadlines, or collect the articles during the conference and have them revised on the spot. If a delay is unavoidable, one should encourage the authors to update their contributions if appropriate. The editors of proceedings are strongly advised to inform contributors about these points at an early stage.

The final manuscript should contain a table of contents and an informative introduction accessible also to readers not particularly familiar with the topic of the conference. The contributions should be in English. The volume's editor(s) should check the contributions for the correct use of language. At Springer-Verlag only the prefaces will be checked by a copy-editor for language and style. Grave linguistic or technical shortcomings may lead to the rejection of contributions by the series editors. A conference report should not exceed a total of 500 pages. Keeping the size within this bound should be achieved by a stricter selection of articles and not by imposing an upper limit to the length of the individual papers. Editors receive jointly 30 complimentary copies of their book. They are entitled to purchase further copies of their book at a reduced rate. As a rule no reprints of individual contributions can be supplied. No royalty is paid on Lecture Notes in Physics volumes. Commitment to publish is made by letter of interest rather than by signing a formal contract. Springer-Verlag secures the copyright for each volume.

The Production Process

The books are hardbound, and the publisher will select quality paper appropriate to the needs of the author(s). Publication time is about ten weeks. More than twenty years of experience guarantee authors the best possible service. To reach the goal of rapid publication at a low price the technique of photographic reproduction from a camera-ready manuscript was chosen. This process shifts the main responsibility for the technical quality considerably from the publisher to the authors. We therefore urge all authors and editors of proceedings to observe very carefully the essentials for the preparation of camera-ready manuscripts, which we will supply on request. This applies especially to the quality of figures and halftones submitted for publication. In addition, it might be useful to look at some of the volumes already published. As a special service, we offer free of charge LaTeX and TeX macro packages to format the text according to Springer-Verlag's quality requirements. We strongly recommend that you make use of this offer, since the result will be a book of considerably improved technical quality. To avoid mistakes and time-consuming correspondence during the production period the conference editors should request special instructions from the publisher well before the beginning of the conference. Manuscripts not meeting the technical standard of the series will have to be returned for improvement.

For further information please contact Springer-Verlag, Physics Editorial Department II, Tiergartenstrasse 17, D-69121 Heidelberg, Germany

Hermann-Josef Röser   Klaus Meisenheimer   (Eds.)

The Radio Galaxy Messier 87

Proceedings of a Workshop Held at Ringberg Castle,
Tegernsee, Germany, 15-19 September 1997

 Springer

Editors

Hermann-Josef Röser
Klaus Meisenheimer
Max-Planck-Institut für Astronomie
Königstuhl 17
D-69117 Heidelberg, Germany

Library of Congress Cataloging-in-Publication Data.

Die Deutsche Bibliothek - CIP-Einheitsaufnahme

The **radio galaxy messier 87** : proceedings of a workshop held at Schloss
Ringberg, Germany, 15 - 19 September 1997 / Hermann-Josef Röser ; Klaus
Meisenheimer (ed.). - Berlin ; Heidelberg ; New York ; Barcelona ; Hong Kong ;
London ; Milan ; Paris ; Singapore ; Tokyo : Springer, 1999
(Lecture notes in physics ; Vol. 530)

ISSN 0075-8450
ISBN 978-3-662-14257-8   ISBN 978-3-540-48667-1 (eBook)
DOI 10.1007/978-3-540-48667-1

The use of general descriptive names, registered names, trademarks, etc. in this publication
does not imply, even in the absence of a specific statement, that such names are exempt
from the relevant protective laws and regulations and therefore free for general use.

Typesetting: Camera-ready by the authors/editors
Cover design: *design & production*, Heidelberg

SPIN: 10644416   55/3144 - 5 4 3 2 1 0 – Printed on acid-free paper

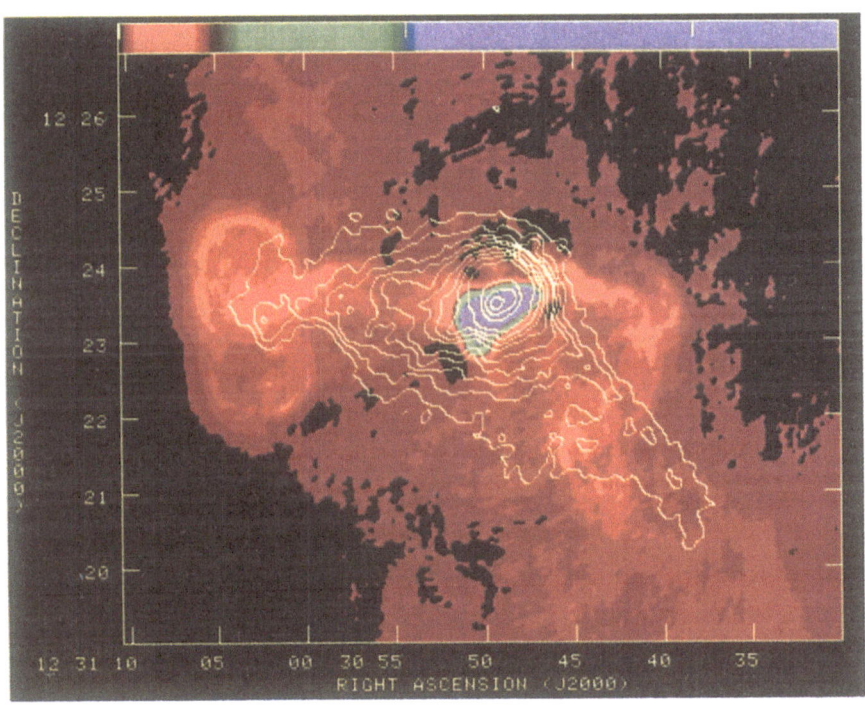

Inner part of the radio galaxy Messier 87

The coloured image represents the central, brighter parts of the radio emission as observed at 330 MHz by the VLA (beamsize ≈ 5″). The well-known radio jet as well as the inner radio lobes are within the blue area at the center. The contours show the residual X-ray structure after subtraction of a power law distribution, which roughly deletes the circularly symmetric emission of the hot gas.

Frazer Owen and Dan Harris

Preface

The idea to have a dedicated workshop on the radio galaxy M 87 was born during the conference dinner of the Cygnus A workshop at Greenbank following a remark by Dan Harris that there are also other very interesting, prominent radio sources, *e.g.* M 87 .... So we decided to make M 87 the topic of our next meeting in our Ringberg workshop series on extragalactic radio sources.

The meeting was organized by Geoff Bicknell (Australian National University Astrophysical Theory Centre), Dan E. Harris (Smithsonian Astrophysical Observatory), Klaus Meisenheimer (MPI für Astronomie), Frazer Owen (NRAO), and Hermann-Josef Röser (MPI für Astronomie).

Again a generous grant from the Max Planck Society made it possible to support invited speakers in an unbureaucratic way. The Max Planck Society also made the Ringberg facilities available to us. The organizers would like to sincerely thank Mr. A. Hörmann and his staff at Ringberg for their excellent support, contributing in an essential and pleasant way to the success of the workshop.

The editors would like to thank Ms. Karin Meissner-Dorn and Ms. Martina Weckauf for their help with the art work.

Heidelberg, April 1999   Hermann-Josef Röser
Klaus Meisenheimer
on behalf of the Scientific Organizing Committee

Contents

1 Böhringer 2 Owen   3 Norman
4 Laing   5 Bicknell   6 Arp
7 Biretta   8 Macchetto 9 Scheuer
10 Binney   11 Axon   12 Matsumoto
13 Blandford 14 Binggeli   15 Eilek
16 Reynolds 17 Sparks   18 Cramphorn
19 Heinz   20 Massaglia 21 Rottmann
22 Harris   23 Crane   24 Klein
25 Röser   26 Tsvetanov 27 Neilsen
28 Dopita

The complete list of participants is found on page 333.

Introducing M 87

Roger Blandford

130-33 Caltech
Pasadena, CA 91125
USA

Abstract. A brief introduction to the study of M 87 is presented. Attention is drawn to the surprisingly low bolometric power of the black hole. It is also argued that recent observations are quite constraining with respect to the origin of the jet power and the cause of its initial collimation. The bright jet knots provide a powerful diagnostic for constraining the character of jet flows as well as the properties of the ambient interstellar medium in the central few kpc of the nucleus. As it has been studied in such great detail throughout the electromagnetic spectrum, M 87 may have much to teach us about active galactic nuclei in general.

1 Historical Background

In 1918, Heber Doust Curtis (1918) , using the Lick Crossly Reflector, observed the galaxy, known as Messier 87 and discovered that it was "exceedingly bright" with a "sharp nucleus". He also reported that "the brighter central portion is about 0.'5 in diameter and the total diameter is about 2'; nearly round". Furthermore, "No spiral structure is discernible" and "A curious straight ray lies in a gap in the nebulosity in pa 20°, apparently connected with the nucleus by a thin line of matter. The ray is brightest at its inner end which is 11'' from the nucleus." This is a remarkable set of observations (by a remarkable observer). Had Curtis been as bold and prescient in his speculations as, for example, Baade and Zwicky were sixteen years later, his observations would have been more widely cited as marking the start of the study of active galactic nuclei and could well have triggered a serious observational program that might have led to the earlier discovery of Seyfert galaxies.

However, it was not until the 1950's when, triggered by new capabilities in the then infant science of radio astronomy, M 87 resurfaced (as Virgo A, Bolton & Stanley 1948, Bolton, Stanley & Slee 1949). Following a suggestion by Shklovsky (1955), Baade (1956) exhibited optical polarization and correctly associated this with synchrotron radiation. It appears to have been Burbidge (1956) who first realized that the minimum magnetic plus cosmic ray energy required to produce this synchrotron emission was impressively large, $\sim 2 \times 10^{55}$ erg for the jet and $\sim 10^{57}$ erg for the halo. (Although we are no longer impressed by these figures, they have stood up well in the light of contemporary observations.) Despite the fact that M 87 was a prototypical example of the burgeoning class of radio galaxies, it was not until the advent of interferometric radio imaging that it became clear that the radio jet looked awfully like its optical counterpart (*e.g.* Turland 1975). By this time, the notion that galactic nuclei contain massive black holes had swung back into favor and a joint imaging and spectroscopic investigation (Young *et al.* (1978), Sargent (1978)) was undertaken to try to find the M 87 black hole. Evidence was found for a central mass of $\sim 3 \times 10^9\,M_\odot$, which, although controversial at the time, has turned out to be pretty close to the contemporary value (Marconi *et al.* 1997). Slightly later Ford & Butcher (1979) discovered the emission line filaments surrounding the nucleus and Schreier *et al.* (1982) were able to distinguish the core, jet and halo at X-ray energy. Since this time, there has been great progress throughout the electromagnetic spectrum and this provides the motivation for this meeting. The presence of a massive black hole, (actually the most massive that we know), has been verified beyond all reasonable doubt (Macchetto *et al.* 1997). The jets have been traced in to small radius, directly implicating, for the first time, general relativity in the explanation of their origin (Junor & Biretta 1995). There have also been conceptual advances and M 87 has once again become a prototypical extragalactic source that brings into unusually sharp focus many of the perplexities that bedevil the general interpretation of active galactic nuclei.

In organizing this introductory talk, I chose to ask a series of rhetorical questions that I had hoped would help guide the discussion. In practice, this did not happen and the workshop followed a more productive path. Nonetheless, the questions remain unanswered and so I feel obligated to repeat them!

2   Why Is M 87 So Dim?

Forty years ago, astronomers marvelled at the prodigious power of M 87. It was a powerful radio galaxy suggesting to a few perceptive theorists that something wonderful was happening in its nucleus. Nowadays, we have become blasé. It is merely a low power, FR I radio source in a giant elliptical galaxy that is mostly distinguished by its proximity. Indeed the major puzzle is "Why isn't it much brighter?". In round numbers, both the core and the knots in the jet radiate a bolometric power $\sim 10^{42}\,\mathrm{erg\,s^{-1}}$, whereas the measured black hole mass is $\sim 3 \times 10^9\,M_\odot$ (Macchetto *et al.* 1997), (the largest current estimate, Richstone 1998), which translates into an Eddington luminosity $L_E \sim 4 \times 10^{47}\,\mathrm{erg\,s^{-1}}$, nearly a million times larger. Even if we make the best guess that we can as to the power carried away by the jets we only get $\sim 10^{43}\,\mathrm{erg\,s^{-1}}$. Now, it can be pointed out that black holes are not obligated to radiate at the Eddington rate; certainly not if they are only accreting slowly. However, in this case, the gas supply ought to be substantial. Probably the most direct estimate is obtained by computing the Bondi accretion rate at the Bondi radius $\sim 100\,\mathrm{pc}$. This gives an accretion rate $\dot{M} \sim 10^{25}\,\mathrm{g\,s^{-1}}$ equivalent to a radiative efficiency $\epsilon \sim 10^{17}\,\mathrm{erg\,g^{-1}}$. If, on the larger scale, we suppose that there is an X-ray cooling flow with $\dot{M} \sim 10^{27}\,\mathrm{g\,s^{-1}}$, some of which may condense into dense clouds of neutral hydrogen (Carter, Johnstone & Fabian (1997)), then $\epsilon \sim 10^{15}$ erg g^{-1}. Another approach is to use submillimeter observations (Despringre & Fraix-Burnet 1996), to try to detect dust, presumably from accreting molecular gas. (This is all reminiscent of the Galactic center, where it can be argued that $\epsilon \sim 10^{14} \sim 10^{-7}c^2$. Similar claims can be made for Centaurus A, NGC 4258 etc. and so the problem is not limited to M 87.)

Now, one popular, contemporary answer to this question is to propose an advection-dominated accretion flow (Narayan & Yi 1995, Reynolds *et al.* 1996). Here the requirements are that the effective viscosity be high (in the language of accretion disk theory $\alpha \sim 1$), that the coupling between ions and electrons be more or less limited to Coulomb scattering (*i.e.* that collisionless plasma processes be excluded) (*cf.* Gruzinov, A. V. 1997, Quataert 1997) and that the magnetic field strength be maintained near an equipartition value. Under these conditions, it is indeed quite plausible that the core bolometric flux not exceed the observations. However, this is not the only option and, as these physical requirements are not necessarily satisfied, it is worth considering alternatives.

One possibility is that most of the gas that is supplied does not actually reach the hole but is, instead driven away in the form of a wind. For example, if the central black hole spins rapidly, it can be coupled magnetically to the inner disk and act as a source of angular momentum which is transported radially outward giving a large local dissipation at all radii. In this case the outflow is likely to be hydromagnetic in character. Another possibility is that the outer disk may be Compton heated ((Ciotti & Ostriker 1997)), (although this would require a far higher Compton temperature than is generally estimated for AGN). A third possibility is that the jet power is even larger than estimated above and that it can provide sufficient mechanical heating of its surroundings to form a pressure driven wind (Binney 1996). It is quite unlikely that any of these outflows is steady and one would, instead, anticipate that some sort of limit cycle behaviour be established with a characteristic timescale ranging from days for the first possibility to millions of years for the next two cases.

3 Does the Jet Power Come from the Disk or the Hole?

Accreting black holes can liberate energy in two quite separate ways. Firstly, the infalling gas may liberate its gravitational binding energy so that, in the limiting case when this happens close to the hole and the radiative cooling time is short compared with the inflow time, the overall radiative efficiency is an appreciable fraction of c^2 per unit mass. However, not all of this binding energy needed be liberated in the form of photons. It may also be extracted as a mechanical outflow from the surface of the disk. A disk efficiency $\sim 10^{18}\,\mathrm{erg\,g^{-1}}$ suffices to power the jet if the accretion rate is $\sim 10^{25}\,\mathrm{g\,s^{-1}}$. Secondly, if the hole itself is spinning, it can be a repository of a large amount of rotational energy; in the case of M 87, this can, quite plausibly, amount to more than $10^{63}\,\mathrm{erg}$, which appears to be ample to power the jets as currently observed for over a hundred Hubble times! How is this energy extracted? I would argue that the only reasonable possibility is through large scale magnetic field threading the black hole event horizon which drive a large scale current flow which can remove both electromagnetic energy and angular momentum in the form of Poynting flux. Presumably, this outwardly directed flow of electromagnetic energy is eventually transformed into outflowing plasma, perhaps via the creation of electron-positron pairs, or through entrainment of surrounding gas. It is important to see where and how jets are powered. This is, naturally, contingent upon black holes having large angular velocity as suggested by ASCA observations of the Seyfert galaxy MCG-6-30-15 (Dabrowski *et al.* 1997, but see Reynolds & Begelman 1997).

Recent observations of M 87 have limited speculation. Firstly, the high resolution VLBI observations of Junor & Biretta (1995) Junor & Biretta 1995 show that the jet is at least polar, and may be even more strongly collimated all the way down to $\sim 2 \times 10^{16}$ cm which is only sixty times the gravitational radius of the hole. Furthermore the observation of apparent, large superluminal motion in the jet knots (Biretta et al. (1999)) indicates that the outflow has an ultrarelativistic speed. (Note, however, that such motions are not observed in the core on VLBI scales (Reid *et al.* 1989). This need not be too surprising if the outflow is mostly steady and neither intermittent nor intrinsically unstable at these small radii.) Both of these observations strongly suggest that the jet either originates from the inner disk or the hole itself. Either way, the problem inevitably involves general relativity.

4   How Is the Jet Collimated and of What Is It Made?

A closely related issue is that of jet collimation. If power is released close to the black hole and allowed to escape freely, then a quasi-spherical outflow should ensue. However, anisotropic density distributions, associated with the disk, or magnetic field, tied to this disk can transform this outflow into the two antiparallel jets that are either observed directly, or, as in the case of M 87, inferred to be present. Jets need not be confined all the way along their length and the opening angles that they subtend at their sources can vary as the external conditions change. In the case of M 87, making a conservative estimate of the jet inclination to the line of sight, we find that the true opening angle subtended by the outer jet is roughly ~ 0.1. If there is no confining pressure then the jet will essentially expand on the Mach cone and this will, in turn, imply that the Mach number satisfies $M > 30$, at least over those radial intervals where the jet is free. (The detection of large perpendicular, optical polarization from the jet walls (Capetti *et al.* (1996)), suggests that the jet is not free on the $\sim$ kpc scale.)

In many ways, magnetic collimation is the most natural mechanism as the spinning disk will naturally wrap magnetic field lines around the jet creating magnetic tension and magnetic pressure gradient which both force the outflowing plasma to flow nearly parallel to the axis of rotation. (There are serious concerns about the stability of magnetic collimation (*cf.* Begelman 1998) that will probably only be fully resolved by three dimensional, MHD simulations. In particular, if the toroidal component of the magnetic field becomes too large relative to the poloidal magnetic field and jet momentum, axisymmetry may be naturally broken and either a precessing structure be formed or the magnetic field lines reconnect across the jet leading to internal dissipation and, presumably, non-thermal emission.) With the coming of VLBI polarization studies, there is some hope that we will be able to measure the polarization structure that might be characteristic of magnetic collimation, although it may be quite hard to remove foreground Faraday rotation, (*cf.* Capetti *et al.* (1996)).

There is an interesting argument that can be applied, in principle, to try to constrain the jet composition. Radio observations of the core of the approaching radio jet are traditionally interpreted as a lower bound on the brightness temperature at a radius $r \sim 2 \times 10^{17}$ cm and an upper bound on the magnetic field strength, and, consequently, a lower bound on the relativistic electron density. It turns out that if the jet contains a proton for every electron, then it must carry a power somewhat in excess of the total inferred jet power $\sim 10^{43}$ erg s^{-1} and so, it is argued, the jet is neutralized by positrons (Reynolds *et al.* 1996). It is probably premature to have much confidence in this argument at present (despite the attraction of the conclusion) because of uncertainties in the above measurements and the additional freedom that is allowed by relativistic effects. However, it is quite possible that it will become much tighter in the future, especially if Faraday depolarisation arguments are used to shore up the estimate of electron density. As we move in closer to the hole, the influence of radiative drag must become important and indicates that something other than electrons and positrons has to carry the jet momentum, presumably electromagnetic field or, conceivably, protons.

5   What's Going on in Knot A?

The prominent sequence of knots clearly visible in radio, optical and now X-ray images of the jet turn out to be a tantalizing diagnostic of physical conditions in the jet and its surroundings. These knots are clearly alive and appear to exhibit X-ray variability and internal motions, that can appear to be superluminal. The observed shape of the knots, after correction for relativistic kinematics, can be used to estimate the jet inclination angle, a crucial, (though controversial), number. It appears to be $\sim 45°$, somewhat different from the complement of the observed disk inclination. This then makes it slightly more puzzling that we don't see the counter-jet directly. Nowadays, it is commonly supposed that the knots are actually strong shocks in the rel-ativistic jet. (Most discussions have assumed gas dynamical shock jump con-ditions, but I expect that hydromagnetic jump conditions are involved and this allows the post shock speed relative to the shock front to be much larger than the values $c/3$ traditionally allowed in purely gas dynamical relativistic shocks.) Furthermore, in a modern application of Burbidge's equipartition argument, we can deduce that the internal pressure within the most promi-nent of these features, Knot A, is $\sim 10^{-8}$ dyne cm^{-2}, roughly ten times the best estimate of the external pressure. However, we still do not know if this suggests that the flow is an expanding transient, or if it is confined by the gas pressure of the cocoon formed by jet waste energy or if we must invoke magnetic stress to inhibit transverse expansion (Bicknell & Begelman 1996).

One of the bigger puzzles is "Why is it so hard to find the counterjet and the counterparts of the brighter knots in M 87?" (Stiavelli, Peletier & Carollo 1997). One possible answer is that it is due to extreme relativistic beaming, but this implies that the jet makes a much smaller angle to the line of sight than is otherwise indicated (Bicknell & Begelman 1996). Another possibility is that surrounding gas on the far side of the source is significantly different from that on the near side, perhaps due to overall motion of M 87 through the Virgo cluster. A very suggestive observation of Herbig-Haro objects as-sociated with young stellar objects (H. Zinnecker, private communication) indicates that, in the case of jets associated with young stellar objects, the shocks are created by the central source and propagate outward as internal shocks in the jet. It is very tempting to suppose that this is also true of relativistic, extragalactic jets associated with black holes. Similar indications of strong, episodic behaviour are found in observations of GRS 1915+105 (Eikenberry, private communication).

The X-ray observations of knots are no less intriguing. If, as suggested by the spectral curvature, the X-rays come from synchrotron emission (Neumann et al. 1997), this implies that relativistic electrons are accelerated in situ to energies ~ 0.1 PeV, just within range of relativistic shock acceleration. The report of X-ray variability from Knot A supports this (Harris, Biretta & Junor 1997).

6   What Can M 87 Teach Us About Other AGN?

In modern language, M 87 is a nearby, giant elliptical galaxy with an active nucleus from which emerges a one-sided optical jet. As it is so close we can examine it in considerable detail and hope to generalize our findings to other objects. Two of its closest cousins are Centaurus A and NGC 6251. However the former lies in the field and the latter is only on the outskirts of a cluster. Comparative observations should be very helpful in trying to determine the relevance of the circumgalactic environment is to the properties of an active nucleus. NGC 1275, by contrast lies at the center of the Perseus cluster, which is reported to possess an even more impressive cooling flow.

M 87 possesses the largest, measured black hole mass, to date. It is very tempting to suppose that, in its day, it was a powerful quasar and accreted gas at a prodigious rate for an extended period. Perhaps the best hope of discovering evidence of this activity is through low frequency radio observations. It is tempting to associate the discovery of the outer halo in these terms.

Finally, returning to the conundrum with which I began, a comparative study of M 87 and its peers may help us understand empirically what it is that determines why some galactic nuclei are so prodigiously powerful and others are some strikingly inefficient and why some are able to produce jets more powerful than the total bolometric luminosity and others show no evidence of bipolar activity. This can then be compared with rapidly developing theoretical notions of how accretion proceeds around massive black holes.

This is a good time to be studying M 87.

Acknowledgements

I am indebted to Geoff Bicknell, Dan Harris, Klaus Meisenheimer, Fraser Owen and especially Hermann-Josef Röser for organising a timely and stimulating meeting and for travel support. I acknowledge NSF support under grant AST 95-29170.

References

Baade, W. (1956): ApJ 123 550
Begelman, M. C. (1998): ApJ 493 291
Bicknell, G. V. & Begelman, M. C. (1996): ApJ 467 597
Binney, J. J. (1994): in *Cosmical Magnetism*, ed. D. Lynden-Bell, Cambridge: Cambridge University Press
Biretta, J. (1993): in *Astrophysical Jets*, ed. D. Burgarella, M. Livio, & C. P. O'Dea, Cambridge: Cambridge University Press, p. 263
Bolton, J. G., Stanley, G. S. (1948): Nature 161 312
Bolton, J. G., Stanley, G. S. & Slee, O. B (1949): Nature 164 101
Burbidge, G. R. (1956): ApJ 124 416
Capetti, A., Macchetto, F. D., Sparks, W. B. & Biretta, J. A. (1997): Astron. Astrophys. 317 637
Carter, D., Johnson, R. M. & Fabian, A. C. (1997): MNRAS 285 L20
Ciotti, L. & Ostriker, J. P. (1997): ApJ 487 L105
Curtis, H. D. (1918): Pub. Lick. Obs. 13 11
Dabrowski, Y., Fabian, A. C., Iwasawa, K., Lasenby, A. N., Reynolds, C. S. (1997): MNRAS 288 L11

DeSpringre, V. & Fraix-Burnet,D. (1996): Proc. IAU Symp. 175, *Extragalactic Radio Sources*, ed. R. Ekers, C. Fanti & L. Padrielli Boston: Kluwer, p. 175

Ford, H. & Butcher, H. (1979): ApJS 41 147

Gruzinov, A. V. (1997): preprint

Harris, D. E., Biretta, J. A. & Junor, W. (1997): MNRAS 284 L21

Junor, W. & Biretta, J. A. (1995): AJ 109 500

Macchetto, F., Marconi, A., Axon, D. J., Capetti, A., Sparks, W. & Crane P. (1997): ApJ 489 579

Marconi, A., Axon, D. J. Macchetto, F. D., Capetti, A., Sparks, W. B. & Crane, P. (1997): MNRAS 289 L21

Meisenheimer, Neumann & Röser (1997): *Jets from Stars and Galactic Nuclei*, ed. W. Kundt Berlin: Springer Verlag

Narayan, R. & Yi, I. (1995): ApJ 444 231

Neumann, M., Meisenheimer, K., Röser, H.-J. & Fink, H. H. (1997): Astron. Astrophys. 318 383

Quataert (1997): preprint

Reid, M. J., Biretta, J. A., Junor, W., Muxlow, T. W. B., & Spencer, R. E. (1989): ApJ 336 112

Reynolds, C. S. & Begelman, M. C. (1997): ApJ 488 109

Reynolds, C. S., DiMatteo, T., Fabian, A. C., Hwang, U, & Canizares, C. R. (1996): MNRAS 283 L111

Reynolds, C. S., Fabian, A. C. Celotti, A. & Rees, M. J. (1996): MNRAS 283 873

Richstone, D. (1998): Proc. IAU Symp. 184. *The Central Region of the Galaxy and Galaxies*, ed. Y. Sofue Berlin: Springer Verlag

Sargent, W. L. W., Young, P. J., Boksenberg, A., Carswell, R. F. & Whelan, J. A. J. (1978): ApJ 229 891

Schreier, E. J., Gorenstein, P., Feigelson, E. D. (1982): ApJ 261 42

Shklovsky, I. S. (1955): Astr. J USSR 32 215

Sparks, W. B., Biretta, J. A. & Macchetto, F. (1996): ApJ 473 254

Stiavelli, M. Peletier, R. F. & Carollo, C. M. (1997): MNRAS 285 181

Turland, B. D. (1975): MNRAS 170 281

Young, P. J., Westphal, J. A., Kristian, J., Wilson, C. P., & Landauer, F. P. (1978): ApJ 221 721

The Virgo Cluster — Home of M 87

Bruno Binggeli

Astronomical Institute of the University of Basel
Venusstrasse 7
CH-4102 Binningen
Switzerland

Abstract. Our current understanding of the structure and dynamics of the Virgo cluster is reviewed. Special emphasis is given to a possible connection between the activity of M 87 and the cluster as a whole. The Virgo cluster is an aggregate of at least three separate subclusters, centered on M 87, M 86, and M 49. The dominant M 87 subclump, with a mass of a few 10^{14} M$_\odot$, is outweighing the other two subclumps by an order of magnitude. There is evidence, from the kinematics of dwarf galaxies and the structure of the X-ray gas, that the M 86 subclump is falling into the M 87 subclump from the back with a relative velocity of $\approx 1\,500$ km s^{-1}. M 87 and M 86 seem to be embedded in a common, cocoon-like swarm of dwarf ellipticals. The orientation of this cocoon, or simply the line connecting M 87 and M 86, is coinciding with the (projected) direction of the jet of M 87. A possible explanation for this apparent coherence between structures on the pc scale of the center of M 87 and the Mpc scale of the Virgo cluster is discussed.

1 Introduction

The purpose of this review is to provide the astrophysicist, who specializes in AGN and in M 87 in particular, with some astronomical background on the physical environment of M 87, which is of course the well-known Virgo cluster of galaxies. I try to convince the reader that such knowledge is not only of cosmographical interest but may be essential for an understanding of certain features of the central engine of M 87. For some, still unknown reason there is a remarkable coherence between the orientation of the jet axis of M 87, which is probably defined by the spin axis of the central black hole on a pc scale, and the orientation of the Virgo cluster on a Mpc scale.

For an astrologer, this micro-macro connection would come as no surprise. Consider the figure of Virgo from Hevel's beautiful *Uranographia* (1690), reproduced here in Fig. 1. The position of M 87, which can be identified with respect to the stars, happens to coincide with the elbow joint of Virgo's left arm, which is hidden behind her left wing. It has always been a mystery what Virgo is pointing at with her left hand. Now we know: this is the direction of the jet of M 87, to within 30 degrees.

So much for astrology in *this* contribution to the present volume.

A mere hundred years after Hevel we find the first mention of the *phenomenon* of the Virgo cluster, still way before its extragalactic nature was

Fig. 1. The figure of Virgo from Hevel's *Uranographia* (1690). North is up, West to the right, as in a modern representation. Everything appears mirrored because the entire sky map was drawn on a globe, to be viewed from the outside. The cross is indicating the position of M 87 and the arrow is the direction of its jet.

known, by Charles Messier in *Connaissance des Temps pour 1784* (see Tammann 1985 for the original passage). Messier noticed an unusual concentration of nebulae in the constellation of Virgo. Fifteen out of the 109 "Messier" objects are, in fact, Virgo cluster members. By identifying them on a conventional sky atlas, one can notice that the Messier's alone nicely trace out the direction of the jet of M 87!

From the rich 20th-century history of the Virgo cluster I mention only the landmark studies of Harlow Shapley and Adelaide Ames in the 30-ties (Shapley, who, ironically, in the "Great Debate" of 1920 had been the opponent of Heber Curtis, discoverer of the jet of M 87) and Gérard de Vaucouleurs and collaborators in the 60-ties and 70-ties (for more history see Tammann 1985).

The modern view of the Virgo cluster presented in the following is essentially based on the Las Campanas photographic survey of the Virgo cluster by Allan Sandage and collaborators (involving the writer), the galaxy redhifts measured by John Huchra, Lyle Hoffman and many others, and the ROSAT imaging of Virgo by Hans Böhringer and colleagues.

2 Global Structure

Figure 2 is a map of the ca. 1300 galaxies in the area of the Las Campanas survey of the Virgo cluster (Binggeli *et al.* 1985) judged to be cluster members. The membership criteria were based on (1), the morphological appearance of the galaxies; *e.g.* dwarf ellipticals, which constitute the dominant population of the cluster, have a characteristically low surface brightness; and/or (2), the measured radial velocities. The velocity criterion works of course only if the cluster is sufficiently isolated in space. Fortunately, this seems to be the case, *i.e.* there is a small void behind the cluster (although not quite so in the case of spirals and irregulars which form a sort of filament that runs through the cluster, *cf.* Sect. 4 below). Velocities are available only for the brightest 400 members. However, morphology is an equally efficient tool to pick up the members; later velocity measurements have nearly always confirmed our morphological judgement (*e.g.* Drinkwater *et al.* 1996). Detailed galaxy morphology is of course limited to the most nearby clusters, such as Virgo.

The magnitude limit of completeness of the Las Campanas survey is around $B_{\mathrm{T}}^{\mathrm{lim}} = 18$, or, if we assume a distance of 20 Mpc, $M_{\mathrm{B_T}}^{\mathrm{lim}} = -13.5$. Undoubtedly, there are hundreds, if not thousands of more, extremely faint members of the Virgo cluster — analogous to the dwarf spheroidal companions of our Galaxy (Phillipps *et al.* 1998). However, these will unlikely alter the structural appearance of the cluster.

Let us now have a look at the structure of the Virgo cluster based on Fig. 2. The primary characteristic is certainly the overall irregularity of the cluster. Although we would not hesitate to call M 87 the "king" of the Virgo cluster (despite the fact that it is not even first-ranked in apparent magnitude; M 49 is slightly brighter), it is not *the* center of *the* cluster. But it is the center of the most massive subcluster, as we shall see below. If one naively draws density contours (isopleths) with a suitable smoothing (number or luminosity-weighted), as in Binggeli *et al.* (1987), M 87 is off the peak density (the cluster "center") by almost one degree. That peak density is closer to the M 84/M 86 lump, to the NW of M 87. However, our smoothing was rather like putting a mattress on the bumpy back of a camel. There is clear evidence for a secondary subcluster around M 86 (see below), so we deal with a double structure, a double conglomerate of galaxies in the central part of the Virgo cluster. The two subclusters, called here the "M 87 subclump" and the "M 86 subclump", seem to be in a state of merging. Although this becomes clear only when we discuss the kinematics and the X-ray properties of the cluster below, the central double structure is quite obvious already from a simple plot of the galaxy positions in the sky (Fig. 2).

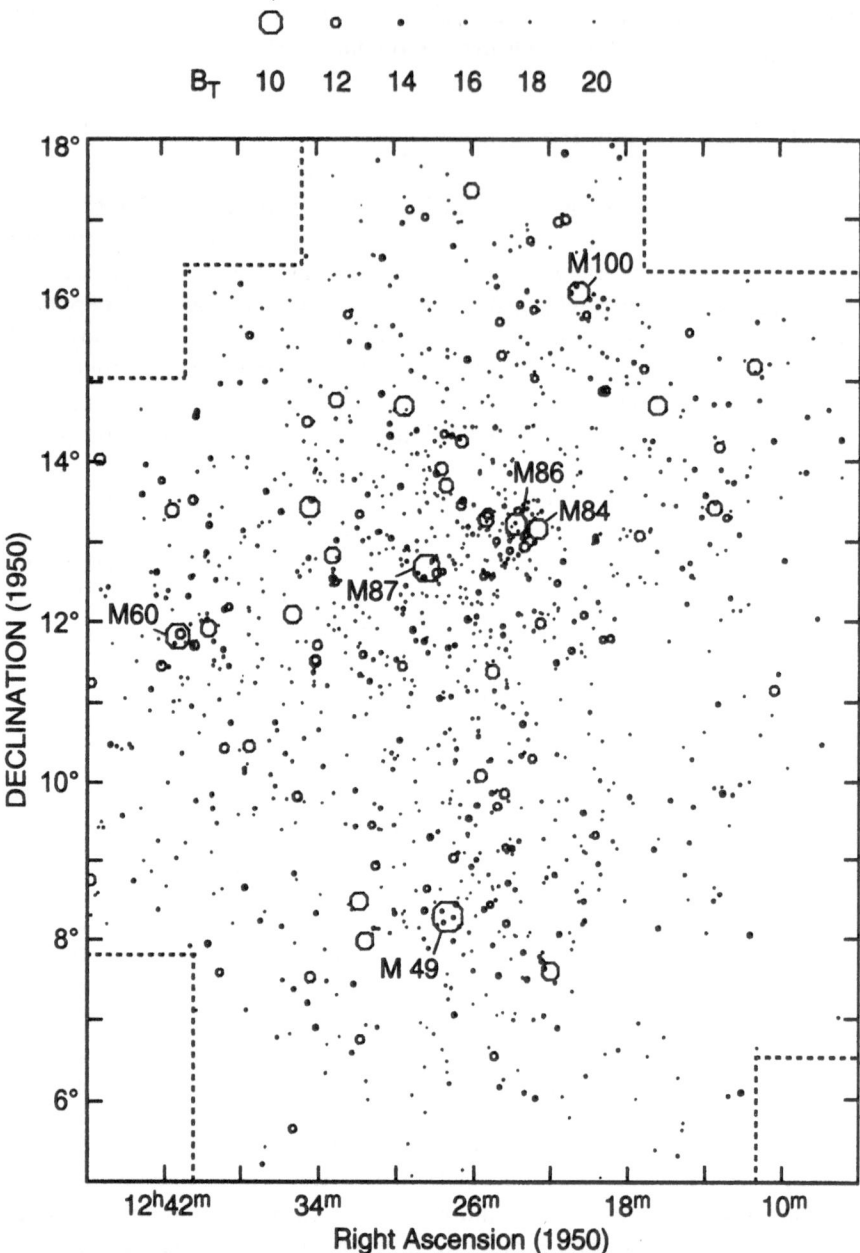

Fig. 2. Map of the Virgo cluster. All cluster members are plotted with luminosity-weighted symbols. The symbol size (area) is proportional to the luminosity of the galaxy. The apparent magnitude scale is given on top of the figure (absolute magnitudes follow from $m - M = 31.5$ if a distance of 20 Mpc is adopted). The most prominent Messier galaxies are indicated. Figure from Binggeli *et al.* (1987).

There is another double structure of the Virgo cluster on a larger scale, along N-S, defined by the northern M 86/M 87 subclump structure (called "cluster A" in Binggeli *et al.* 1987) on the one hand, and the southern galaxy concentration around M 49, called here the "M 49 subclump" (= "cluster B") on the other hand. Possibly, there is a small subclump around M60 (= "cluster C"). A number of very small bound subsystems, essentially groups of galaxies consisting of one bright galaxy plus a swarm of dwarf galaxies (as, *e.g.*, M 100 + satellites in the far North), are likely to exist, but these are difficult to identify even with kinematic data (for the general question of bound companions in the Virgo cluster, see Ferguson 1992, and Binggeli 1993).

So there are two main axes of the Virgo cluster: one N-S, *i.e.* M 100–M 86/M 87–M 49, and one E-NW, *i.e.* M 60–M 87–M 84/M 86. Remarkably, the former axis is nearly perfectly aligned with the position angle of the outer isophotes of M 87 (*e.g.*, Weil *et al.* 1997), while the latter is perfectly aligned with the jet axis of M 87. The two axes appear also very prominently in the X-ray image of the Virgo cluster (Fig. 5).

3   Galaxy Contents and Kinematics

The distribution and clustering properties of galaxies in clusters are known to depend strongly on the Hubble type (Dressler 1980). The Virgo cluster is no exception of this. Fig. 3 shows the projected distribution of Virgo cluster members divided into the main morphological types: giant early types (E + S0), dwarf early types (dE + dS0), spirals and smooth, magellanic-type irregulars (S + Im), and clumpy irregulars (blue compact dwarfs, BCD). Only galaxies with known velocities are depicted, although this is a restriction only for the dEs, as all other types are essentially complete with respect to kinematic data (all dEs are shown in Fig. 9, Sect. 6, in a different context).

From Fig. 3 we note: (1) late-type galaxies are much more dispersed than early types — independent of luminosity, (2) The southern M 49 subclump is very spiral and irregular-rich, unlike the northern M 87/M 86 subclump structure, (3) the E-NW axis of the cluster, which is aligned with the projected jet direction of M 87, is nicely traced out by the Es and S0s, (4) there is a prominent asymmetry in the distribution of dEs with respect to M 87 (of which more in Sect. 6 below). For more details, including the Dresslerian morphology-density relation for the Virgo cluster, the reader is referred to Binggeli *et al.* (1987).

These morphological differences are even more pronounced in the kinematic space. In Binggeli *et al.* (1993) we have collected and statistically analysed the radial velocities of ca. 400 Virgo cluster members. Most measurements are from the optical *Center of Astrophysics Redshift Survey* (Huchra *et al.* 1983, Geller & Huchra 1989) and the *Arecibo* H I survey of late-type Virgo members by Hoffman *et al.* (1987, 1989). Special efforts to get the velocities of a number of dE galaxies, which play a key role in our analysis,

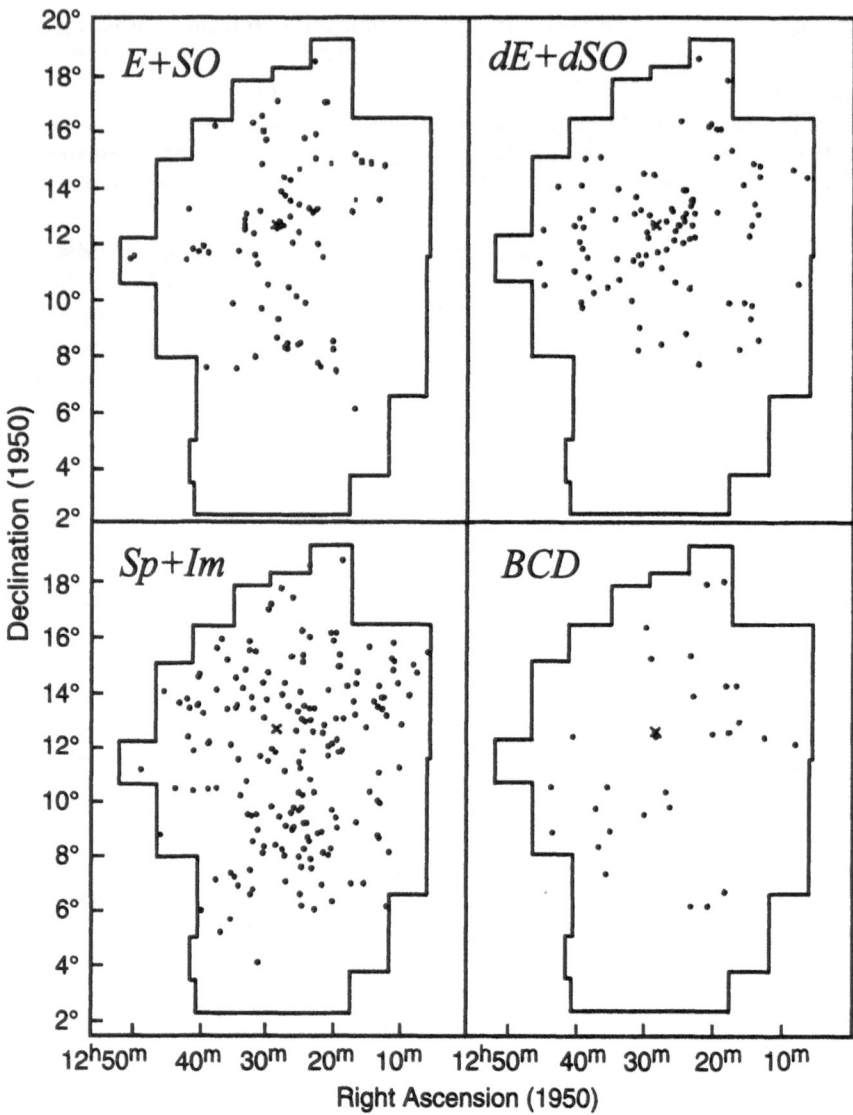

Fig. 3. The distribution of the main morphological classes of galaxies in the Virgo cluster: E + S0, dE + dS0, spirals + magellanic irregulars, and clumpy irregulars (BCDs), shown in four panels. Only galaxies with known radial velocities are shown. The irregular-shaped contour is indicating the area of the Las Campanas Survey of the Virgo cluster (Binggeli *et al.* 1985). The crosses mark the position of M 87. Figure from Binggeli *et al.* (1993).

are due to Bothun & Mould (1988). In fact, all 800 odd Virgo members still lacking a velocity are dEs: their notoriously low surface brightness renders spectroscopy essentially unfeasible, at least at present.

The basic kinematic data for the main morphological types, reproduced from Binggeli *et al.* (1993), are listed as velocity means and (r.m.s. or 1σ) dispersions in Table 1, and shown as distributions in Fig. 4. From these we note the following: (1) the velocity distribution of late-type members (spirals and irregulars) is significantly broader than that of early types (giant and dwarf E + S0s), *i.e.* late types are more dispersed in space *and* velocity, (2) the velocity distribution of spirals and irregulars is distinctly non-Gaussian, though it is fairly symmetric with a low-velocity and a high-velocity wing, (3) the velocity distribution of dwarf ellipticals is non-Gaussian *and* non-symmetric, being skewed towards low velocities, (4) the velocity of M 87 is off the cluster mean by $+200\,\mathrm{km\,s^{-1}}$ (as its projected position is also off a naively determined global cluster center), but is coinciding with the peak (median) of the velocity distribution of dEs, (5) the velocity of M 49 is not significantly different form the cluster mean, while the velocity of M 86 is even negative, coinciding with the low-velocity tail of the velocity distribution of dEs.

Table 1. Heliocentric velocities of Virgo cluster members

sample	N	$\langle v \rangle$ $(\mathrm{km\,s^{-1}})$	σ_v $(\mathrm{km\,s^{-1}})$
E + S0	75	1017	589
dE + dS0	93	1139	649
S + Im	188	1031	737
BCD	29	1110	795
All	**399**	**1064**	**699**
M 87		*1258*	
M 86		*−227*	
M 49		*969*	

These kinematic features, in connection with the projected spatial distributions discussed before, have been interpreted in the following way (*cf.* Huchra 1985, Binggeli *et al.* 1987, 1993). The broad velocity distribution of spirals and irregulars likely means that most of these galaxies are not yet relaxed (virialized); if they are only bound to the cluster, one indeed expects a velocity dispersion that is higher by $\approx \sqrt{2}$ than the dispersion of the presumably older, relaxed E + S0 population. The existence of low and high-velocity wings in the S + Irr distribution is a strong indication for infalling/expanding shells of late-type galaxies around the core of the cluster.

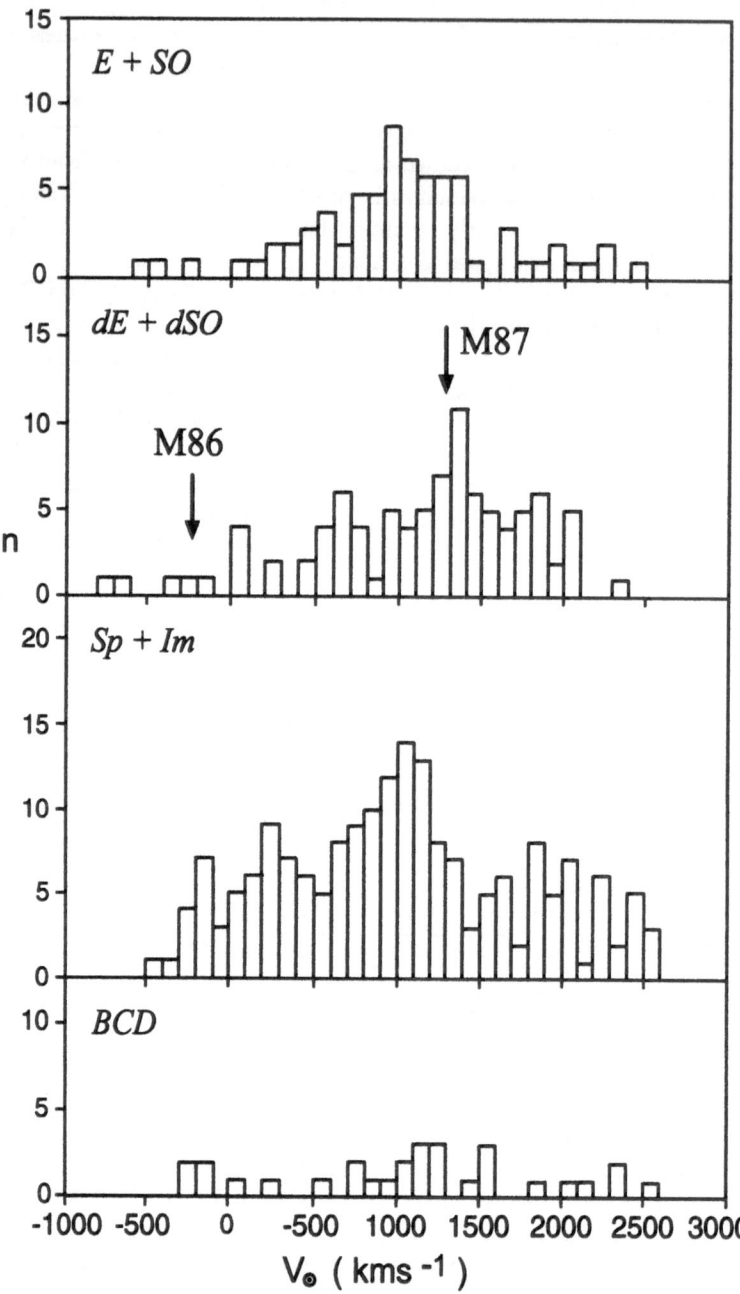

Fig. 4. The heliocentric velocity distributions of Virgo cluster members divided into the four main morphological classes shown in Fig. 3. The velocities of M 86 and M 87 are indicated in the panel for early-type dwarf galaxies.

It is quite plausible that nearly all spirals and irregulars are late, or even future arrivals, and hence are not yet virialized. The surrounding low-density field of the cluster, where these types of galaxies predominate, is subject to a global clustercentric velocity perturbation with a characteristic infall pattern (Rivolo & Yahil 1983, Tully & Shaya 1984). Field late-type galaxies are constantly fed into the cluster. Based on H I properties, it seems even possible to discriminate beteween spirals that have already fallen through the cluster core and spirals that are still in approach: the former, which are typically found in the central cluster area, are naturally identified with those spirals that are strongly H I-deficient (Haynes & Gionavelli 1986) and have very small H I disks (Cayette *et al.* 1990).

A clear asymmetry in the velocity distribution of a cluster of galaxies is almost certainly an indication of ongoing subcluster merging (*e.g.* Schindler & Böhringer 1993). The present asymmetric velocity distribution of Virgo dEs is taken as sign of the merging between the M 87 and M 86 subclumps (or rather, the infall of the M 86 subclump into the more massive M 87 subclump), for which there is additional evidence from X-ray observations, as will be discussed in Sect. 6. Both giant galaxies are obviously the centers of huge swarms of dwarf ellipticals, which is why the velocity of M 87 is coinciding with the peak of the dE velocity distribution and not with the cluster mean, while M 86 is apparently falling into, or through the M 87 subclump from the back, hence with a high relative (negative) velocity, dragging along a smaller swarm of dwarfs, some of which are the most blueshifted galaxies in the sky (*cf.* Sect. 6). Finally, the well-behaved velocity of M 49, coinciding more or less with the cluster mean, would suggest that the M 49 subclump is approximately at the same distance as the M 87/M 86 core structure, and that their supposed future merging will take place in the plane of the sky. This view is supported by the lack of a significant difference in the distance moduli of the M 49 and M 87/86 subclumps based on a host of different distance indicators (*cf.* Federspiel *et al.* 1998).

4   Cluster Depth and Environment

The absolute distance of the Virgo cluster is still a matter of debate. Distances quoted range from 15 to 22 Mpc. In general, the most reliable extragalactic distance indicators are of course the Cepheids. Cepheids at (and slightly beyond) the Virgo cluster distance are now within the reach of *HST*. This achievement was so long awaited that the first Cepheid-based distance determination of a Virgo cluster spiral (M 100) by Freedman *et al.* (1994) had an enormous impact. The resulting distance of ca. 17 Mpc was simply taken as *the* distance of the Virgo cluster. But we know (*cf.* above) that Virgo spirals avoid the cluster core and may be in the field far off the cluster. M 100 is probably lying at the near side of the cluster. Indeed, as more spiral distances are nailed down by *HST*-observed Cepheids, the average distance of the spirals is growing (Tammann & Federspiel 1997). Recently, Böhringer *et al.* (1997) have made the clever suggestion to use spirals as *HST*-Cepheid targets that show clear signs of ram pressure stripping, thereby ensuring proximity to the cluster core.

The safest would be to use only elliptical and S0 cluster members (or, ideally, M 87 alone!) for a distance determination. Unfortunately, the primary RR Lyrae stars are much too faint at the distance of Virgo even for *HST*. The secondary distance indicators which can be applied to Virgo ellipticals give controversial results: globular clusters, $D_n - \sigma$, and novae tend to give large distances ($D \approx 20\,\mathrm{Mpc}$), surface brightness fluctuations (SBF) and planetary nebulae (PN) lead to a small $D \approx 16\,\mathrm{Mpc}$ (an overview of the methods can be found in Jacoby et al. 1992). Great efforts are spent in the application of the SBF method (Tonry *et al.* 1997) because its claimed distance uncertainty for an individual galaxy is almost as small as with Cepheids ($\leq 0.2\,\mathrm{mag}$). However, Tammann (*e.g.* Tammann 1996, Tammann & Federspiel 1997) argues that the method is not yet mature for use, as long as the variations of the stellar populations among ellipticals are not really understood. A different problem might also undermine the PN method (Tammann 1996).

Fortunately, the absolute distance of the Virgo cluster is not very relevant for our discussion, and for definitiveness I continue to use $D = 20\,\mathrm{Mpc}$. More interesting is the question of the depth of the cluster. Can we resolve this depth with any of the distance indicators in use? What accuracy in the distance modulus would be required? According to Fig. 2, the angular width of the Virgo cluster in the sky is $\approx 8°$. Assuming the cluster is as deep as it appears wide (*i.e.* approximate spherical symmetry), and with a mean distance of 20 Mpc, the front-to-back depth is 2.8 Mpc, or $\lesssim 0.3\,\mathrm{mag}$ in distance modulus. So the cluster could just barely be resolved with the most accurate distance indicators.

But again: this assumes spherical symmetry. Should the cluster be deeper than wide, in the form of a cigar or finger pointing towards us, we might resolve (or claim to resolve) the cluster depth with even the worst distance indicator. This happened in the early days of the SBF method (Tonry *et al.* 1988) when the cluster literaly exploded. Since then, people have become more cautious, and an apparent dispersion of the SBF distances among Virgo ellipticals beyond of what can be expected from spherical symmetry is usually ascribed to unaccounted-for variations in the stellar populations (Pahre & Mould 1994, Jensen *et al.* 1996). A recent claim by Young & Currie (1995) that *dwarf* ellipticals are distributed in a prolate structure pointing towards us, based on the shape of the luminosity profile of these galaxies, has been shown to be flawed (Binggeli & Jerjen 1998). There is presently no indication that early-type galaxies in the Virgo cluster are *not* as strongly clustered in space as they are observed to be in sky projection.

Remains the Tully-Fisher method for spiral galaxies (and of course also the Cepheids, which, however, are too costly for a gross application) to map the outskirts and the large-scale environment of the Virgo cluster. There is consistent evidence that Virgo spirals are distributed in a prolate cloud, or filament, stretching essentially from the cluster backwards to the so-called "W cloud" — again roughly along our line of sight (Fukugita *et al.* 1993, Yasuda *et al.* 1997, Federspiel *et al.* 1998). There is no doubt about the reality of the feature: spiral and irregular galaxies in the Local Supercluster are known to be gathered in filamentary "clouds" of galaxies (*cf.* Tully & Fisher 1987). The Virgo spiral filament is probably part of a very long filament that runs from Virgo way back to the "Great Wall" at the distance of the Coma cluster (Hoffman *et al.* 1995), and it might even be connected, on the near side, with the "Coma-Sculptor cloud" that runs through, *i.e.* includes the Local Group (*cf.* Tully & Fisher 1987). If so, we should not be surprised to observe a "finger of God" — because we *live* in a finger of God.

5   The M 87 Subclump

The absolute dominance, in terms of mass, of the M 87 subclump in the Virgo cluster is not well seen in the optical (Fig. 2, which shows the luminosity distribution of the cluster), but it becomes strikingly clear from an X-ray image of the cluster. The reason for this is that, on the assumption of hydrostatic equilibrium, the hot X-ray gas is directly tracing the gravitational potential, *i.e.* the dark mass of the system in which it is embedded. In Fig. 5 we reproduce a *ROSAT All-Sky-Survey* image of the Virgo cluster, which is identical in scale to the optical image in Fig. 2. As mentioned by Böhringer et al. (1994), who did the definitive study of the X-ray properties of the Virgo cluster, there is good overall match between the optical and the X-ray, even in detail, note *e.g.* the relatively sharp Western edge seen in the optical as well as in the X-ray. This means that the galaxies and the hot gas are more or less in equilibrium. However, there is this striking difference in appearance of the M 87/M 86 core structure. In the optical, we note a swarm of galaxies around M 86/M 84, which of course is responsible for shifting the optical center of the cluster away from M 87 (see Fig. 2), while in the X-ray, M 87 is clearly *the* center of the cluster, or more precisely: it is clearly lying at the bottom of the most massive subclump. This is evidence that at least the core of the Virgo cluster is not in dynamical equilibrium. Our interpretation of what is going on in the core has been mentioned before: there is a smaller subclump around M 86 which is falling into the dominating M 87 subclump from the back (Binggeli *et al.* 1993, Böhringer *et al.* 1994). From the X-ray halos of M 87, M 86, and M 49, Böhringer *et al.* (1994) estimate several times $10^{14}\,M_\odot$ for the mass of the M 87 subclump, and an order of magnitude less for the mass of each of the M 86 and M 49 subclumps.

The M 87/M 86 subclump interaction will be further discussed below. Here we concentrate on the M 87 subclump. Nulsen & Böhringer (1995) have used *ROSAT* data to calculate the mass profile of the M 87 subclump (superceding,

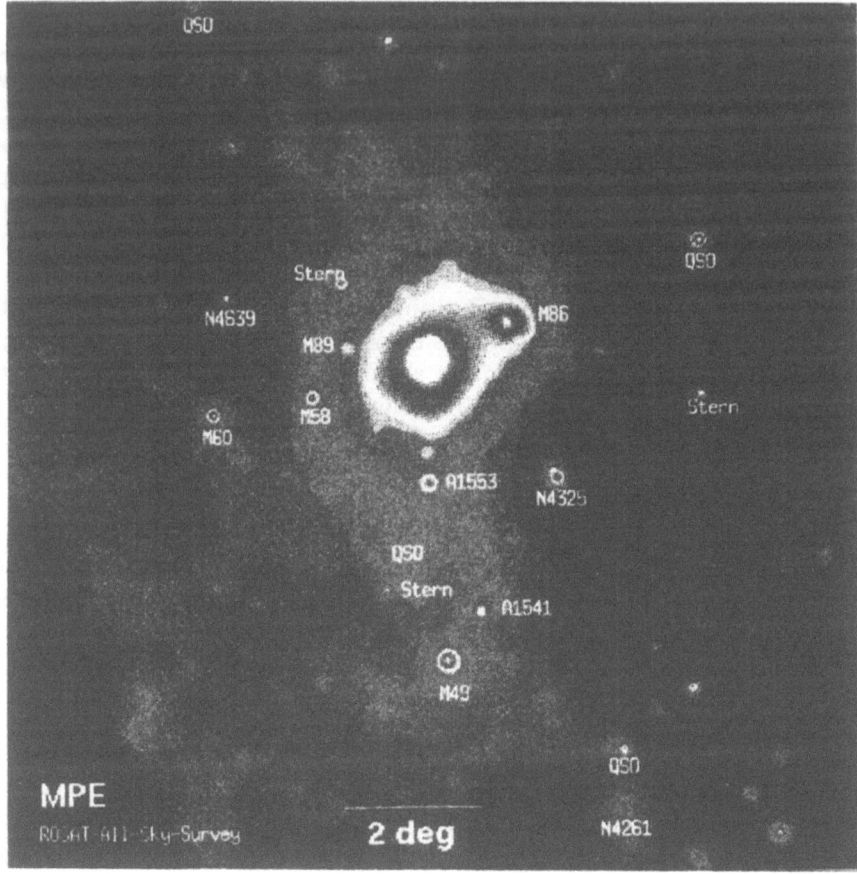

Fig. 5. X-ray image of The Virgo cluster from the *ROSAT All-Sky-Survey.* Various foreground stars, quasars, Abell clusters, and Virgo cluster members (NGCs and Messiers) are indicated. The large, bright spot is centered on M 87. The scale is identical to that of Fig. 2. Note the bright halos of M 49, M 86, and — very massive and dominating — M 87. Courtesy of Dr. H. Böhringer and the *MPE*, Garching.

but more or less agreeing with, the classic study by Fabricant & Gorenstein, 1983, based on *Einstein* data). Two components could be distinguished in the mass profile, one belonging to the galaxy M 87 itself, which is in good accord with the masses indicated by the stars and the globular clusters of M 87 (refs. given in Nulsen & Böhringer 1995), and one belonging to the subclump in which M 87 is embedded. However, the mass (dark matter) profile is really continuous, and it would seem to be difficult to discriminate between "M 87 subclump" galaxies that are gravitationally bound to M 87 proper and others that are merely bound by the subclump as a whole. Nevertheless, this is what I have tried to do several years ago (Binggeli 1993), and I show some

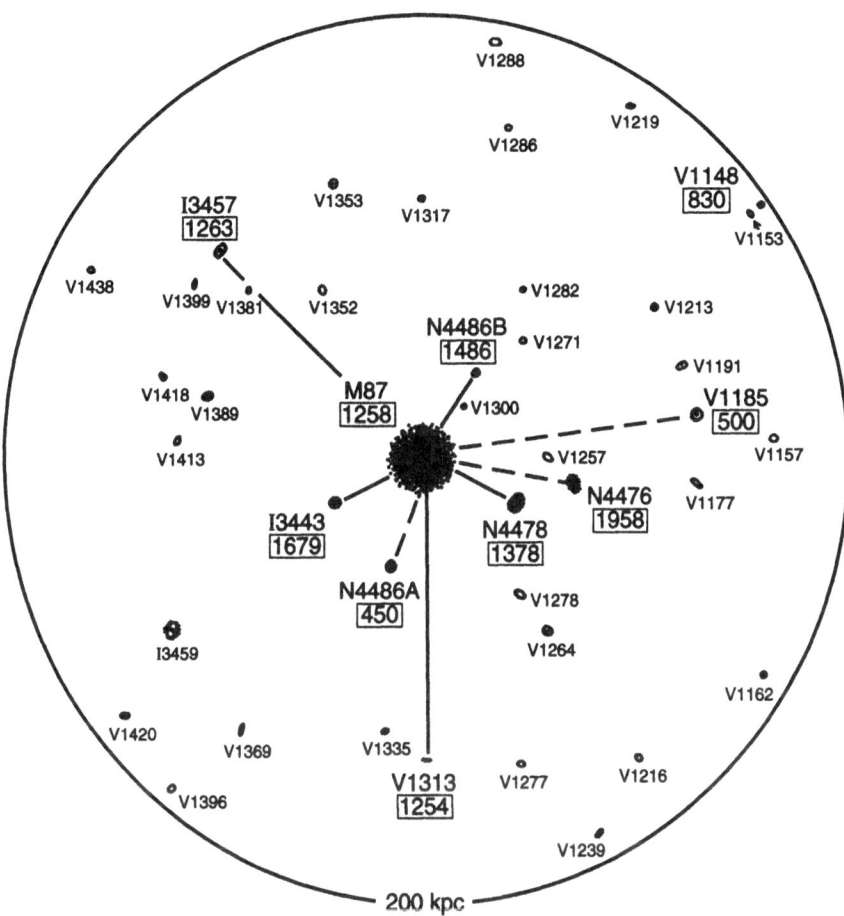

Fig. 6. Detailed map of the neighbourhood of M 87 within a projected radius of 200 kpc. The size and morphology of the galaxies is schematically indicated (filled image = high surface brightness, open = low surface brightness, central dot = nucleus, irregular contour = irregular). Faint galaxies are identified by their VCC number (Binggeli *et al.* 1985). Possible satellites are connected to the center by lines. Figure from Binggeli (1993).

of this here because it gives me the opportunity to show the closer galaxy environment of M 87.

Figure 6 is a schematic of this environment, within a projected radius of 200 kpc from M 87. There are 41 Virgo cluster members in this area, only 10 of which have known velocities (including M 87). Most of these must be bound to the M 87 subclump; some might also belong to the M 86 subclump. But what we can try to do is to find those galaxies which are likely bound companions to the *galaxy* M 87, in the following way. For a candidate companion we require, in addition to the (pre-selected) small projected distance, d, a sufficiently small velocity difference, $|v|$, to M 87. These two quantities can now be combined to give a "projected mass", $q = |v|^2 \cdot d/G$. By averaging q for N candidates, and multiplying the result with a suitable constant f that accounts for the distribution and projection of orbits, one gets a "projected mass" estimate for the central object (Bahcall & Tremaine 1981). Since we know the mass profile of M 87 from the X-ray halo, we can now add up possible companions with growing q until the upper-limit X-ray mass from Nulsen & Böhringer (1995) is exceeded, *i.e.* until the mass of M 87, at that average projected distance, needed to bind the would-be companions becomes too large. With $f = 24/\pi$ (*cf.* Bahcall & Tremaine 1981), I found five candidates in this way, which in combination give a total "projected mass" for M 87 at $d \approx 85$ kpc of $\approx 4 \cdot 10^{12}\,M_\odot$, which is (made to be) in accord with the X-ray mass of M 87 at that radius. These five candidates are listed with their relevant data in Table 2. Note also the smallness of their relative luminosities, which is a working condition for the method of Bahcall & Tremaine (1981). All five galaxies are, in fact, fairly compact in their appearance (see Fig. 6); NGC 4486B, *e.g.*, is legendary as a M 32-type compact elliptical; so they are very probably companions to M 87 indeed. However, two other *apparently* good candidates, NGC 4486A and NGC 4476, are certainly not real companions: the "projected mass" of each of these galaxies alone far exceeds the X-ray mass limit. Many more of the remaining 30 dwarf galaxies with unknown galaxies might be M 87 satellites, of course.

6   The M 87/M 86 Subclump Interaction ...and the Jet

We now enter the most interesting and perhaps most important issue of the Virgo cluster structure: the interaction between the M 87 and M 86 subclumps. The *galaxy* M 86 has long been known to interact with the intracluster medium (ICM) of the Virgo cluster: Forman *et al.* (1979) discovered

Table 2. Possible satellites of M 87

galaxy	type	$L/L_{M\,87}$	$|\Delta v|$ (km s^{-1})	d (kpc)	q ($10^{12}\,M_\odot$)
VCC 1313	BCD	0.001	4	130	≪0.01
IC 3457	dE3,N	0.013	5	139	≪0.01
NGC 4478	E2	0.094	120	55	0.18
NGC 4486B	E1	0.013	228	46	0.55
IC 3443	dE0,N$_{pec}$	0.003	421	47	1.93

d = projected distance to M 87
q = "projected mass" (see text)

a plume in the X-ray structure of M 86, which they interpreted as being due to ram pressure stripping   of the hot gas of M 86 by the Virgo ICM. This general picture has recently been confirmed by Rangarajan *et al.* (1995) based on *ROSAT* data, although these authors find no evidence for a cooling flow associated with an apparent optical emission seen on very deep photographs (Nulsen & Carter 1987). The surprisingly high metal abundance in the X-ray plume is attributed to the destruction of dust in the stripped material. From the morphology of the X-ray structure of M 86, Rangarajan *et al.* (1995) estimate a southward velocity of M 86 in the plane of the sky of ca. $500\,\mathrm{km\,s^{-1}}$, in addition to the radial velocity, relative to M 87, of almost $-1500\,\mathrm{km\,s^{-1}}$ (*cf.* Table 1).

However, we probably do not deal with a single galaxy, M 86, that is plunging through the Virgo ICM at a high relative speed and thereby suffering ram pressure stripping. The large gas fraction of M 86 is more typical for a whole group of galaxies (Böhringer *et al.* 1994). A first hint at the existence of such a group, or rather the "M 86 subclump" of galaxies, was found in the velocity distribution of dwarf ellipticals (Binggeli *et al.* 1993). So we have a small X-ray cluster, the M 86 subclump, which is banging into a big X-ray cluster, the M 87 subclump (note that this is more specific than the "Virgo ICM"), from the back.

Let us have a closer look at the velocities of dwarf ellipticals just mentioned. The velocity distribution of dEs was shown in Fig. 4, and attention was drawn to the fact that this distribution is asymmetric, with M 87 coinciding with the peak at higher-than-average velocities, and with a long tail of low velocities around the velocity of M 86. This asymmetry was taken as evidence for the ongoing merging of the M 87 and M 86 subclumps.

That the strange distribution of dE velocities is indeed coupled with M 87 and M 86 becomes more obvious by plotting the projected positions of the dwarfs in the cluster for different velocity ranges. This is shown in Fig. 7. Two features are striking: (1) there is a strong clustering of low-velocity dEs, in particular *negative*-velocity dEs, to the NW of M 87, around M 86, (2) there is a concentration of high-velocity dEs to the W of M 87. The dEs with negative velocities are interesting in themselves, because two of them are, in fact, the most *blue*-shifted galaxies known in the sky! Although there is good reason to hunt for the most *red*-shifted galaxies in the universe, it is only just that these blue-record holders are finally portrayed — which I do in Fig. 8.

How can we interpret the concentration of dwarfs to the W of M 87? The dEs with the most negative velocities ($v \approx -700\,\mathrm{km\,s^{-1}}$) are ca. $500\,\mathrm{km\,s^{-1}}$ off M 86, which we suppose is at rest with respect to the M 86 subclump as a whole. Thus, these dEs could simply constitute the low-velocity tail of the M 86 subclump, even with a velocity dispersion that is significantly smaller than $500\,\mathrm{km\,s^{-1}}$. On the other hand, one could envision a rather small intrinsic (original) velocity dispersion of the M 86 subclump, in which case the negative-velocity dwarfs would have been accelerated by the tidal

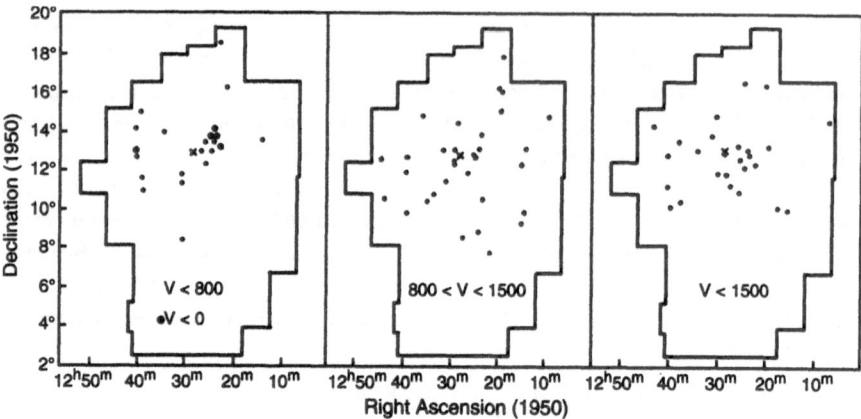

Fig. 7. The distribution of dwarf ellipticals in the Virgo cluster shown for three different velocity ranges: low velocities ($v < 800$) in the left panel, velocities close to the cluster mean ($800 < v < 1500$) in the central panel, and high velocities ($v > 1500$) in the right panel. Velocities are heliocentric and in km s^{-1}. The irregular contour is the boundary of the Las Campanas Virgo cluster survey (Binggeli *et al.* 1985). The position of M 87 is indicated by a cross. Note the concentration of high-velocity and low-velocity dEs to the right of M 87. In particular, there is a clustering of dEs with negative velocities around M 86 (drawn as fat dots). Figure taken from Binggeli *et al.* (1993).

influence of the M 87 subclump. Conversely, the concentration of high-velocity dEs to the W of M 87 could be due to M 87 subclump members that have been tidally accelerated towards M 86.

At this point we have, finally, to look at the distribution of *all* 800 odd dEs in the cluster, not only those for which we have a velocity. This is presented in Fig. 9. We note a very striking clump of dEs in the central region defined by M 87 and M 86, which is best seen with the very faintest dEs ($B_T > 18$, in the right panel). The feature looks somewhat like a cocoon that is embedding M 87 and M 86 at its extreme ends. Without modelling, *i.e.* numerical simulation of the Virgo cluster dynamics with a full account of galaxy morphology, it remains unclear what this dE cocoon, including its kinematic oddities discussed above, exactly means. However, its very existence strongly suggests that these dwarf galaxies are not in equilibrium with the rest of the cluster. We take it as best evidence for the strong ongoing interaction between the M 87 and M 86 subclumps.

But perhaps the most remarkable feature of all is the perfect alignment of the symmetry axis of this interaction, *i.e.* the elongation of the dE cocoon, or simply the direction from M 87 to M 86, with the (projected) jet axis of M 87. The merging of the two subclumps is taking place on a Mpc scale, while the jet is likely originating from the central pc of M 87 — this is a factor of

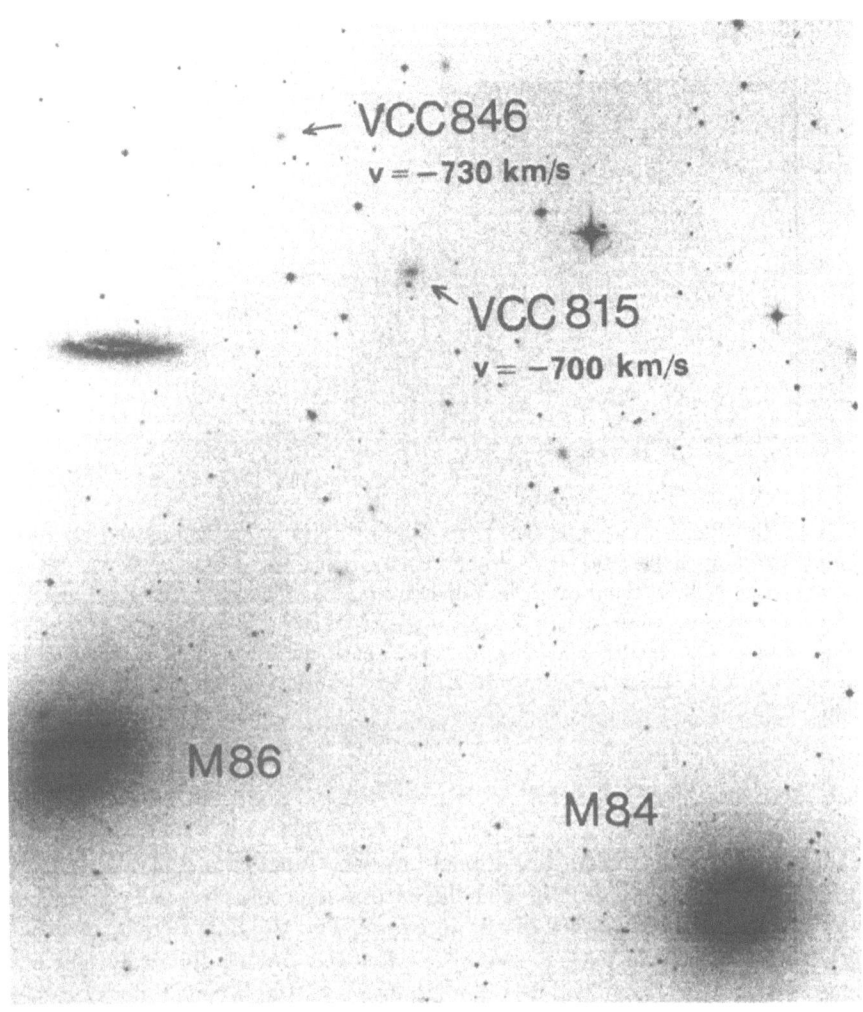

Fig. 8. A small piece of the Virgo cluster reproduced from one of the 67 Las Campanas photographic plates used for the Virgo cluster survey (Binggeli *et al.* 1985), showing M 84, M 86, NGC 4402, and a number of inconspicuous dwarf galaxies, two of which, however, are the most blue-shifted galaxies in the whole sky known to date: VCC 815 and VCC 846 with heliocentric velocities of -700 and $-730\,\mathrm{km\,s^{-1}}$, respectively. The scale of the photograph is given by the angular distance between M 84 and M 86, which is 18 arcminutes.

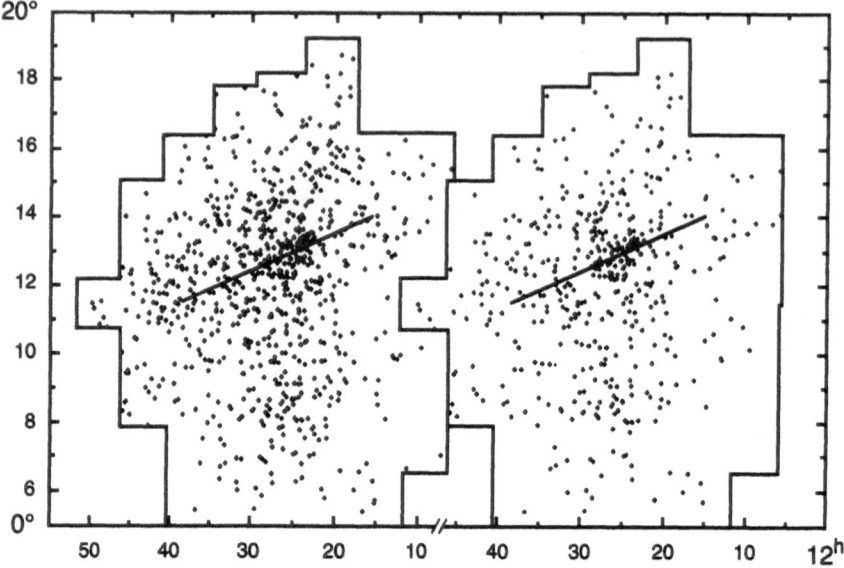

Fig. 9. The distribution of all 828 dwarf ellipticals (left panel) and of the 476 dwarf ellipticals fainter than $B_T = 18$ (right panel) in the Virgo cluster. The positions of M 87 and M 86 are indicated by filled circles. The jet axis of M 87 is shown as line. The irregular contour marks the boundary of the Las Campanas survey of the Virgo cluster (Binggeli *et al.* 1985). Note the dense, elliptical-shaped concentration of dEs which fills the area between M 87 to M 86 along the jet axis of M 87.

one million difference in scale! To put it bluntly: — *how does the jet know of the position of M 86?*

Recently, a very interesting model has been put forward by West (1994) which might explain the observed coherence of structures from the AGN scale to the large-scale structure of the universe. The top-down chain of events evoked by West (1994) goes roughly as follows. During the early epoch of cluster formation, a central dominant (often a cD) galaxy is built via mergers of smaller galaxies. These mergers proceed in a coherent manner along preferred directions which are related to the large-scale cluster surroundings. As a consequence of this anisotropic formation process, the central galaxy is prolate in shape and has a built-in memory of the shape of its parent cluster. Cold gas falling into the prolate potential well will settle into a disk whose angular momentum vector is aligned with the major axis of the central galaxy. Angular momentum loss results in an inward flow of gas, leading to the creation and feeding of a black hole whose spin axis will be aligned with the angular momentum vector. Radio jets are expected to emanate along the black hole spin axis, and thus will be aligned with the major axis of the central dominant cluster galaxy.

This model explains two observed alignment effects: (1) the shape of the central cluster galaxy tends to be aligned with the surrounding distribution of matter on a scale of ca. 20 Mpc, (2) the radio major axis of strong radio galaxies, defined by the radio lobes, tends to be aligned with the optical major axis. The first effect was found for relatively low redhifts, but it will probably also hold for high-redshift clusters. The second effect is only for high-redshift radio galaxies. At lower redshifts, *i.e.* later epochs, the radio-optical alignment might be washed out by the relaxation of the stellar component, though West (1994) does give low-redshift examples with a strong alignment, such as *Cygnus A*. In combination, the two alignment effects lead to a coherence of structures from parsecs to Megaparsecs.

Could the model of West (1994) explain the alignment of the jet axis of M 87 with the M 87-M 86 axis? First, we should note that the major axis of the galaxy M 87 is not pointing in this direction, but is rather oriented NNW, ca. 50° off the jet axis (*e.g.*, Weil *et al.* 1997). However, as we have seen in Sect. 2, the position angle of the galaxy M 87 is in accord with a second cluster axis, viz. that defined by the line connecting M 87 and M 49. In fact, it is in perfect accord with "Supergalactic Equator" (*cf.* Fig. 4 in Binggeli *et al.* 1987), which is the symmetry axis of the Local Supercluster. Hence the shape of the galaxy M 87 is, after all, aligned with its surroundings on a 20 Mpc scale.

But the jet is pointing to M 86, or rather: the projection of the jet is. In 3D-space, the jet is pointing towards us at an angle of 30° – 40° to the line of sight (Bicknell & Begelman 1996, also Bicknell in this volume). So — is it just a coincidence? Without taking resort to an extremist scenario, such as Arp's (1986, also Arp in this volume), I find this hard to believe. Not only M 86 and M 84 are lying "in the way"; there are also M 59 and M 60 on the other side, and there is a whole chain of ellipticals tracing out the projected jet axis (*cf.* again Figs. 2 and 3 here, and Figs. 4 and 7 in Binggeli *et al.* 1987). This is suggesting *some* kind of causal connection between the orientation of the jet and the distribution of matter on a Mpc scale, be it along the lines of West's (1994) model, or by some yet unknown processes.

7   Concluding Remarks

The Virgo cluster is a typical cluster in every sense, and we would not expect it to be otherwise. Not even the fact that it is made of (at least) three smaller clusters (one of which is grossly dominating in mass, however), which are on the verge to merge, turns out to be very special. Many other clusters, at a closer look, show almost exactly the same features. A good example is the somewhat more distant Centaurus cluster, where we distinguish between a massive, X-ray emitting and dE-rich subclump (centered on the cD NGC 4696) and at least one less massive, spiral-rich unit, which is falling into the dominant structure with a high relative velocity (Jerjen 1995, Stein *et al.*

1997). Even the Coma cluster, formerly the prototype of a hypthetical class of "regular", "relaxed" clusters, has given way to this picture: it also is an aggregate of three subunits, two of which must have merged long ago, but still show traces (as the respective central dominant galaxies, NGC 4874 and NGC 4489, are not yet merged), and the third of which (centered on NGC 4839, of Virgo cluster size and mass!) is in the process of merging (White *et al.* 1993, Colless & Dunn 1996). We seem to live in the epoch of rich cluster formation.

What *is* special about the Virgo cluster, for us, is of course its proximity. There is no other cluster of comparative richness lying that close (Fornax, Ursa Major, and Coma I are all much less rich than Virgo). That the cluster is harbouring an active galaxy is not unusual (in fact, as Blandford emphasizes in this volume, M 87 is a fairly lousy, *i.e.* inactive AGN). But again: it is the proximity which may render M 87 a kind of Rosetta stone for AGN astrophysics. Indeed, enormous efforts are spent in the attempt to unveil the secrets of the center of M 87, of which this volume bears ample witness. The efforts spent to investigate the extragalactic *environment* of M 87 are very small in comparison. However, as I tried to show here, the central pc of M 87 may be intimately connected with the structure and dynamics of the Virgo cluster as a whole.

Of the many features of the Virgo cluster yet to be studied, I mention two which I regard as especially relevant for the present discussion. (1) As mentioned before, we still lack radial velocity data for ca. 800 Virgo members. Among them are the hundreds of dEs which form the cocoon around M 87/M 86. It would be highly desirable to know the velocities of as many of these dwarfs as possible, to get a more complete picture of the cluster kinematics. Present-day technology should allow the measurement of at least the brighter of these objects. (2) Based on such data, with the projected positions of all, and the radial velocities of nearly all Virgo galaxies, plus (possibly) the X-ray gas distribution taken as model input (initial conditions), one could "run" the whole Virgo cluster. In particular, one should be able to simulate the M 87/M 86 subclump interaction, or at least put useful constraints on its dynamics. For a full 3D-simulation, the computing time might be prohibitively large. However, even with very simplifying assumptions to save computing time, such a simulation might be very rewarding. A cluster simulation with realistic, *i.e.* observed quantities as input parameters (not only in the statistical sense, but galaxy-by-galaxy) has not yet been carried out. The Virgo cluster would be the obvious first choice for such a project (for a first, crude attempt, see Schindler & Binggeli 1994).

Acknowledgements: My involvement with the Virgo cluster was initiated by an illuminating and pleasant collaboration with Dr. Allan Sandage and Prof. G. A. Tammann, to whom I'm very grateful. I thank Dr. Hans Böhringer for the nice X-ray image of the Virgo cluster (Fig. 5). This work was supported by the Swiss National Science Foundation.

References

Arp, H. 1986, JA&A, 7, 71

Bahcall, J.N., Tremaine, S. 1981, ApJ, 244, 805

Bicknell, G.V., Begelman, M.C. 1996, ApJ, 467, 597

Binggeli, B. 1993, Habilitationsschrift, University of Basel

Binggeli, B., Jerjen, H. 1998, A&A, in press

Binggeli, B., Popescu, C.C., Tammann, G.A. 1993, A&AS, 98, 275

Binggeli, B., Sandage, A., Tammann, G.A. 1985, AJ, 90, 1681

Binggeli, B., Tammann, G.A., Sandage, A. 1987, AJ, 94, 251

Böhringer, H., Briel, U.G., Schwarz, R.A., Voges, W., Hartner, G., Trümper, J. 1994, Nat, 368, 828

Böhringer, H., Neumann, D.M., Schindler, S., Huchra, J.P. 1997, ApJ, 485, 439

Bothun, G.D., Mould, J.R. 1988, ApJ, 324, 12

Cayette, V., van Gorkom, J.H., Balkowski, C., Kotanyi, C. 1990, AJ, 100, 604

Colless, M., Dunn, A.M. 1996, ApJ, 458, 435

Dressler, A. 1980, ApJ, 236, 351

Drinkwater, M.J., Currie, M.J., Young, C.K., Hardy, E., Yearsley, J.M. 1996, MNRAS, 279, 595

Fabricant, D., Gorenstein, P. 1983, ApJ, 267, 535

Federspiel, M., Tammann, G.A., Sandage, A. 1998, ApJ, in press

Ferguson, H.C. 1992, MNRAS, 255, 389

Forman, W., Schwarz, J., Jones, C., Liller, W., Fabian, A.C. 1979, ApJ, 234, L27

Freedman, W.L. et al. 1994, Nat, 371, 757

Fukugita, M., Okamura, S., Yasuda, N. 1993, ApJ, 412, L13

Geller, M., Huchra, J. 1989, Sci, 246, 857

Haynes, M.P., Giovanelli, R. 1986, ApJ, 306, 466

Hoffman, G.L., Helou, G., Salpeter, E.E., Glosson, J., Sandage, A. 1987, ApJS, 63, 247

Hoffman, G.L., Lewis, B., Helou, G., Salpeter, E.E., Williams, H. 1989, ApJS, 69, 65

Hoffman, G.L., Lewis, B., Salpeter, E.E. 1995, ApJ, 441, 28

Huchra, J. 1985, in The Virgo Cluster, ESO Workshop proceedings No. 20, ed. O.-G. Richter and B. Binggeli, ESO, Garching, p. 181

Huchra, J., Davis, M., Latham, D., Tonry, J. 1983, ApJS, 52, 89

Jacoby, G.H. et al. 1992, PASP, 104, 599

Jensen, J.B., Luppino, G.A., Tonry, J.L. 1996, ApJ, 468, 519

Jerjen, H. 1995, PhD dissertation, University of Basel

Nulsen, P.E.J., Böhringer, H. 1995, MNRAS, 274, 1093

Nulsen, P.E.J., Carter, D. 1987, MNRAS, 225, 939

Pahre, M.A., Mould, J.R. 1994, ApJ, 433, 567

Phillipps, S., Parker, Q.A., Schwartzenberg, J.M., Lones, J.B. 1998, ApJ Lett., in press

Rangarajan, F.V.N., White, D.A., Ebeling, H., Fabian, A.C. 1995, MNRAS, 277, 1047

Rivolo, A.R., Yahil, A. 1983, ApJ, 274, 474

Schindler, S., Binggeli, B. 1994, in Cosmological Aspects of X-Ray Clusters of Galaxies, ed. W.C. Seitter, Kluwer, Dordrecht, p. 155

Schindler, S., Böhringer, H. 1993, A&A, 269, 83

Stein, P., Jerjen, H., Federspiel, M. 1997, A&A, 327, 952

Tammann, G.A. 1985, in The Virgo Cluster, ESO Workshop proceedings No. 20, ed. O.-G. Richter and B. Binggeli, ESO, Garching, p. 3

Tammann, G.A. 1996, PASP, 108, 1083

Tammann, G.A., Federspiel, M. 1997, in The Extragalactic Distance Scale, STScI Symposium, ed. M. Livio *et al.* Cambridge University Press, Cambridge, p. 137

Tonry, J.L., Ajhar, E.A., Luppino, G.A. 1988, ApJ, 100, 1416

Tonry, J.L., Blakeslee, J.P., Ajhar, E.A., Dressler, A. 1997, ApJ, 475, 413

Tully, R.B., Fisher, J.R. 1987, Nearby Galaxies Atlas, Cambridge University Press

Tully, R.B., Shaya, E. 1984, ApJ, 281, 31

Weil, M.L., Bland-Hawthorn, J., Malin, D.F. 1997, ApJ, 490, 664

West, M.J. 1994, MNRAS, 268, 79

White, S.D., Briel, U.G., Henry, J.P. 1993, MNRAS, 261, L8

Yasuda, N., Fukugita, M., Okamura, S. 1997, ApJS, 108, 417

Young, C.K., Currie, M.J. 1995, MNRAS, 273, 1141

M 87 as a Galaxy

Walter Dehnen

Theoretical Physics
1 Keble Road
Oxford OX1 3NP
United Kingdom

Abstract. I review recent studies about the gravitational potential and stellar dynamics of M 87 in particular, and the dynamics of the stars in the presence of a super-massive central black hole, in general.

At large radii, investigations of both the X-ray emitting gas and the velocity distribution of globular clusters indicate the presence of large amounts of non-luminous matter, possibly belonging to the inner parts of the Virgo cluster.

At small radii, there is no evidence from the stellar kinematics, at most a hint, for the existence of a central point mass, whereas the gas dynamics reveal the presence of a highly concentrated mass in the centre of M 87, possibly a super-massive black hole (BH). Given the existence of such a central mass, the stellar kinematics indicate a strong tangential anisotropy of the stellar motion inside a few arcseconds. The implications of this result for the evolution and formation history of M 87 and its central BH are discussed. I also discuss in more general terms the structural changes that a highly concentrated central mass can induce in its parent galaxy.

1 Introduction

According to their observable properties, elliptical galaxies can be divided into two classes. This dichotomy is most clearly revealed in the central brightness distribution (*cf.* Fig. 3 of Gebhardt *et al.* 1996). Consequently, the two classes are commonly called 'power-law galaxies' and 'core galaxies' (though — as so often in astronomy — the latter term is highly misleading: these galaxies actually do not have a core of constant density). Core galaxies have shallow central luminosity density profiles with $j \propto r^{-\gamma}$, $\gamma \lesssim 1.3$, ellipticities E0 to E3–4, elliptic to boxy isophotes, and negligible rotation $v_{\rm rot} \ll \sigma$. They are on average bright ($M_V \lesssim -19.5$) and often radio-loud and X-ray-active, possess extended stellar envelopes and rich ($N \gtrsim 2000$), extended globular cluster (GC) systems which are multi-modal in their properties. These galaxies are thought to be of round to triaxial shape supported by anisotropy in the stellar motions.

Power-law galaxies, on the other hand, have steep central density cusps with $\gamma \gtrsim -1.5$, ellipticities up to E7, elliptic to disky isophotes, and significant rotation velocities $v_{\rm rot} \sim \sigma$. They are on average fainter ($M_V \gtrsim -21.5$) and show no radio or X-ray activity; their surface density follows a de Vaucouleurs profile and their GC systems are poor ($N \lesssim 1500$) and with a profile following that of the stellar light. These galaxies are believed to be of near-oblate shape supported by rotational motions.

Clearly, M 87 having a shallow density cusp ($\gamma \approx 1.2$), round isophotes, negligible rotation, radio and X-ray activity, an extended stellar envelope, and a rich and bi-modal globular cluster system is a generic representative of the class of core galaxies. It is generally believed, that the core galaxies are formed by one or more major merger events. In face of this hypothesis, it is important to ask whether M 87 is consistent with being a merger remnant.

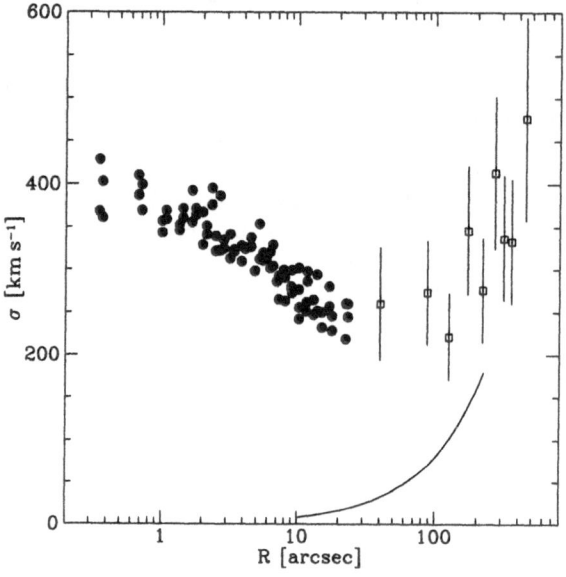

Fig. 1. Velocity dispersion for M 87 as measured for stars (filled circles, van der Marel 1994) and planetary nebulæ (open squares, Cohen & Ryzhov 1997). The full line is the rotational velocity measured by the latter authors.

2   The Matter Distribution at Large Radii

The presence of dark matter around spiral galaxies was established by studies of the motions of their gaseous disk, whose emission lines can be traced to large radii. Elliptical galaxies in general have very little cold gas and one has traditionally used stars as dynamical tracer population. However, at large radii the stellar absorption-line spectra are very hard, if not impossible, to observe with useful signal-to-noise ratio. Thus, in order to probe the potential at large radii of elliptical galaxies, one needs different kinematical tracers.

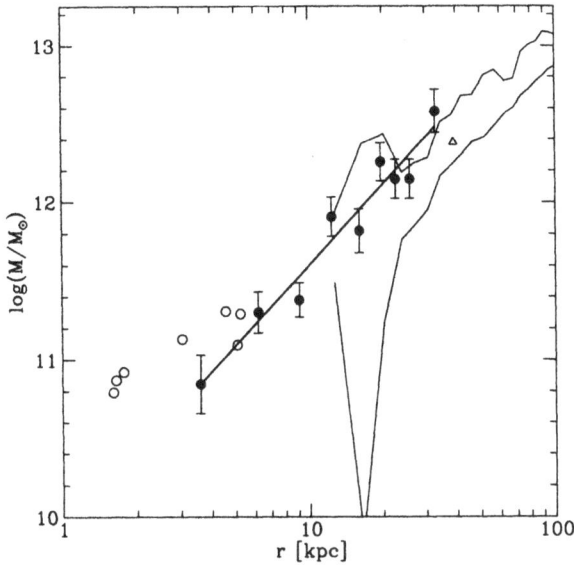

Fig. 2. Mass enclosed in radius r as derived from studies of stellar kinematics (open circles, Sargent *et al.* 1978), globular clusters (triangle, Mould *et al.* 1990, and filled circles, Cohen & Ryzhov 1997), and X-ray gas (thin lines: lower and upper limits, Nulsen & Böhringer 1995, corrected for the difference in the adopted distance to M 87). The solid line represents a power-law fit to the filled circles: $M \propto r^{1.7}$.

Possible candidates are gas rings, planetary nebulæ (PN), and globular clusters (GC). Gas rings are rather rare among elliptical galaxies, and in general one is left with PNs or GCs. The problem with using these or stars as tracers is that they do not move on circular orbits like the gas, but form dynamically hot systems, which complicates the interpretation of the measured kinematics.

Recently, Cohen & Ryzhov (1997) have studied the GC system of M 87. The derived rotation velocity v_{rot} and velocity dispersion σ are displayed in Fig. 1 together with stellar kinematical data for the central parts. There is a clear change in the kinematic properties at about 100″ from the centre: in the inner parts $v_{rot} \approx 0$ and σ decreases from $\approx 400\,\mathrm{km\,s^{-1}}$ to $\approx 200\,\mathrm{km\,s^{-1}}$; in the outer parts σ increases reaching $\approx 400\,\mathrm{km\,s^{-1}}$ at 400″ and v_{rot} becomes significant. This kinematical behaviour is very similar to that observed for NGC 1399 (cD galaxy in the Fornax cluster) by Arnaboldi & Freeman (1996) using PNs as tracers (their Fig. 2). An obvious explanation is that the change in kinematical properties constitutes the transition from the highly concentrated galaxy to the less concentrated and dark-matter-dominated Virgo cluster.

Cohen & Ryzhov have analysed their data in terms of the mass distribution and found that in the range $3\text{kpc} \le r \le 30\text{kpc}$ (using a distance to M 87 of 15 Mpc) their data imply a mass density $\rho \propto r^{-1.3}$ and a mass of $\approx 3 \times 10^{12} M_\odot$ inside 44kpc. These findings are in agreement with estimates derived from X-ray observations (Fabricant & Gorenstein 1983, Nulsen & Böhringer 1995), see Fig. 2. It is intriguing that the derived density slope of $\gamma = 1.3$ agrees well with 1.4 recently predicted for the inner parts of dark matter halos by high-resolution simulations of CDM cosmogonies (Moore *et al.* 1997).

3   Small Radii: Stellar Kinematics and the Black Hole

Because of its stellar kinematics, M 87 has for a long time been suspected to harbour at its centre a super-massive black hole[1] (BH), though this conclusion was always controversial (*cf.* Sargent *et al.* 1978, Binney & Mamon 1982, Merritt 1987). The best currently available photometry (Lauer *et al.* 1992) and stellar kinematical data (van der Marel 1994) show a weak density cusp and a slightly centrally peaked velocity dispersion σ (see Fig. 1). As van der Marel's analysis showed, these data can easily be interpreted by a pure stellar model with constant mass-to-light ratio and anisotropy[2] of $\beta \approx 0.5$ (van der Marel 1994), *i.e.* radially anisotropic as expected for a galaxy formed by violent relaxation. Alternatively, one can explain the centrally peaked σ by a central BH of a few billion solar masses. However, such a model does not work with radial anisotropy near the BH, which would give too high a central σ, and requires the opposite: isotropy or tangential anisotropy, depending on the mass of the BH.

Bender *et al.* (1994) found a change in the profile of the stellar absorption lines: the coefficient h_4 of the Gauss-Hermite expansion of the profile changes sign at $r \approx 3''$, such that $h_4 < 0$ inside and $h_4 > 0$ outside. This implies a change in the underlying dynamical properties in the sense that $\beta_{r<3''} < \beta_{r>3''}$, as predicted by models with central BH (though it would be very hard to quantify this).

---

[1] Clearly, from stellar dynamical and similar arguments, one cannot infer that the dark object is of the size of its Schwarzschild radius. However, lacking other plausible explanations for the high mass concentrations at the centres of many galaxies, it is generally believed that these are actually black holes, and I adopt this hypothesis throughout this paper.

[2] The anisotropy parameter is defined to be

$$\beta \equiv 1 - \sigma_t^2/\sigma_r^2,$$

where σ_t and σ_r denote the tangential and radial velocity dispersions. For an isotropic system $\beta = 0$; for radial anisotropy $\beta > 0$ reaching $\beta = 1$ for pure radial orbits; and for tangential anisotropy $\beta < 0$ with $\beta = -\infty$ for pure circular orbits.

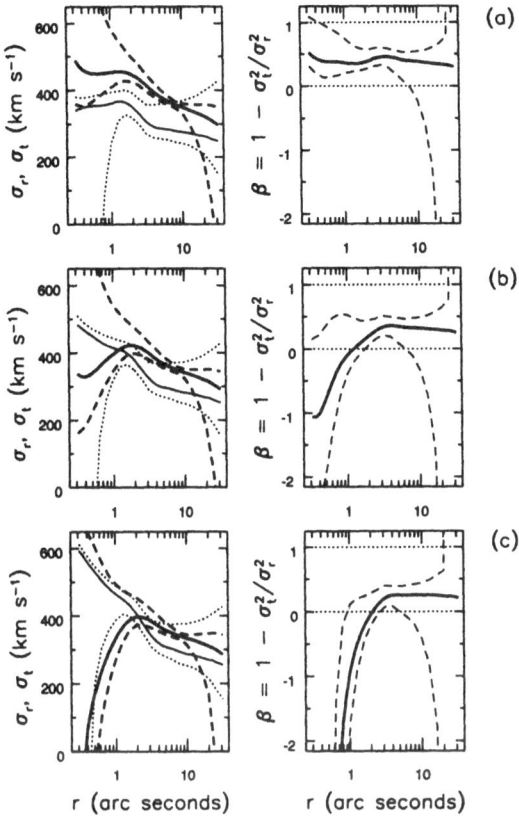

Fig. 3. Velocity dispersions and anisotropy vs. radius for M 87 as inferred by Merritt & Oh (1997) from the stellar kinematics for a BH with assumed mass of (a) 1, (b) 2.4, and (c) 3.8 billion solar masses. Dotted lines give 95% confidence bands.

Actually, since 1994 we *know* from HST observations of a central gas-disk that M 87 hosts a central BH of $M_\bullet = (3.2 \pm 0.9) \times 10^9 M_\odot$ (Harm *et al.* 1994, Ford *et al.* 1994, Marconi *et al.* 1997 and this volume). Starting from this fact and assuming spherical symmetry and a constant stellar mass-to-light ratio, Merritt & Oh (1997) solved simultaneously the projection equation for the stellar kinematics and the Jeans equation describing dynamical equilibrium. This gave them a non-parametric estimate for the anisotropy implied by the observed stellar kinematics and the BH. Their results for a BH of 1, 2.4, and 3.8 billion solar masses are displayed in Fig. 3. For a sufficiently massive BH, the central stellar motions are strongly tangentially anisotropic ($\beta \lesssim -1$). The more massive the BH, the stronger the inferred anisotropy and the larger is the radius at which β changes sign. Outside of this radius, the stellar motions are slightly radially anisotropic in agreement with van der Marel's (1994) study. For the most recent estimate of $M_\bullet$, the effect is very significant leading to highly tangential anisotropy. In the following section, I discuss possible explanations for this effect.

4   Stellar Dynamics Around a Black Hole

Can we understand the abrupt change in the stellar motions from radial to tangential anisotropy near the BH? Let us consider various possible scenarios for the formation of the BH at the centre of a galaxy.

Adiabatic Growth of the Black Hole
If the BH grows slowly over a long period of time, the actions of the stellar orbits are conserved. However, the mapping between actions and ordinary phase-space coordinates changes as the potential evolves from harmonic (for an initially isothermal core) to Keplerian. This change results in a stellar density cusp with $\rho \propto r^{-3/2}$ and a mildly tangential anisotropy of at most $\beta = -0.3$ (Young 1980, Goodman & Binney 1984, Quinlan *et al.* 1995) inside the sphere of influence of the BH, which has radius

$$r_h = \frac{G M_\bullet}{\sigma^2} \left(\simeq 1'' \text{ for M 87} \right). \tag{1}$$

While this radius is of the correct size, the effect is much too weak in order to explain Merritt & Oh's result of $\beta \lesssim -1$. Furthermore, M 87 has a central density slope $\gamma \approx 1.2 < 3/2$.

Growth by the Capture of Stars
A BH may grow by tidally disrupting and capturing stars that happen to come too close to it. Such stars will predominantly have low angular momenta, while near-circular orbits are hardly affected. Hence, this process gives rise to $\beta < 0$. However, the distance from the BH that a star must reach before it gets destroyed is of the order of the BH's Schwarzschild radius, which is smaller than r_h by several orders of magnitude.

Growth by the Accretion of Black Holes
If two galaxies, each hosting a massive BH at its centre, merge, the BHs will sink to the centre of the remnant and finally merge as well (Ebisuzaki *et al.* 1991). Begelman *et al.* (1980) have outlined the stages of this process:
(1) Due to dynamical friction with the background stars, the BHs sink into the centre and form a BH binary.
(2) The binary looses energy and angular momentum (it "hardens") due to three-body interactions with passing stars.
(3) When the separation between the BHs has become sufficiently small, gravitational radiation becomes very efficient in hardening the binary until it finally merges to a single BH.

Depending on the time-scale of the whole process, repeated events of this kind are possible[3], in particular for a central-cluster elliptical, such as M 87, which over its lifetime has likely cannibalized many minor companions. The least understood mechanism here is (2), which is also the process most relevant to the possible effect on the stellar dynamics. One problem, for instance, is whether or not the eccentricity of the binary BH increases, which in turn is relevant for (3), since for a highly eccentric binary gravitational radiation can take over at higher energies, *i.e.* earlier, than for a less eccentric binary. Another problem is loss-cone depletion: three-body interactions that harden the binary eject stars out of the centre, and the number of interaction candidates diminishes. Eventually, this may even halt the hardening before gravitational radiation can take over (the re-fueling of the loss cone due to two-body relaxation is much too slow).

It is clear that process (2) will create tangential anisotropy, since, as for the tidal disruption, low-angular-momentum stars are more likely to interact with the BHs – they are, however, not eaten by the BH but ejected via a slingshot. Evidently, this process reaches out to the radius r_h, where the circular motion around the BHs equals the mean stellar velocity dispersion.

Recently, Quinlan (1996) and Quinlan & Hernquist (1997) have studied processes (1) and (2) by detailed simulations. They developed an N-body code in which the BH-BH interaction is computed exactly (Newtonianly; relativistic effects are unimportant before stage 3), the BH-star interactions by a softened Keplerian force, and the star-star interactions via an expansion of the stellar potential in basis functions, *i.e.* essentially collisionless. Their code thus follows the dynamical evolution of the BHs and stars in a self-consistent way. In their simulations, Quinlan & Hernquist started with the stars being in spherical isotropic equilibrium following a Jaffe model (which has $\gamma = 2$) and various choices for the masses and initial positions of the BHs. Their general conclusions relevant here are as follows. (i) The total mass of the stars ejected from the centre by three-body interactions is about twice the mass of the BH binary. (ii) Wandering of the BH pair significantly increases the loss cone and mitigates the problem of its depletion. (iii) Inside about r_h, the density profile has become much shallower, almost flat, the velocity dispersion increases in a Keplerian fashion, and the stellar motions are significantly tangentially anisotropic with $\beta \approx -1$ (see Fig. 4).

Thus, at least qualitatively, this process can explain the presence of a shallow stellar cusp with significant tangential anisotropy around a central black hole. However, there are some points where M 87 does not fit smoothly into this picture. For example, the radius inside which the density of M 87 becomes shallow ($\approx 10''$) is clearly larger than the radius inside which $\beta < 0$ ($\approx 3''$), while the simulations indicate that these radii should be similar. Also, the

---

[3] One BH-merger has to be finished before a third BH arrives, since more than two BHs cannot co-exist for long in a galaxy's centre, as all but two of them will quickly be ejected by sling shots.

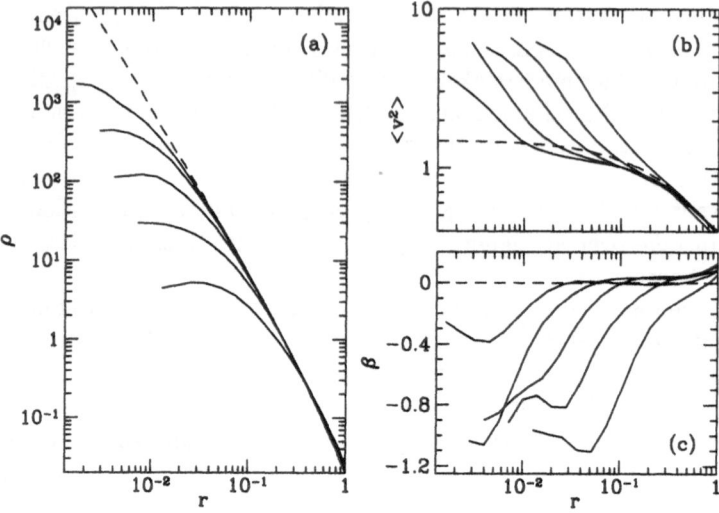

Fig. 4. Response of a Jaffe model to the hardening of a BH binary as computed by Quinlan & Hernquist (1997). The dashed lines show the initial (a) density, (b) velocity dispersion, and (c) anisotropy. The solid lines show these quantities when the binary reaches a certain hardness. The binary's components have equal mass of 0.04, 0.02, 0.01, 0.005, and 0.0025 (leftmost) in units of the total mass in stars. Reprint from New Astronomy, Vol. 2, Quinlan & Hernquist, "The dynamical evolution of massive black hole binaries - II. Self-consistent N-body integrations", p. 533-554, 1997 with kind permission of Elsvier Science - NL, Sara Burgerhartstraat 25, 1055 KV Amsterdam, The Netherlands.

central cusp of M 87 is not completely dug out as in Quinlan & Hernquist's simulations. These discrepancies indicate that reality is more complicated, possibly involving several such accretion events with "small" secondary BHs and/or dissipational processes (*e.g.* star formation) as indicated by the presence of the central gas-disk.

5   Influence of a Central Black Hole on Larger Scales

The processes discussed in the last section influence the structure and dynamics of the stellar system only in the immediate neighbourhood of the BH, *i.e.* inside r_h. A central BH, however, will influence all those stars that ever come near the centre. Many stars in triaxial galaxies are on box orbits, which pass arbitrarily close to the centre after a sufficiently long time. It was argued by Gerhard & Binney (1985) that scattering and re-distribution of box orbits by a central BH would cause at least the inner parts of a triaxial galaxy to become rounder or more axisymmetric.

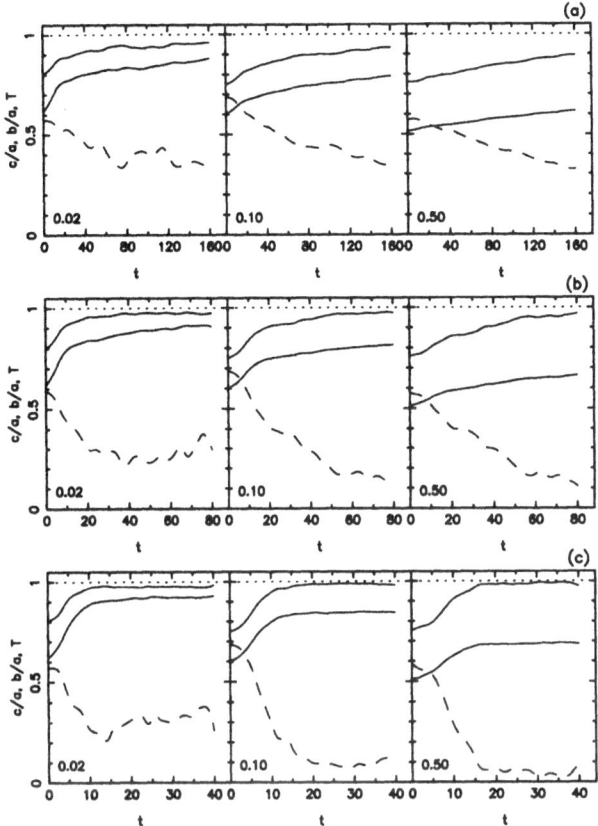

Fig. 5. Time-evolution of the axis ratios b/a and c/a (full lines) and the triaxiality index $T = (a^2 - b^2)/(a^2 - c^2)$ (dashed) in the simulations of Merritt & Quinlan (1998). The mass $M_\bullet$ of the BH grown in $t_{\mathrm{grow}} = 15$ is 0.003 (a), 0.01 (b), and 0.03 (c) of the total mass in stars. The numbers in the left corner of each frame are the fraction of particles, ranked by binding energy, that were used in computing the shape parameters.

Using a modification of the N-body code employed by Quinlan & Hernquist, Merritt & Quinlan (1998) have recently studied this process numerically. They created a stable triaxial equilibrium model by the simulated collapse of a non-equilibrium configuration, and followed the time-evolution of this model when at its centre a point mass was slowly grown. Fig. 5 shows, for three different BH masses, the time-evolution of the shape of the model. In all simulations, the galaxy model tends to become axisymmetric, even at the half mass radius. For the lightest BH with $M_\bullet/M_g = 0.003$ (M_g denotes the mass of the galaxy), this process is still ongoing at the end of the simulation, whereas for $M_\bullet/M_g = 0.01$ it is nearly finished. The most interesting result, however, is that for $M_\bullet/M_g = 0.03$ the process is almost finished at the time

t_{grow} when the BH has reached its final mass. Further simulation of the authors with varying t_{grow} confirmed this result: for $M_\bullet/M_g = 0.03$ the shape becomes axisymmetric as soon as possible, *i.e.* at $t = \max\{t_{grow}, t_{dyn}(r)\}$, where $t_{dyn}(r)$ denotes the dynamical or crossing time at radius r.

This behaviour can be explained by the BH inducing chaos in the orbital motions of the triaxial galaxy. For small $M_\bullet$, the orbits become weakly chaotic, *i.e.* they behave like a box orbit over some time, but after a sufficiently long time they fill their energy surfaces. For ever larger $M_\bullet$, the orbits become ever more strongly chaotic until at some critical $M_\bullet$ the Liapunov time (the time in which two neighbouring trajectories diverge) equals the dynamical time, *i.e.* the orbits no longer resemble box orbits at all. In the simulations of Merritt & Quinlan this critical $M_\bullet$ is about 2.5% of M_g. In general, this number should depend on details of the triaxial configuration, but is likely to be of the same order, *i.e.* $\sim 10^{-2} M_g$.

From the existence of this critical $M_\bullet$, Merritt & Quinlan draw a very interesting conclusion. In order for BHs to grow by gas accretion (the standard model for AGN), the gas has to reach the BH from large radii, *i.e.* it has to lose its angular momentum. In axisymmetric galaxies, the conservation of angular momentum along ballistic orbits renders gas-fueling of the centre very difficult. Thus, a BH may cut off its own gas supply by changing the shape of its host galaxy. If this picture actually applies, the BH mass should be no larger than the critical $M_\bullet \sim 10^{-2} M_g$. A correlation in this sense is indeed observed: $M_\bullet$ inferred from the dynamics of several early-type galaxies is always of this order (cf. Kormendy & Richstone 1995). In M 87, $M_\bullet/M_g$ is only $\sim 0.5\%$, which might be a consequence of mergers that convert disks into spheroids and hence increasing M_g (Merritt, private communications).

Several BHs, however, are hosted by barred spiral galaxies (*e.g.* the Galaxy, NGC 1068). Tumbling bars are mainly made of stars on so-called x_1 orbits, which avoid the very centre. Hence, the mechanism working on box orbits for triaxial bulges may not (or not as well) work for barred spirals.

6 Summary

The kinematics of M 87 are well studied, which make this galaxy a good test case for the theories of galaxy formation. Outside $\sim 100''$, the velocity dispersion profile rises indicating the presence of large amounts of non-luminous matter. The inferred density profile $\rho \propto r^{-1.3}$ is consistent with predictions from CDM cosmogony for the inner parts of dark-matter halos.

The massive black hole (BH), detected in the very centre of M 87 by gas motions, together with the observed stellar kinematics implies a significant tangential anisotropy of the stellar motions. Among the formation histories discussed for a BH in a galactic centre, only the model of accretion of other massive BHs, originating from the centres of cannibalized companions, can explain such a strong anisotropy. This scenario also predicts a shallow stellar density cusp as observed for M 87. (Quantitatively, there are some discrepancies, which may well be due to over-simplification in the simulations of this process.)

A massive BH at the centre of a triaxial galaxy renders, by the destruction of box orbits, the shape of its host axisymmetric. This mechanism becomes very fast once the BH mass reaches a critical value, which is of the order of 1% of its host's mass. Since the conservation of angular momentum along ballistic orbits in an axisymmetric galaxy obstructs gas-fueling of the centre, this process may pose an upper limit for the mass a BH can reach by gas-accretion. An upper limit of this order is indeed observed among BH masses inferred from the dynamics of early-type galaxies.

Acknowledgements

I am grateful to the organizers for inviting me to this wonderful workshop. Special thanks to David Merritt, who helped improving on an early version and made Figs. 3 and 5 available in electronic format.

References

Arnaboldi M., Freeman K., 1996, in Arnaboldi, Da Costa, Saha, eds., *The Nature of Elliptical Galaxies* A.S.P. Conf.Ser. 116, 54

Begelman, M.C., Blandford, R.D., Rees, M.J., 1980, Nature, 287, 307

Bender, R., Saglia, R.P., Gerhard, O.E., 1994, MNRAS, 269, 785

Binney, J.J., Mamon, G.A., 1982, MNRAS, 200, 361

Cohen, J.G., Ryzhov, A., 1997, ApJ, 486, 230

Ebisuzaki, T., Makino, J., Okumura, S.K., Nature, 354, 212

Fabricant, D., Gorenstein, P., 1983, ApJ, 267, 535

Ford, H.C., et al., 1994, ApJ, 435, L27

Gebhardt, K., et al., 1996, AJ, 112, 105

Gerhard, O.E., Binney, J.J., 1985, MNRAS, 216, 467

Goodman, J., Binney, J.J., 1984, MNRAS, 207, 511

Harm, R.J., et al., 1994, ApJ, 435, L35

Kormendy, J., Richstone, D., 1995, ARA&A, 33, 581

Lauer, T.R. et al., 1992, AJ, 103, 703

Marconi, A., et al., 1997, MNRAS, 289, L21

Merritt, D., 1987, ApJ, 319, 55

Merritt, D., Oh, S-P., 1997, AJ, 113, 1279

Merritt, D., Quinlan, G.D., 1998, ApJ, in press

Moore, B., et al., 1997, submitted to ApJ Letters (astro-ph/9709051)

Mould, J.R. et al., 1990, AJ, 99, 1823

Nulsen, P.E.J., Böhringer, H., 1995, MNRAS, 274, 1093

Quinlan, G.D., 1996, NewA, 1, 35

Quinlan, G.D., et al., 1995, ApJ, 440, 554

Quinlan, G.D., Hernquist L., 1997, NewA, 2, 533

Sargent, W.L.W. et al., 1978, ApJ, 221, 731

van der Marel, R.P., 1994, MNRAS, 270, 271
Young, P., 1980, ApJ, 242, 1232

M 87 as a Younger Progenitor Galaxy in the Virgo Cluster

Halton Arp

Max-Planck-Institute für Astrophysik
D85740 Garching
Germany

Abstract. The structure of the Virgo cluster with the brighter, redder galaxy M 49 at its center argues that the rest of the cluster, including M 87, originated from M 49 and is younger. M 87 (Vir A), like most other bright radio galaxies, *e.g.* Cen A, Per A, For A, shows current ejection activity as well as conspicuous, lines of galaxies originating from its center. It is argued that M 87 is showing second generation ejection of objects which are evolving into younger galaxies.

Observations show that in general quasars are ejected along the minor axes of active galaxies and then evolve into alignments of low redshift, companion galaxies. In M 87, it is argued that the knots in the jet are decelerating outward, evolving into quasars, BL Lac objects and finally lower redshift, aligned companions. If this is true the knots must consist of a low-particle mass plasma and the physics of the jet would have to be recalculated with this new assumption.

1 Structure of Virgo Cluster

The most informative picture of the Virgo cluster is a sample plot of the bright galaxies over the Virgo region. The galaxies are distributed in a rough spiral with the brightest E galaxy, M 49 at the center (Arp 1967). In Fig. 1 we show a plot by Kotyani (1981) of the spiral galaxies and next to it a plot of the strongest radio sources. The "S" shape of the cluster is very conspicuous in these younger, active galaxies and suggests that they were ejected from the central M 49 with a slight rotation of the ejection axis during the process.

The fact that M 49 is slightly brighter and redder than M 87 indicates it is more massive and hence the dominant galaxy in the cluster. M 87 (3C 274) and 3C 273 are then indicated to be currently more active, second generation galaxies. Both have 20″ jets and are aligned exactly across M 49 (Arp and Burbidge 1990). M 49 has also the lowest redshift ($cz_o = 822$ km/sec) which, as will be discussed later, is evidence of it being the original parent galaxy.

2 M 87 as the Origin of Secondary Galaxy Ejection

Figure 2 shows the startling fact that essentially all the E galaxies in the vicinity of M 87 are aligned through it *and exactly along the line of the famous blue jet*. This clearly is not an accident and in any case is supported by a

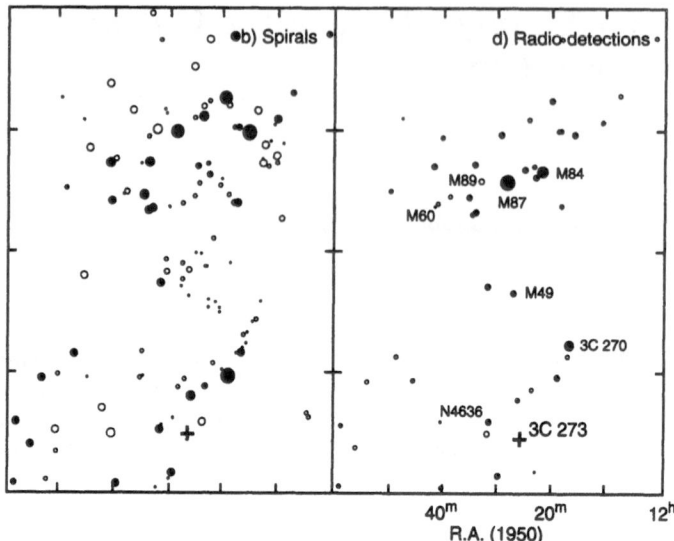

Fig. 1. b) The distribution of galaxies with young stars (spiral) and d) energized plasma (radio galaxies). To these plots by Kotyani (1985) the quasar 3C 273 has been added to indicate that it is probably a primary ejecta from the largest, oldest galaxy, in the center of the cluster, M 49.

study of the 14 brightest radio emitters in the sky which showed 13 to have lines of galaxies associated with them. Where radio jets were conspicuous, they were aligned with the galaxies (Arp 1968).

In Fig. 2c all the bright galaxies around Cen A are plotted and it is seen they form an alignment coincident with the outer lobes of the radio source. The inner radio and X-ray jets of Cen A are younger and have rotated some degrees counterclockwise. As in M 87 there is some "S" shape to the outer radio lobes and there generally may be some precession within a narrow cone angle for these kinds of ejections.

In Fig. 1b & d, it is seen there is a hollow oval of spiral galaxies surrounding the line of E galaxies through M 87. Since spirals contain generally younger stars and have generally higher redshifts, these could only be third generation galaxies arising from ejection in various directions from the second generation line of E's (Arp 1987, p 141).

In this respect it is interesting to note the bright X-ray sources in Fig. 2b which lie along the line of the jet. The X-ray source in the NW direction of the jet is identified as the quasar PG 1211+143. But the quasar definition is arbitrary, based on a luminosity derived from an assumed redshift distance. Actually PG 1211+143 is low redshift, $z = .085$, and more nearly resembles an active compact galaxy (N galaxy) or a BL Lac object (Arp 1995). It resembles very closely Mark 205 at $z = .070$ (Arp 1996). PG 1211+143 also

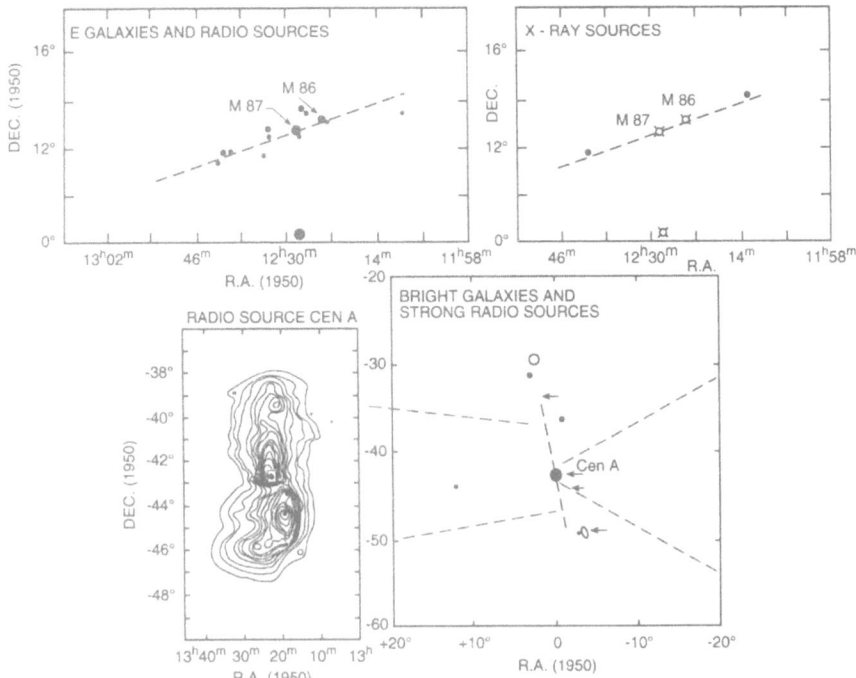

Fig. 2. a) shows the line of E galaxies along the direction of the jet in M 87 (Arp, 1968; 1987). b) the same line of the M 87 jet is outlined by the strongest X-ray sources in the Virgo Cluster. c) shows a similar line of younger galaxies along the line of the jet in Centaurus A.

shows ejection of radio material and, along the same line, a pair of quasars of $z = 1.01$ and 1.28.

3 Quasars Ejected from Active Galaxies

For 30 years evidence has accumulated that active galaxies eject quasars in opposite directions (Arp 1987; 1997a). These results are shown in schematic form in Fig. 3. Most recently this empirical evolutionary schema has been confirmed in detail by a single active Seyfert galaxy, NGC 3516 (Chu *et al.* 1997). This is shown in Fig. 4. X-ray sources in general, including those coming out along the minor axis of NGC 3516 had been previously shown to be physically associated with bright Seyfert galaxies (Radecke 1997). The important point about this evidence is that it shows empirically that quasars evolve into low redshift companion galaxies. The clinching evidence then comes from the co-alignment of companion galaxies and quasars.

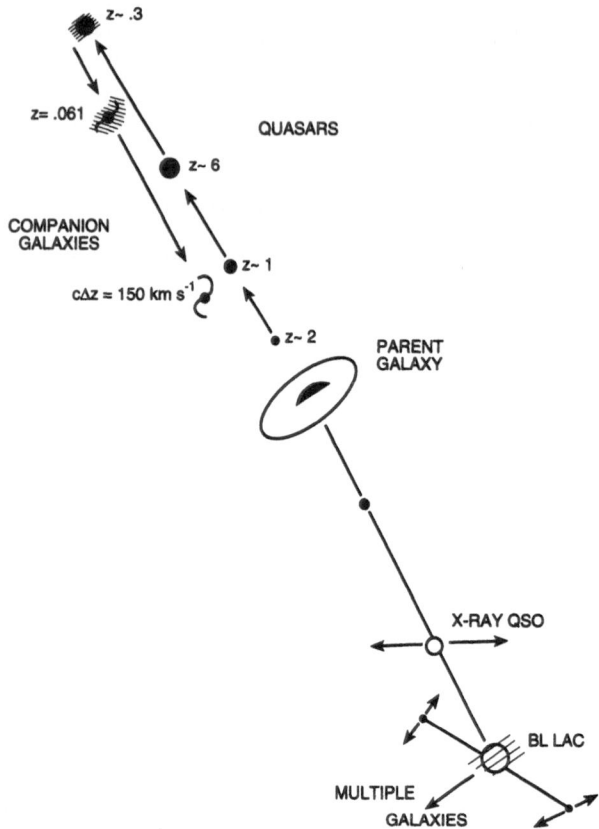

Fig. 3. The evidence from about 30 years study of associations with active galaxies has been summarized here. It appears to represent an empirical sequence of evolution from quasars to normal companion galaxies.

4   Companion Galaxies and Quasars Aligned Along the Minor Axes of Disk Galaxies

Erik Holmberg (1969) showed companion galaxies were preferentially aligned along the minor axes of disk galaxies and concluded: "...the satellites have been formed from gas ejected from the central galaxies." The latest confirmation of the alignment of companion galaxies along the minor axis comes from Zaritsky *et al.* (1997). But as Figs. 3 and 4 show, *quasars are also aligned along the minor axis of disk galaxies.* The most important evidence presented so far is their Fig. 5 which shows that the quasars and the companion galaxies *are distributed in the same volume of space* along the minor axes. Along the minor axis the ejecta would have only small amounts of angular momentum and hence move on plunging orbits. Since the galaxies are older they can

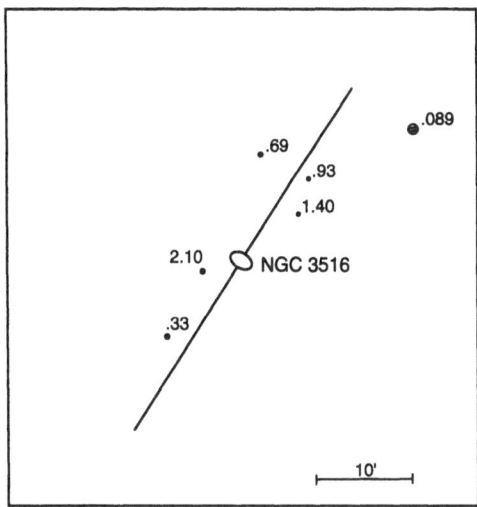

Fig. 4. Recent observations by Yaoquan Chu *et al.* (1997) show that the brightest X-ray sources in the vicinity of the very active Seyfert galaxy NGC 3516 are being ejected out along the minor axis and decrease in redshift as they age.

have drifted somewhat further from the minor axis (to ±35 from the ±20 degrees of the quasars) or can represent some minor axis procession with time. Note that recent HST exposures of the M 87 nucleus (shown in the present workshop) reveal an inner disk which has a minor axis closely along the jet direction.

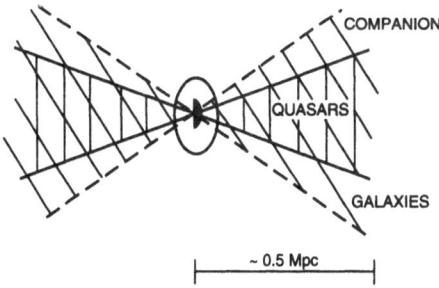

Fig. 5. Schematic representation of companion objects along minor axes of disk galaxies (±35 degrees from Holmberg 1969; Sulentic *et al.* 1978; Zaritsky *et al.* 1997). Quasars are observed ±20 degrees from minor axis (Arp 1997b; 1998)

The unavoidable conclusion is that quasars evolve into companion galaxies. This is in a way conventionally believed, but here the evidence requires the high redshift quasars to be at distances like the Virgo Cluster and to decrease their redshifts and increase their luminosities as they evolve.

5   M 87 as a Progenitor Galaxy

The evidence summarized in Figs. 3, 4 and 5 show that the alignment of companion galaxies along the direction of the jet in M 87 must have evolved from ejections of quasars along the jet. Is there supporting evidence for this? There is in the sense that the only material objects which we see coming out along the line of the jet are the blue synchrotron knots. The essential properties are:

5.1   Knots in the Jet as Proto BL Lac Objects

a) The knots have strong radio, X-ray, high polarization, continuous spectra.
b) Only BL Lac objects and knots in jets share these properties.
c) In 1979 Sulentic *et al.* found BL Lac's associated $\sim 1°$ from bright galaxies.
d) A very conspicuous excess of BL Lac's has now been found around bright Seyferts (Arp 1997a).
e) The only empirical sequence that can be inferred is that BL Lac activity represents periodic outbursts of energy in material which is evolving from proto galactic knots through quasars to galaxies as the knots travel outward.

5.2   Knots in the Jet as Superfluid

The absolutely key point is that in order to explain the high redshifts of the younger ejected material one must invoke particle masses in the ejected matter which start out near zero and grow with time (Narlikar and Arp 1993). Low mass electrons transitioning in an atom emit and absorb redshifted photons. The initially created plasma will consist of low mass particles with large cross sections — a definition of a fluid. This reminds us that Ambarzumian in 1958 described photographs of galaxies forming in ejected jots as condensing from a "superfluid".

Another key point is that the low mass particles in order to conserve momentum must slow both their translational and random motions as they age and gain mass. This means that a low particle mass plasma has to cool and become more massive, leading in the direction of a self gravitating body and star formation. Of course, the plasmoid is radiating continuum radiation in the beginning before it forms atoms. But even after it has formed some stars it is liable to intermittent, secondary out bursts which give the typical BL Lac and energetic galaxy nuclei continuum spectra.

Perhaps the most telling point is that at the observed ejection velocities of 5 to 6 c in the M 87 jet discussed in this workshop (deprojecting to bulk velocities very close to c) there would have to be an extraordinary amount of energy pumped into the knots in order to achieve this acceleration. Even if this could be achieved they would then blow apart and be unable to condense into the companion galaxies which are ultimately observed. On the other hand particles with near zero mass are almost completely energy and naturally travel with signal velocity, c, and thus emerge with the observed properties of the knots. They then can then decelerate, cool and condense.

6   Summary

Because of the large intrinsic redshifts which are observed during the evolution of small, energetic objects into near normal redshift companion galaxies, it would seem necessary to reconsider the assumptions underlying the conventional computations on the nature of the plasma knots in the jet in M 87.

References

Ambarzumian, V.A. (1958): *Onzieme Conseil de Physique Solvay* (ed. R. Stoops, Bruxelles)

Arp, H. (1967): Ap. J. **148**, 321

Arp, H. (1968): P. A. S. P. **80**, 129

Arp, H. and Sulentic, J. W. (1979): Ap.J. **233**, 44

Arp, H. (1987): *Quasars, Redshifts and Controversies* (Interstellar Media, Berkeley)

Arp, H. (1995): *A&A* **296**, L5

Arp, H. (1996): *A&A* **316**, 57.

Arp, H. and Burbidge, G. (1990): Ap.J. **353**, L1

Arp, H. (1997a): *A&A* **319**, 33

Arp, H. (1997b): IAU Symposium 183, Kyoto Aug. 1997, ed. K. Sato (Kluwer, Dortrecht).

Arp, H. (1998): Ap.J. **496**, 661.

Kotyani, C. (1985): *ESO Workshop on the Virgo Cluster of Galaxies* (ed. O.G. Richter and B. Binggeli), p 13

Chu, Y., Wei, J. Hu, J., Zhu, X. and Arp, H. (1997): Ap.J. **500**, 596.

Holmberg, E. (1969): Arkiv of Astron., Band 5, 305.

Narlikar, J.V. and Arp, H. (1993), Ap.J. **405**, 51.

Radecke, H.-D. (1997): *A&A* 319, 18.

Sulentic, J.W., Arp, H. and Lorre, J. (1979): Ap.J. **233**, 44

Zaritsky, D., Smith, R., Frenk, C.S. and White, S.D.M. (1997): Ap.J. **478**, L53.

The Surface Brightness Fluctuations and Globular Cluster Populations of M 87 and Its Companions

Eric H. Neilsen Jr., Zlatan I. Tsvetanov and Holland C. Ford

Johns Hopkins University
Baltimore, MD
USA

Abstract. Using the surface brightness fluctuations in HST WFPC-2 images, we determine that M 87, NGC 4486B, and NGC 4478 are all at a distance of $\sim$ 16 Mpc, while NGC 4476 lies in the background at $\sim$ 21 Mpc. We also examine the globular clusters of M 87 using archived HST fields. We detect the bimodal color distribution, and find that the amplitude of the red peak relative to the blue peak is greatest near the center. This feature is in good agreement with the merger model of elliptical galaxy formation, where some of the clusters originated in progenitor galaxies while other formed during mergers.

1 Introduction

An accurate estimate of the distance to M 87 is important for the study of its other properties, in particular for determining the correct physical scales and luminosities from measured angular separations and apparent magnitudes. As the brightest, central galaxy in its portion of the Virgo cluster, M 87 has a rich environment, including several companions and exceptionally large population of globular clusters (GC). By measuring accurate distances we can separate true neighbours from close projections, while mapping global properties of the GC population can give important clues for the merger history of M 87 and the Virgo cluster in general.

Because of its smooth morphology, the surface brightness fluctuation (SBF) method of distance determination is ideally suited for application to M 87. The method works by measuring the statistical effects of the galaxy being made up of a large number individual stars.

The basis for the method is clearest if one considers the idealized case of a galaxy with uniform surface brightness, whose stars all have the same luminosity, l. In an image of this galaxy, each pixel will contain the light from some number of stars, $n(x, y)$. The number of stars per pixel will have an average $\bar{n}$, and a standard deviation $\sqrt{\bar{n}}$. By examining the image, one can measure the mean flux in a pixel, $\bar{f} = \bar{n}l$, and its standard deviation, $\sigma_f = \sqrt{\bar{n}}l$. Finally, one can use these to calculate the flux of a star in the galaxy: $l = \sigma_f^2/\bar{f}$. If we can estimate the absolute luminosity from this star, it may be used as a "standard candle," and the distance can be measured.

In practice, there are a number of complications. The morphology of the galaxy is not uniform, the variance due to stellar statistics must be separated from variance due to noise and contaminating objects, and galaxies are made up of stars with a variety of luminosities. All of these difficulties can be overcome to a large extent. A complete description of the process of SBF measurement can be found in the original paper by Tonry & Schneider (1988), or in Neilsen, Tsvetanov, & Ford (1997).

A second method of distance determination uses the globular cluster luminosity function (GCLF). Assuming that the luminosity function of globular clusters can be estimated, one may use it as a standard candle. It is particularly convenient to measure the GCLF on data for which one has already measured SBF, because it is necessary to locate the globular clusters in the image to obtain an SBF measurement. The creation of a catalog of globular clusters also allows for the study of the spatial and color distributions of the globular cluster population.

2   The Data

The Hubble Space Telescope archive provided all of the images used in this study. Figure 1 shows the placement of the fields around M 87. For each

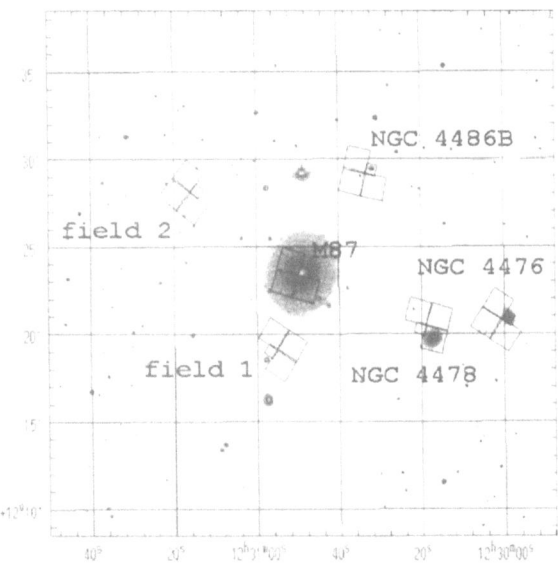

Fig. 1. The placement of the fields around M 87. The image on which our fields are shown was created using an image from the GASP plates and an image taken by Hintzen *et al.* (1993), supplied courtesy the NCSA Astronomy Digital Image Library.

field, we used HST WFPC-2 images with a filter approximating the I band (F814W) and a filter approximating the V band (either F555W or F606W). Table 1 presents, for each field, the projected separation from the center of M 87 to the center of the field, the filters used, the total exposure times, and the number of exposures in each filter.

Table 1. The data used in this study.

Field	Sep. (arcmin)	Bands	Filters	t_{exp} (sec)	# exp.
M 87 center	0.4	V, I	F555W, F814W	2400, 2400	4, 4
field 1	4.5	V, I	F606W, F814W	3380, 4200	5, 4
NGC 4486B	6.7	V, I	F555W, F814W	1800, 2000	3, 4
field 2	8.2	V, I	F606W, F814W	1800, 13400	3, 5
NGC 4478	8.2	V, I	F606W, F814W	16800, 16500	6, 6
NGC 4476	12.0	V, I	F555W, F814W	2400, 2400	4, 4

3   The Distances

In table 2, we present the distances determined through our SBF and GCLF measurements. We also report the distance as determined by the planetary nebula luminosity function, as given by Ciardullo, Jacoby, & Tonry (1993).

Table 2. The distances to M 87 and its companions in Mpc, by three methods.

Method	M 87	NGC 4486B	NGC 4478	NGC 4476
SBF	15.8 ± 1.0	16.2 ± 1.0	15.1 ± 1.0	21.1 ± 1.1
GCLF	14.2 ± 1.7		16.4 ± 3.7	19.2 ± 3.4
PNLF	14.9 ± 0.7			

To calculate the SBF distances, we measured the mean color in the utilized region of the galaxy using the two filters and converted to $V - I$ using the calibration of Holtzman et al. (1995), and the F814W SBF calibration of Ajhar et al. (1997). To calculate the GCLF distances, we estimate the peak of the GCLF using Ashman, Conti, & Zepf (1995), whose theoretical results agree well with the observational calibration of Whitmore (1996). The distances measured by the different methods agree well. From these distances, it appears that NGC 4486B and NGC 4478 are genuine companions of M 87, while NGC 4476 lies in the background.

4   The Globular Cluster System of M 87

In four of our fields, (the center of M 87, field 1, the NGC 4486B field, and field 2) we expect our globular cluster candidate catalog to be dominated by clusters from M 87; this allows us to study the color distribution in several locations. It has been known for some time that the cluster population becomes bluer with distance from the center (see Strom *et al.* 1981), and a variety of explanations have been proposed. Strom et al. (1981) suggests that the clusters all formed at the same time, probably preceding the formation of the galaxy itself, and that the gradient in colors has the same origin as the gradient in the color of the integrated halo. In contrast, Ashman & Zepf (1992) claim that M 87 was formed as the result of mergers, which play a critical role in the formation of the globular cluster population. In this model, during mergers the red clusters are formed in the center of M 87 from the gases in the merging galaxies, while the blue clusters, originally associated with the progenitors, form the more extended cluster halo.

Figure 2 presents the color distributions in each field. In every field, there is a peak near $V - I = 0.95$. The center field clearly shows a distinct peak near $V - I = 1.2$ as well. There are traces of this second peak in field 1 and in the NGC 4486B field as well, but with smaller amplitude relative to the blue peak. It appears that there are two populations of globular clusters, the bluer of the two populations having a larger spatial extent.

A Kolmogorov-Smirnov (K-S) test on the 3 fields with reasonably good statistics rejects with a confidence of better then 97% that the data in each field arise from the same parent population, even if the different fields are shifted to have the same mean color. The data are consistent, however, with clusters in each of the fields arising from a double Gaussian distribution, supporting the a model where the central, red clusters are formed during mergers.

5   Conclusions

The surface brightness fluctuations and globular cluster luminosity functions allow us to measure the distances to M 87 and several of its apparent neighbours, and determine which were true companions. Our globular cluster catalogs provide valuable information on the variation of the color distribution of globular clusters with the distance from the center. However, our statistics were insufficient to provide a more detailed understanding. In particular:

- M 87, NGC 4486B, and NGC 4478 appear to be true companions, all about 16 Mpc away. NGC 4476, on the other hand, is in the background, at approximately 21 Mpc.
- The K-S test on the globular cluster color distributions indicates that the samples detected at the different distances are unlikely to have arisen from the same population. We therefore conclude that the color distribution of the globular cluster population in M 87 varies with distance.

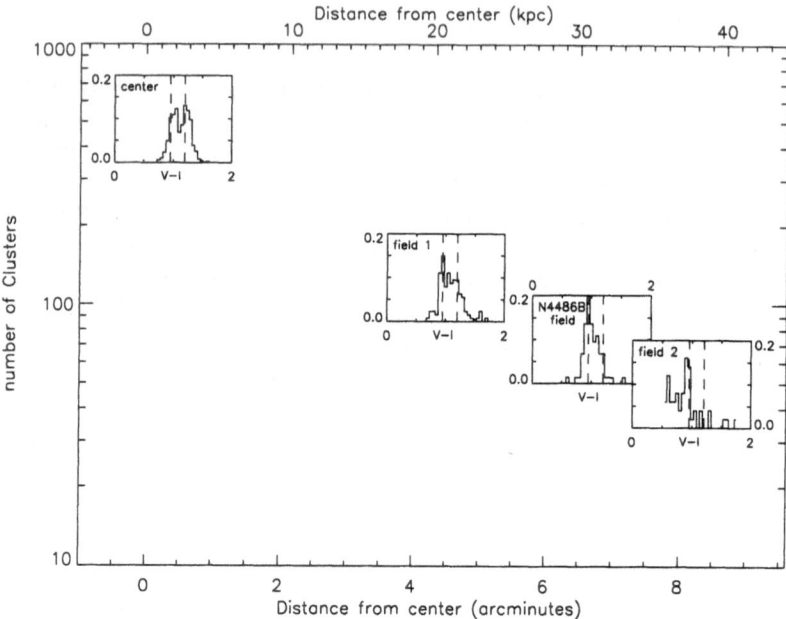

Fig. 2. The color distribution of M 87's globular cluster in several fields. The place-ment of each plot represents the distance from the center of the field to the center of M 87, and the total number of clusters with good $V - I$ colors data. The dashed lines are placed at $V - I = 0.95$ and $V - I = 1.20$.

- Even when the cluster populations are artificially shifted so that they have the same mean color, the K-S test still indicates that the different samples are unlikely to have arisen from the same population. This casts doubt on models where all clusters have the same origin.
- The color distribution in the central field is significantly better fit by two Gaussian distributions than one. Double peaked distributions with the peak widths and the difference in peak colors constrained to match the best fit double peaks of the center fit the other fields well. This result is consistent with current merger models.

References

Ajhar, E. A., Lauer, Tod R., Tonry, J. L., Blakeslee, J. P., Dressler, A., Holtzman, J. A., & Postman, M., 1997, AJ, 114, 626.

Ashman, K. M., Conti, A., & Zepf, S. E., 1995, AJ, 110, 1164.

Ashman, K. M., & Zepf, S. E., 1992, ApJ, 384, 50.

Ciardullo, R., Jacoby, G. H., & Tonry, J. L., 1993, ApJ, 419, 479.

Hintzen, P., Angione, R., Talbert, F., Cheng, K. P., Smith, E., & Strecher, T. P., 1993, in NASA Ames Research Center, The Evolution of Glaxies and Their Environment.

Holtzman, J. A., Burrows, C. J., Casertano, C., Hester, J. J., Trauger, J. T., Watson, A. M., & Worthey, G., 1995, PASP, 107, 1065.

NCSA Astronomy Digital Image Library, http://imagelib.ncsa.uiuc.edu/imagelib

Neilsen, E. H., Jr., Tsvetanov, Z. I., & Ford, H. C., 1997, ApJ, 483, 745.

Strom, S. E., Forte, J. C., Harris, W. E, Strom, K. M., Wells, D. C., & Smith, M. G., 1981, AJ, 245, 416.

Tonry, J. L., & Schneider, D. P., 1988, AJ, 96, 807.

Whitmore, B. C., in The Extragalactic Distance Scale, ed. M. Livio, M. Donahue, & N. Panagiua (Cambridge: Cambridge Univ. Press).

The Large-Scale Structure of Virgo A

Uli Klein

Radioastronomisches Institut der Universität Bonn
Auf dem Hügel 71
D 53121 Bonn
Germany

Abstract. The large-scale structure of the nearby radio galaxy Vir A is reviewed. It is argued that epochs of varying activity are also manifest on large scales. The radio sources exhibits three basic components: the inner bright lobes (projected size ~5 kpc), the intermediate lobes (~30 kpc), and the outer diffuse lobes (~80 kpc). It seems difficult to reconcile these features, which have rather different brightnesses, energy densities and spectra, with a jet that had constant thrust and orientation over the past 10^8 yrs.

1 Introduction

Many active galactic nuclei (AGN) that exhibit jets on pc scales reveal some kind of continuation of this transport of matter and momentum on larger scales (a few to several 100 kpc), and the radio structures frequently indicate a good memory of the sources' orientations for at least a couple of 10^6 yrs. Owing to its proximity Vir A is one of the best studied radio galaxies. Its jet, which has been studied from radio waves to X-rays (see *e.g.* Meisenheimer *et al.* 1996), is often referred to as a textbook template to study jet physics and AGN properties.

Bolton *et al.* (1949) reported the first tentative identification of a radio source with an extragalactic object, namely the radio source in the constellation of Virgo with the elliptical galaxy M 87. Mills (1953), using his transportable radio link interferometer at 101 MHz, was the first to make any assessment of the extent of the radio source (4ʹ6) associated with M 87. Biraud *et al.* (1960) first mentioned the core-halo structure; the halo was subsequently studied by Maltby and Moffet (1962), and a first estimate of the age of the radio source was made by Lequeux (1962) who employed the spectral steepening of the radio halo to derive $\tau \approx 10^7$ yrs. The first notion that the radio halo of Vir A is polarized came from Morris *et al.* (1964). This was followed by a more detailed study of the linear polarization by Seielstad and Weiler (1969) who already conjectured that the observed depolarization must occur within the radio source itself. In the meantime, M 87 had also become the first X-ray source identified with an extragalactic object (Byram *et al.* 1966). It was especially the impact by improved observations with interferometers (Graham 1970; Cameron 1971; Turland 1975) which facilitated a

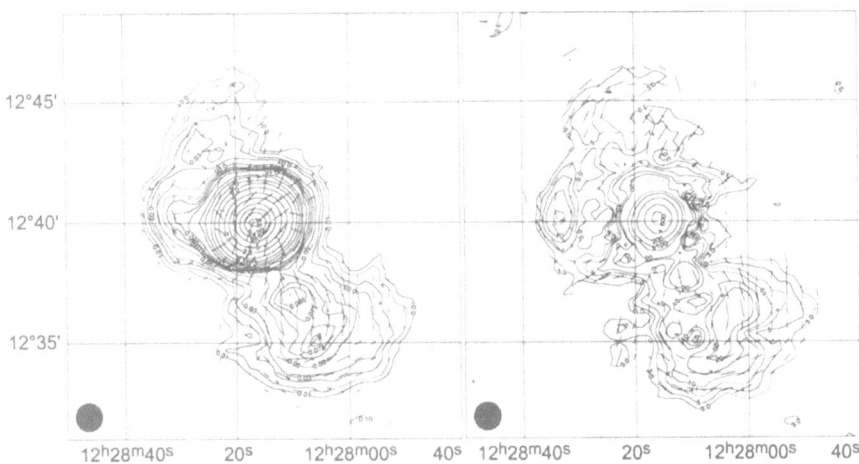

Fig. 1. Maps of Vir A at 10.6 GHz (from Rottmann *et al.* 1996). Left: contours of total intensity (in Jy/b.a.) are superimposed by vectors of linear polarization, rotated by 90° to mimic the structure of the magnetic field, their lengths being proportional to polarized intensity. Right: contours of polarized intensity (in mJy/b.a.) are superimposed by vectors of fractional polarization.

better decomposition of the central radio source and the halo, and such studies led Turland (1975) to conclude that the radio halo of Vir A is "probably a relic of earlier activity". The first single-dish study of the halo of Vir A was presented by Andernach *et al.* (1979) who analyzed its radio spectrum and linear polarization, emphasizing a conspicuous north-south asymmetry. They in particular reported rather extreme values for the radio spectrum between 4.8 and 10.7 GHz ($\alpha = 3 \cdots 4$, $S_\nu \sim \nu^{-\alpha}$). Kotanyi (1980) pointed out the morphological difference between Cen A and Vir A, which he attributed to the presence of the confining medium around the latter.

Observations with higher resolution which were subsequently carried out with the VLA (Feigelson *et al.* 1987; Kassim *et al.* 1993; Böhringer *et al.* 1995) clearly demonstrate that the so-called radio halo of Vir A is in fact made up by large lobes. The radio spectrum and linear polarization of these lobes were recently revised by Rottmann *et al.* (1996). They found a less extreme steepening at high radio frequencies than Andernach *et al.* (1979), and about equally polarized lobes at 10.6 GHz (see Fig. 1). This strongly suggests that the depolarization asymmetry observed at longer wavelengths is an orientation ("Laing-Garrington") effect (Laing 1988; Garrington *et al.* 1991).

This brief outline of the large-scale radio continuum structure of Vir A shows that we are not dealing here with a standard source such as Cyg A. It is quite obvious that the structure of Vir A is most likely influenced by the environment, and it is well known that the interstellar medium in the host galaxy, M 87, is relatively dense (*e.g.* Sparks *et al.* 1993; Böhringer *et al.*

1993). The caveat then is that Vir A may not be typical at all, being located at the centre of a galaxy cluster.

2   The Morphological Zoo of Radio Galaxies

The morphological appearance and radio luminosity of extragalactic radio sources is manifold and certainly depends on the intrinsic power of the AGN feeding the extended lobes, but also on the environment (ISM or IGM). By virtue of this, the classification schemes which have been established are not simply morphological, but also bear importance to the main physical properties of the radio sources. Before we might try and classify Vir A as a radio source, the morphological variety of extragalactic radio sources shall be briefly summarized here.

2.1   The FR I/FR II Dichotomy

Most radio galaxies can be subdivided into two classes distinguished by their morphology and luminosity. They are termed FR I and FR II, which is due to the work of Fanaroff and Riley (1974) who were the first to realize the basic categories of powerful edge-brightened and less powerful edge-darkened extragalactic radio sources.

Radio sources of type FR I have monochromatic radio luminosities $P <$ 10^{25} W Hz^{-1} at 178 MHz. They are edge-darkened, in the sense that the ratio of the separation of their brightest components to their total size is < 0.5. Their jets are mostly one-sided close to the cores (typically $< 10\%$ of the jet lengths), and become two-sided beyond a couple of kpc. They exhibit a strong flaring between 1 and 10 kpc from the cores, with $d\phi/d\theta$ changing from ~ 0.1 to ~ 0.5. There is often a distinct gap between the core and the commencement of the jet. The magnetic fields are initially parallel to jets, and become oriented perpendicular further out. The jets finally disperse into a plume of diffuse emission, the radio spectrum of which steepens with increasing distance from the source. FR I's are commonly found in galaxy groups and clusters. In the optical regime, they exhibit absorption- or weak emission-line spectra.

Radio sources of type FR II have monochromatic radio luminosities $P >$ 10^{25} W Hz^{-1} at 178 MHz. They are edge-brightened: their brightest components — usually the hot-spots — are found at their outer edges, so that their ratio of the separation of their brightest components to their total size is about unity. Their cores and jets are generally much less prominent than those in FR I's, with one-sided jets throughout, if any. Their opening angles are small ($d\phi/d\theta \approx 0.05$), and the magnetic field in the jets is parallel except in the frequently visible knots. The particles propagating from their hot-spots form extended diffuse lobes, which usually bend backwards, connecting to the cores with bridges of emission. The radio spectra steepen along this bridge emission from the hot-spots towards the cores. Local FR II's are not seen in clustered environments, their surroundings show normal field galaxy densities. Their optical spectra are featured by strong emission lines.

2.2   Variants

While the majority of extragalactic radio sources fit into the above FR I/II scheme, there is a huge variety of variants to this simple classification, which has mostly to do with the different environments that radio sources find themselves in. The influence of a surrounding medium is most obvious in case of tailed radio sources. These are subdivided into so-called narrow-angle tailed (NAT) and wide-angle tailed (NAT) sources.

The prototype of the former is NGC 1265 (see *e.g.* O'Dea and Owen 1986). These sources, which are found in galaxy clusters, exhibit strongly bent twin jets that are deflected by the ram pressure of the external medium as the galaxy moves through the intra-cluster medium. Although their luminosity ranks them among the FR I category, their magnetic fields are always seen to be parallel to the jets, and remain parallel and extremely well-ordered in the extended radio tails that are trailing behind them (see O'Dea and Owen 1986; Mack *et al.* 1993).

WATs are more powerful than NATs, their radio luminosity being in the transition zone between FR Is and FR IIs. The prototype here is 3C 465 (*e.g.* Eilek *et al.* 1984). This class has first been established by Owen and Rudnick (1976). Their jets end up in hot-spots, beyond which the radio structure flares up into plumes, giving these sources an overall FR I appearance. There is a distinct change in position angle between the inner jets and the outer plumes. WATs are found to never bend before decollimation (O'Donoghue *et al.* 1990). Like with the NATs, it seems obvious at first glance that they must be shaped by the surrounding gas of the rich cluster environment that they usually find themselves in. However, it turned out that even the very symmetric structure of the prototype, 3C 465, is difficult to explain in terms of relative motion or buoyancy (Eilek *et al.* 1984), and more complex motion of the ICM as a result of sub-cluster merging has to be invoked (*e.g.* Pinkney *et al.* 1994).

Radio sources with *"X"*-shapes (also termed "winged sources") have stimulated a discussion whether or not some AGN and the beams they produce precess. It is the structure of the prototype, NGC 326, which was employed to unveil the effect of precession on the structure of a radio source (Ekers *et al.* 1978). This was corroborated by the numerical simulations of Gower *et al.* (1982) who successfully reproduced the radio structures of winged sources with precessing beams. Later, Leahy and Parma (1992) invoked the combination of intermittent activity and changes in the ejection orientation to explain the wings. Klein *et al.* (1995) proposed a model that still hinges upon continuous precession, but with the possibility of exhausting jets. They showed that in case of the winged source 0828+32 there is a continuous spectral steepening from the young lobes into the secondary, presumably old, lobe system. According to Worrall *et al.* (1995) buoyancy forces exerted by the surrounding X-ray gas are responsible for the shapes of the dumbbell galaxy NGC 326.

Flux density limited radio source samples are known to contain a large portion of compact sources, with angular sizes below $2''$. The corresponding projected linear sizes are ≤ 15 kpc, assuming $H_0 = 75$ km s^{-1} Mpc^{-1}. They show steep radio continuum spectra ($\alpha \geq 0.5$). Fanti *et al.* (1990) have shown that the majority of these compact steep-spectrum sources (CSS) cannot be larger sources foreshortened by projection effects, which means that their radio emission originates on sub-galactic scales. Two scenarios have been proposed that would naturally explain their observed small sizes. They could reflect an early stage in the evolution of radio sources; this is the youth scenario (Phillips and Mutel 1982; Carvahlo 1985). Alternatively, they are physically distinct, owing to unusual conditions in the ISM of their host galaxies, such that there is a higher density and/or turbulence, inhibiting the radio source from growing to larger sizes. This is the frustration scenario (van Breugel *et al.* 1984). Most CSS exhibit double-lobed structures such as seen in classical radio galaxies. Quite a few of these are symmetric, which gave rise to the terms CSO (compact symmetric objects, with sizes ≤ 0.5 kpc) and MSO (medium-size symmetric objects, with sizes > 0.5 kpc). Such symmetric compact sources are considered as scaled-down versions of larger-sized double radio sources (Fanti *et al.* 1990), which could still be evolving from "baby Cyg A's" (Begelmann 1996) into powerful FR II's (see also Fanti and Spencer 1996).

3   Virgo A on Three Scales

In the introduction it was emphasized that Vir A may serve as a nice template to understand the physics of radio sources, such as AGN or jet physics (fluid mechanics). Would this also hold on larger scales? Would we be able to classify this nearby radio galaxy using one of the above schemes?

The jet which connects the core to the north-preceding inner lobe is undisturbed out to $\sim 25''$ (2 kpc); it is seen to expand at almost constant rate ($d\phi/d\theta \approx 0.07$) over its first half, along position angle 291°. At about 2 kpc projected distance from the nucleus, it appears to undergo a strong oblique shock rendering a turbulent and disrupted flow beyond this point, followed by a sudden backward bend (see Hines *et al.* 1989). Something similar must be happening on the opposite (south-following) side, at almost the same distance. It looks like the jet and (unseen) counterjet impinge on an invisible obstacle in the ISM of M 87. This inner 5 kpc region which has a high relativistic energy density and strong magnetic fields (Hines *et al.* 1989; Owen *et al.* 1990) is clearly overpressured as compared to the ambient X-ray gas ($P_{rel} \approx 10^{-9}$ dyn cm^{-2}, $P_{ext} \approx 5 \cdot 10^{-10}$ dyn cm^{-2}). The equipartition magnetic field strength is $B_{eq} \lesssim 90\mu G$ (Hines *et al.* 1989). The spectrum is steep, with $\alpha_{east} = 0.80$ and $\alpha_{west} = 0.84$ between 5 and 230 GHz (Salter *et al.* 1989). Owen *et al.* (1990) find larger rotation measures in the eastern lobe than in the western one.

Out of this inner brightest region which is discussed in more detail by Owen (1997) the intermediate (30 kpc) lobe system emerges. This has considerably lower brightness ($\sim 1/70$ of the inner lobes) and a different orientation, with a position angle of 270°. This lobe system exhibits very abrupt terminations on either side, at ~ 15 kpc distance from the nucleus. It is fed by a flow that is obviously much less collimated. The sudden termination of this structure which is accompanied by bow-shock phenomena (an "ear"-shaped structure in the east, and a very strong bend in the west; see *e.g.* Fig. 2 of Böhringer *et al.* 1995) suggests that on this larger scale, too, some obstacle exists which strongly confines the radio structure. This intermediate-scale structure has been little studied so far. Comparing the 10.6 GHz data of Rottmann *et al.* (1996) with those of Kassim *et al.* (1993) yields a spectral index $\alpha > 1$. A slight asymmetry is indicated in the degree of linear polarization, $p_{east} < 1\%$ and $p_{west} \approx 1\%$ at 1.4 GHz (Feigelson *et al.* 1987). The intermediate lobes are clearly underpressured ($P_{rel} \approx 6 \cdot 10^{-12}$ dyn cm^{-2}, and $P_{ext} \approx 10^{-10}$ dyn cm^{-2}). The equipartition magnetic field strength is $B_{eq} \lesssim 8\mu G$.

Finally, the largest structure appears to emerge out of the intermediate lobes, one towards the north, the other towards the south, giving Vir A an overall "*S*"-shaped appearance (Fig. 1). Again, there is a marked drop in brightness between the intermediate and outer lobes (by a factor of ~ 3). Andernach *et al.* (1979) reported two remarkable properties of this "halo": an extremely steep spectrum, and a strong north-south asymmetry of this spectrum as well as of the fractional polarization measured at 2.7 and 4.8 GHz. While the recent study at 10.6 GHz by Rottmann *et al.* (1996) qualitatively confirms the spectral characteristics, the actual numbers become more moderate. Maximum values are $\alpha_{south} = 1.6$ and $\alpha_{north} = 2.6$ between 4.8 and 10.6 GHz (instead of ~ 2 and > 3, respectively) [1]. This strong asymmetry is also seen at lower frequencies; using the data of Kassim *et al.* (1993) we obtain $\alpha_{south} = 1.3$ and $\alpha_{north} = 1.5$ between 327 MHz and 10.6 GHz. So this asymmetry is much less pronounced at low frequencies, and hence not likely to be an asymmetry in the magnetic field strength.

A clearer picture of the magnetic field of the outer lobes now emerges from the high-frequency polarization map of Rottmann *et al.* (1996; see also Fig. 1). The degrees of polarization of both lobes are about equal at 10.6 GHz, with values in the range of 30 $\cdots$ 60 %. The highest degrees ($> 60\%$) are seen at the edges of the lobes, with a circumferential structure of the (well-aligned) B-field. Andernach *et al.* (1979) speculated that the lack of polarization in the

---

[1] The differences are clearly due to the greatly improved receivers and data analysis, in particular cleaning of single-dish data (see Klein and Mack 1995).

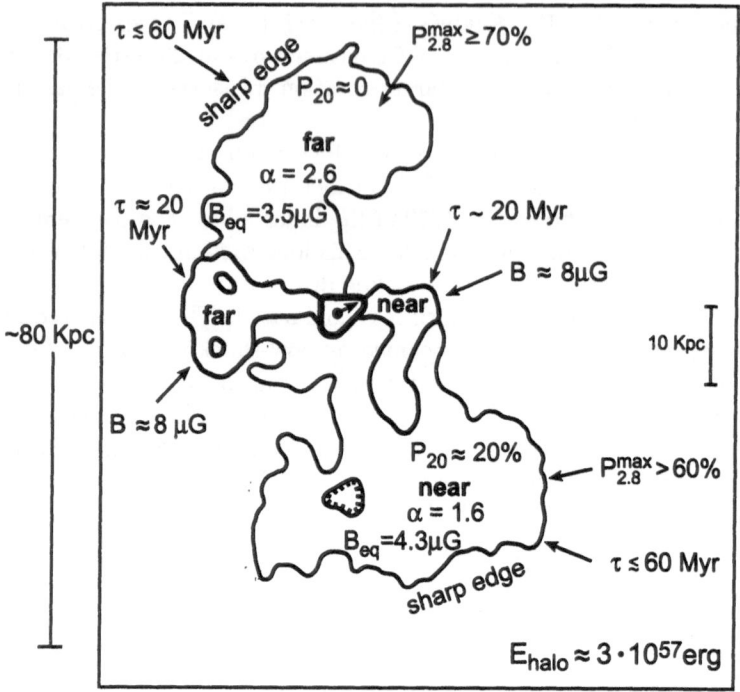

Fig. 2. Sketch of the large-scale radio continuum structure of Vir A.

northern lobe could be due to a greater turbulence there. The new results are now clearly in favour of an orientation ("Laing-Garrington") effect, with the southern lobe being the nearer one and the northern one pointing away from us.

The outer lobes appear to be underpressured by a factor of ~ 30 ($P_{rel} \approx 1.5 \cdot 10^{-12}$ dyn cm^{-2}, and $P_{ext} \approx 4 \cdot 10^{-11}$ dyn cm^{-2}). The equipartition magnetic field strength is $B_{eq} \sim 4\,\mu$G. Note that these numbers rest upon equipartition and a homogeneous medium (see Eilek 1997).

In Figure 2 the three distinct structures (5, 30, 80 kpc) are sketched, along with some of the physical parameters. Some particle ages are also indicated at different locations. These were computed according to the model of Jaffe and Perola (1973), which assumes permanent isotropization of pitch angles.

4   FR I or II, or What?

So what kind of shelf could we put Vir A into? Its total radio power ($P_{178} \simeq 4 \cdot 10^{25}$ W Hz^{-1})[2] places it just above the transition between FR I's and II's.

---

[2] A distance of 17 Mpc is assumed.

While its morphology is complex in comparison to standard representatives of the two classes, it is clearly FR I-like on large scales: the brightest structures are concentrated towards the inner region, and the extended lobes mimic the outer plumes of FR I sources.

However, if we look at the individual radio components (5, 30, 80 kpc scales), Vir A becomes a real puzzle. They are distinguished by pronounced differences in brightness ($\sim$ 200:3:1) and position angle ($\sim$ 290°, $\sim$ 270°, $\sim$ 0°). Both, the innermost (5 kpc) and intermediate (30 kpc) structure are edge-brightened, so looking at them as separate entities they would be FR II-like. The youngest (innermost) structure is also reminiscent of CSS sources, and could still be growing in size. The easiest interpretation would require a change in the orientation of the ejection during different epochs of activity. This would also naturally explain the fact that the jet suddenly decollimates and bends backwards, which it wouldn't need to if it still followed the flow channel from the previous activity. Feigelson *et al.* (1987) offered two alternative interpretations of the structure of Vir A: (i) ejection of inhomogeneous jets by a precessing nucleus, and (ii) continuous ejection of jets that bend and twist in varying directions. Vir A may have had a history similar to that of "*X*"-shaped sources. Models for restarting jets or alternating ejection ("flip-flop") have been discussed long ago (Christiansen 1973; Rees 1976), and support came with refined VLA images of sources like Her A (Dreher and Feigelson 1984) or 3C 219 (Clarke *et al.* 1992), which clearly show intermittent activity of their AGN. Turland (1975) already speculated about multiple outbursts in M 87, and more and more evidence is accumulating that the activity of AGN has not been continuous.

Blandford (1997) draws attention to the puzzle why M 87 does not host a quasar. With a central black hole of $3 \cdot 10^9 M_\odot$ it should easily produce a quasar, or should have had one before, which for some reason is currently defunct. Such a phase of "appropriate" activity should have left behind a giant (*i.e.* Mpc sized) radio continuum halo, which is definitely not seen! In fact, an inspection of the various existing radio maps shows that the edges of the radio source are everywhere very sharp and well-defined, even at the lowest frequencies. A more extended aged, steep-spectrum halo should be readily disclosed in the 74 and 327 MHz maps of Kassim *et al.* (1993), which is not seen though.

5   Future Work

As outlined above, the picture that we currently have for the large-scale structure of Vir A is still rather puzzling. In order to refine it we need to get some handle on its three-dimensional structure, and, notwithstanding the caveats against using synchrotron spectra to derive ages of radio source components, spectral analysis is the only means to date them back to different epochs of activity. In case of Vir A, this is not an easy task in view of the required dynamic range of the data. The high-frequency observations that have been started with the Effelsberg 100 m dish at 10.6 GHz (Rottmann *et al.* 1995) are being continued at 32 GHz, while the low-frequency regime is being covered with the VLA (Kassim *et al.* 1993; Owen 1997).

To explore the 3-D structure, a thorough RM analysis is indispensable. This has so far been done for the innermost bright region (Owen *et al.* 1990), while for the larger structures collection of the necessary data has commenced only recently. Understanding the 3-D structure is intimately connected with solving the problem of the possible relation between the radio continuum and excess X-ray emission (Harris 1997; Böhringer 1997). To make the outer radio lobes of Vir A a megaparsec large, they would have to be projected by $\sim 85°$ against the sky plane; a rough estimate of the true orientation of the outer lobes should be made on the basis of a Faraday rotation and depolarization study. The necessary data to perform such an analysis may soon be available.

References

Andernach, H., Baker, J.R., von Kap–herr, A., Wielebinski, R.: (1979) A&A **74**, 93 – 99

Begelman, M.C.: (1996) *Cygnus A – Study of a Radio Galaxy* eds. C.L. Carilli and D.E. Harris, (Cambridge University Press), 209 – 214

Biraud, F., Lequeux, J., Le Roux, E.: (1960) Observatory **80**, 116 – 118

Blandford, R.D.: (1999), these proceedings

Böhringer, H.: (1999), these proceedings

Böhringer, H., Voges, W., Fabian, A.C., Edge, A.C., Neumann, D.M.: (1993) MNRAS **264**, L25 – L28

Bolton, J.G., Stanley, G.J., Slee, O.B.: (1949) Nature **164**, 101 – 102

van Breugel, W.J.M., Miley, G.K., Heckman, T.A.: (1984) AJ **89**, 5 – 22

Byram, E., Chubb, T., Friedman, H.: (1966) Science **152**, 66 –

Cameron, M.J.: (1971) MNRAS **152**, 439 – 460

Carvahlo, J.C.: (1985) MNRAS **215**, 463 – 471

Christiansen, W.A.: (1973) MNRAS **164**, 211 – 221

Clarke, D.A., Bridle, A.H., Burns, J.O., Perley, R.A., Norman, M.L.: (1992) ApJ **385**, 173 – 187

Dreher, J.W., Feigelson, E.D.: (1989) Nature **308**, 43 – 45

Eilek, J.A.: (1999), these proceedings

Eilek, J.A., Burns, J.O., O'Dea, C.P., Owen, F.N.: (1984) ApJ **278**, 37 – 50

Fanaroff, B.L., Riley, J.M.: (1974) MNRAS **167**, 31p – 35p

Fanti, R., Fanti, C., Schilizzi, R.T., Spencer, R.E., Nan Rendong, Parma, P., van Breugel, W.J.M.: (1990) A&A **231**, 333 – 346

Fanti, R., Spencer, R.E.: (1996) IAU Symp. 175 *Extragalactic Radio Sources*, eds. R. Ekers *et al.* p. 63 – 66

Feigelson, E.D., Wood, P.A.D., Schreier, E.J., Harris, D.E., Reid, M.J.: (1987) ApJ **312**, 101 – 110

Garrington, S.T., Leahy, J.P., Conway, R.G., Laing, R.A.: (1988) Nature **331**, 147 – 149

Graham, I.: (1971) MNRAS **149**, 319 – 339

Harris, D.E: (1999), these proceedings

Hines, D.C., Owen, F.N., Eilek, J.A.: (1989) ApJ **347**, 713 – 726

Jaffe, W.J., Perola, G.C.: (1973) A&A **26**, 423 – 435

Kassim, N.E., Perley, R.A., Erickson, W.C., Dwarakanath, K.S.: (1993) AJ **106**, 2218 – 2228

Klein, U., Mack, K.-H.: (1995) Workshop on *Multi-feed Systems for Radio Telescopes* ed. D.T. Emerson, ASP Conf. Ser., 318 – 326

Klein, U., Mack, K.-H., Gregorini, L., Parma, P.: (1995) A&A **303**, 427 – 439

Kotanyi, C.: (1980) A&A **83**, 224 – 248

Laing, R.A.: (1989) Nature **331**, 149 – 151

Leahy, J.P., Parma, P.: (1992) *Extragalactic Radio Sources – From Beams to Jets* eds. J. Roland, H. Sol, and G. Pelletier, (Cambridge University Press), 307 – 308

Lequeux, J.: (1972) Ann. Astroph. **25**, 221 – 260

Mack, K.-H., Feretti, L., Giovannini, G., Klein, U.: (1993) A&A **280**, 63 – 75

Maltby, P., Moffet, A.T.: (1962) ApJS **7**, 141 – 163

Meisenheimer, K., Röser, H.-J., Schlötelburg, M.: (1996) A&A **307**, 61 – 79

Mills, B.Y.: (1953) Aust. J. Phys. **6**, 452 – 470

Morris, D., Radhakrishnan, V., Seielstad, G.A.: (1964) ApJ **139**, 560 – 569

O'Dea, C.P., Owen, F.N.: (1986) ApJ **301**, 841 – 859

O'Donoghue, C.P., Owen, F.N., Eilek, J.A.: (1990) ApJS **72**, 75 – 131

Owen, F.N.: (1999), these proceedings

Owen, F.N., Rudnick, L.: (1976) ApJ **205**, L1 – L4

Owen, F.N., Eilek, J.A., Keel, W.C.: (1990) ApJ **362**, 449 – 454

Phillips, T.J., Mutel, R.L.: (1982) A&A **106**, 21 – 24

Pinkney, J., Burns, J.O., Hill, J.M.: (1994) AJ **108**, 2031 – 2045

Rees, M.J.: (1976) *The Physics of Nonthermal radio Sources* ed. G. Setti (Dordrecht, Reidel), 107 – 120

Rottmann, H., Mack, K.-H., Klein, U., Wielebinski, R.: (1996) A&A **309**, L19 – L22

Salter, C.J., Chini, R., Haslam, C.G.T.: (1989) A&A **220**, 42 – 48

Seielstad, G.A., Weiler, K.W.: (1969) ApJS **18**, 85 – 126

Sparks, W.B., Ford, H.C., Kinney, A.L.: (1993) ApJ **413**, 531 – 541

Turland, B.D.: (1975) MNRAS **170**, 281 – 294

Spectral Analysis of the Large-Scale Radio Emission of M 87

H. Rottmann[1,2], N. Kassim[3], K.-H. Mack[4,2], U. Klein[2], and R. Perley[5]

[1] Max-Planck-Institut für Radioastronomie
Auf dem Hügel 69
D 53121 Bonn
Germany

[2] Radioastronomisches Institut der Universität Bonn
Auf dem Hügel 71
D 53121 Bonn
Germany

[3] Code 7213, Remote Sensing Division
Naval Research Laboratory
Washington, DC 20375-5351
USA

[4] Istituto di Radioastronomia del CNR
Via Gobetti 101
I-40129 Bologna
Italy

[5] National Radio Astronomy Observatory
Socorro, NM 87801
USA

Abstract. We present first results from a spectral analysis of the outer radio halo of M 87. Radio observations at 4 frequencies (333 MHz, 1.6 GHz, 4.8 GHz and 10.55 GHz) have been used to construct spectral index and break frequency distributions. Implications on the possible mechanisms for the formation of the halo are discussed.

1   Introduction

Radio observations of M 87 on different scales have revealed that the source morphology is complex and that it cannot be easily categorized in the classical FR I / FR II scheme. In particular the source can be divided into at least two components, the inner and the outer radio source without any obvious transition between them. East of the nucleus the inner jet extends to a distance of about 5 kpc, where it bends, disrupts and forms the inner radio lobes. On larger scales we can observe the outer jets (we refer to them as jets, even though they are not very well collimated) which extend to a distance of about 30 kpc where they seem to impinge on the ambient medium. In the east, one can see the so-called "ear", a peculiar, shell-like feature, in shape very different from a classical cocoon observed in other radio galaxies. On high-resolution images (Owen, these proceedings) even a fainter and more extended second shell can be seen surrounding the "ear".

2   Spectral analysis

A careful study of the synchrotron spectra provides an excellent tool to shed light on some interesting aspects of the source history of M 87. Spectral aging calculations allow a rough estimate of the source age. Gradients in the break frequency distribution can reveal, whether precession can account for the formation of the halo, or whether other processes (*e.g.* expanding bubbles) are to be favored. Also, a comparison of the spectral properties of the inner and outer source — in particular in the transition zone — will allow to decide whether the extended structure is a relic of a previous outburst or if it is still being fed by the inner active source.

2.1   General spectral properties of the halo

The only comprehensive spectral study of the halo of M 87 has been performed by Andernach *et al.* (1979) including observations at 4.8 GHz and 10.7 GHz with the 100 m telescope. They have noted the overall steep spectrum of the halo, with spectral indices generally between 1.6 and 2.5 ($I \propto \nu^{-\alpha}$). Extreme values reaching of $3 - 5$ (with large errors) were derived for the northern part of the halo, where the signal/noise-ratio was poor due to the low sensitivity of the receiver systems. We have re-observed M 87 at 4.8 GHz (ongoing) and 10.55 GHz (Rottmann *et al.* 1996) with the new and more sensitive receivers of the 100 m telescope. A preliminary analysis of the spectral index distribution shows good agreement with the values given by Andernach *et al.* (1979) for the southern part of the halo. However, we derive more moderate values of $\alpha <$ 2.5 for the northern lobe. The spectrum — with spectral indices exceeding 1.6 everywhere in the halo — is exceptionally steep which can probably be attributed to greater synchrotron losses in the dense environment.

The rather low resolution of the 4.8 GHz single-dish map ($157''$) does not allow us to study the spectral properties of the various features like the "ear" and the outer shell visible in the halo of M 87. In a second step we have performed a spectral analysis at $69''$ resolution including observations at 10.55 GHz (Effelsberg), 1.6 GHz (VLA) and 333 MHz (VLA). The spectral index map between 1.6 GHz and 10.55 GHz is displayed in Fig. 1a. At this resolution the spectral index distributions already reveal quite a lot of fine structure. The break frequency can be determined by fitting a spectral model to the spectral index distribution. We have used a JP-model (Jaffe & Perola 1973). Best results were obtained with an injection index of $\alpha_i = 0.85$. The corresponding break frequency distribution is displayed in Fig. 1b.

The spectral north/south asymmetry reported by Andernach *et al.* (1979) can be seen nicely. However, at lower frequencies it becomes less pronounced.

Between 1.6 GHz and 10.55 GHz we find values of $\alpha \sim 1.7$ in the northern part and $\alpha \sim 1.5$ in the southern part of the halo ($\Delta\alpha = 0.2$). Between 333 MHz and 1.6 GHz the spectra in the north and south are almost equally steep with $\Delta\alpha = 0.05$.

The southern lobe is much more homogeneous in spectral behaviour than the northern lobe. In the south an extended flat region coincides with the location where the western jet finally disrupts and forms the southern lobe. This region gradually steepens towards the edges of the source. In the northern halo several areas of flat spectral index can be made out. On the eastern rim of the "ear", where the jet seems to interact with the ambient medium a region of flat spectrum can be seen. Also the "outer ring" surrounding the "ear" is traced very nicely by emission with a considerably flatter spectrum then the surrounding lobe material.

2.2   The origin of the large-scale structure

Precession has been proposed as the reason for the roughly S-shaped morphology of the extended structure of M 87 (*e.g.* Feigelson *et al.* 1987). The displacement of the inner jet in respect to the outer jets in a counterclockwise direction is supporting this idea. Such a precession with constant injection of material should be visible in the break frequency distribution. Assuming that the involved magnetic fields are fairly homogeneous and re-acceleration does not occur a gradient from high to low break frequencies should be visible along the path of precession. From our analysis there is no clear evidence for such a gradient in break frequency and therefore precession with constant injection seems to be unlikely. However, precession with several independent outbursts is consistent with our findings. In such a case the material from more recent active periods can mix with older material, blurring out differences in break frequency.

Alternatively it has been proposed that the large scale structure of M 87 is produced by outflow from the inner lobes (Eilek these proceedings). In this scenario the "ear" and the shell surrounding it can be interpreted as expanding bubbles. In the northern halo support can be given to this idea by the spectral analysis. Both, the rim of the shell and the "ear", are characterized by a fairly flat spectrum as compared to the ambient material. The young age and/or re-acceleration through shocks on the edges of the expanding bubble can account for the flat spectral indices. Also the magnetic field is aligned in a tangential direction at the edge of the shell with high degrees of polarization (Rottmann *et al.* 1996) indicative of interacting magnetized plasma. In this picture the halo could have formed by a series of expanding bubbles. In the southern halo things are less clear. Neither in total (or polarized) intensity nor in break frequency there are any obvious shell-like structures.

2.3   The age of the halo

The break frequency ν_b is related to the particle age as given by Alexander & Leahy (1987):

$$t = 1.59 \cdot 10^3 \cdot \frac{\sqrt{B}}{B^2 + B_R^2} \cdot \frac{1}{\sqrt{(1+z)\nu_b}} \qquad (1)$$

The biggest uncertainty is introduced by our inability to directly measure the source magnetic field strength. Generally, the B-field derived from equipartition is an underestimate of the true field strength. In the case of M 87, due to interactions with the dense environment, the geometry of the source is probably very amorphous which also adds to the uncertainty of deriving a reliable equipartition value. However, to give a rough estimate of particle age we will use $B_{eq} = 0.4\mu G$ as given by Feigelson *et al.* (1987). The lowest break frequencies observed in the halo is $\nu_b \sim 6\,\mathrm{GHz}$. This corresponds to a particle age of 48.5 Myr. The maximum break frequencies derived are $\sim 12\,\mathrm{GHz}$ (southern lobe) and $\sim 9\,\mathrm{GHz}$ (northern lobe, outer bubble) with corresponding ages of 34 Myr and 39.5 Myr respectively. How do these particle ages compare to the true source ages ? As Eilek (1996) has pointed out, the simple equation *particle age = source age* very often leads to an underestimate of the true source age, due to bad estimates of the magnetic fields involved, inhomogeneities in the fields and possible re-acceleration. This can be especially true for M 87, where interaction with the dense medium has resulted in a very amorphous source with possibly very inhomogeneous magnetic fields. The polarization data have shown that the regions of maximum break frequencies coincide with regions of high degrees of polarization, so very likely the magnetic field strength is locally enhanced to values much greater than the equipartition value. If shocks occur in these areas also re-acceleration has to be taken into account, which will result in higher true source ages than derived from the particle ages. However, the particle age can serve as a lower limit for the ages of the various components.

2.4   Future work — connection of the inner and the outer source

An important question is whether the inner and outer radio sources are intrinsically connected. In other words: Is the extended structure still being fed with energy and matter or is it merely a relic from a previous outburst? The inner lobes seem to be well confined with sharp boundaries and no clear continuation from the inner to the outer jets. The inner lobes are overpressured while the extended outer structure seems to be in pressure equilibrium with the ambient medium. Possibly material from the inner lobes can "leak" into the outer jets. Again information from the spectra can contribute to answering this question. If the outer lobes are supplied with radiating particles from the inner source the initial electron energy distribution should be identical. The spectrum of the outer lobes is reproduced nicely by spectral models with an initial power law slope of $\alpha_{inj}=0.85$. Neumann *et al.* (1995) have performed a spectral analysis of a bright filament belonging to the inner eastern lobe. They find a straight power law with a slope of $\alpha=0.85$ indicating that indeed the energy distribution of the radiating particles is identical on both scales. Of course this is no definite proof for a connection of the inner and outer lobes as it is possible that different outbursts will produce the same initial particle energy distribution. As future work we will perform a spectral analysis with higher resolution of the transition region where the outer jets connects to the inner lobe. If there is no supply of young material from the central source to the outer lobes there should be an abrupt change in spectral behaviour at the boundary area with high break frequencies (few hundred THz) in the inner lobes and low ones (few GHz) along the outer jets. With the new high resolution images available now (see Owen these proceedings) such an analysis is possible.

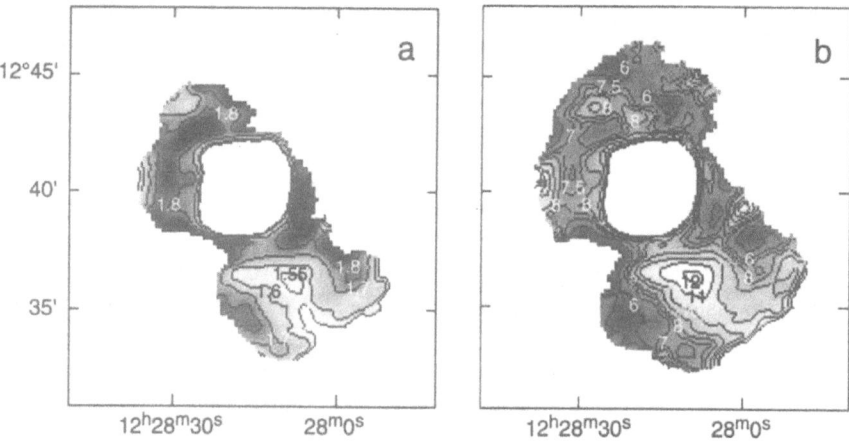

Fig. 1. (a) The spectral index distribution between 1.6 GHz and 10.55 GHz. (b) The derived break frequency distribution including frequencies of 333 MHz, 1.6 GHz and 10.55 GHz. Contours are labeled in GHz.

References

Alexander, P., Leahy, J.P.: (1987) MNRAS, 255, 1

Andernach, H., Baker, J.R., von Kap-herr, A., Wielebinski, R.: (1979) A&A, 74, 93–99

Eilek, J.: (1996) *Energy Transport in Radio Galaxies and Quasars*, eds. P.E. Hardee, A.H. Bridle, J.A. Zensus, ASP Conference Series Vol. 100., 281 –286

Feigelson, E.D., Wood, P.A.D., Schreier, E.J., Harris, D.E., Reid, M.J.: (1987) ApJ ,312, 101 – 110
Jaffe, W.J., Perola, G.C.: (1973) A&A, 26, 423 – 435
Neumann, M., Meisenheimer, K., Röser, H.-J., Stickel, M.: (1995) A&A, 296, 662
Rottmann, H., Mack, K.-H., Klein, U., Wielebinski, R.: (1996) A&A, 309, L19 – L22

The Structure of the Radio Halo

Frazer Owen

NRAO
Socorro, NM 87801
USA

Abstract. VLA imaging of the most extended structure in M 87 at 20 and 90 cm reveals a complex structure extending to about 40 kpc in radius. The images suggest a complicated, turbulent, transonic flow beginning at the boundary with the inner lobes and ending at the boundary of the bubble-like structure which is seen on the largest scales of the radio emission. The scale is interestingly similar to the "cooling core" of the X-ray source.

Models of the jet argue for kinetic luminosities of $\sim 10^{44}$ ergs s^{-1}. Most of this energy must ultimately be dissipated in heat. The morphology of the outer lobes suggests that this is occurring on the scale of the cooling core. This input energy rate exceeds the current luminosity of the cooling core in X-rays ($\sim 10^{43}$ ergs s^{-1}). Thus this suggests that the jet is providing the energy to power to the M 87 cooling core and that currently the excess energy is being absorbed in the expansion of the region rather than a contraction as envisioned in "cooling flow" theory.

1 Previous Work and the New VLA Image

It has been well known since the 1950's that most of the radio emission from M 87 below 1 GHz comes a region with a scale much larger the the region of the inner lobes (Mills 1952; Baade & Minkowski 1954). However, until recently observations of this scale usually have had a resolution of one arcminute or worse. Böhringer *et al.* 1995 have recently published an 18 cm VLA image of this structure which begins to show the structure of the region at 12″ resolution. I have followed this up with VLA images at 20 and 90 cm which show the structures more clearly and raise some interesting questions.

In Figure 1 I show an old 20 cm image with one arcminute resolution and also a new one with 5″ beam. Figure 2 contains a 90 cm image and a 20 cm blowup of the field center is shown in Figure 3. All the high resolution images have a set of radially symmetric artifacts centered on the core and inner lobes. However, on top of that one can see a complex pattern of filaments and apparent flow patterns leading out into the diffuse halo. The 90 cm image contains data from all four arrays, unlike the Böhringer *et al.* 1995 image and thus more correctly describes the structure on the largest scales. The 20 cm image was made from data from the B, C, and D arrays and thus also describes the full structure of the halo. The images show that on the largest scales M 87 can be described as two almost symmetric bubbles. However, a

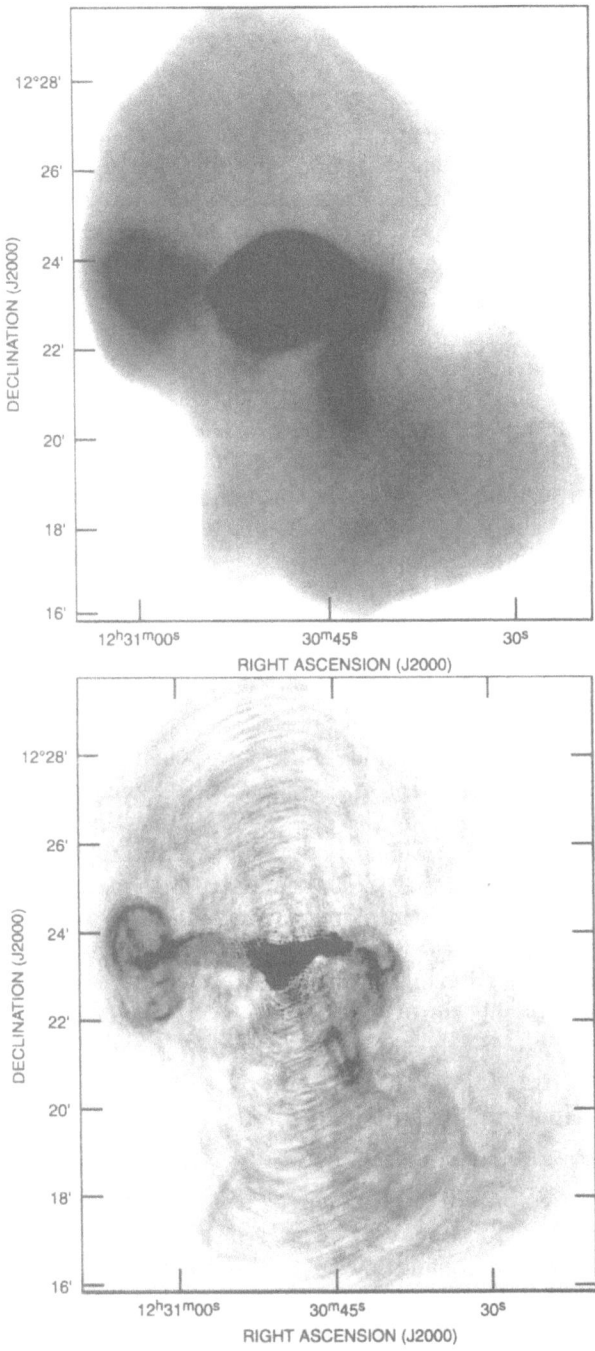

Fig. 1. 20 cm images of M 87 at 1′ and 5″ resolution

quite complex set of structures is embedded in these lobes. These suggest a bent, ordered flow out from the inner lobes which is made up of a great deal of complicated fine structure. This fine structure seems consistent with a transonic flow which is brighter at places due to shocks.

Other observational evidence concerning this region also supports a transonic environment in this general region. Transonic velocities are seen in the velocity field of the Hα filaments to the north of the inner lobes (Keel, Owen & Eilek 1996). Also the Faraday rotation results suggest the existence of magnetic fields in this region that are consistent with being generated in a transonic environment (Eilek, Owen, & Zhou 1999). The image of the M 87 halo suggests that the collimated flow from the nucleus is driving the turbulence.

The analysis of the radio jet in the inner lobes suggests an average kinetic energy flux of 10^{44} erg s^{-1}, much in excess of its radio luminosity (Bicknell, Cameron & Gingold (1990)). Supposedly most of this energy eventually is deposited in the surrounding gas in the form of heat. It seems likely that the apparent flow and shocks seen in the radio images of the halo are part of this process. If this interpretation is correct we are seeing the energy produced by the jet, coming from the central black hole, being deposited in the external medium on a scale of 40 kpc (80 kpc in diameter). The structure of the radio source suggests that the collimated flow is depositing most of it energy 10's of kpc from the nucleus rather than very near the center of the galaxy.

This scale is approximately the same one which the excess X-ray emission from the cooling core is being radiated. However, the total X-ray luminosity of this region is $\sim 1.0 \times 10^{43}$ ergs s^{-1} (Fabricant & Gorenstein 1983). At this radius, the cooling time is about 5×10^9 years (Stewart et al (1984)) which is somewhat less than the cooling time usually assumed for the arbitrary definition of the cooling radius. Thus it appears that more energy is being deposited in the hot thermal medium on this scale than is currently being radiated away. The rest of the energy is probably going into to the continued expansion of the radio bubble. Thus instead of slowly flowing inward, the gas in this region is probably slowly expanding.

2   What Powers Cooling Flows ?

Thus in M 87 there appears to be an energy source in the core of the central galaxy which is exceeding the excess energy output from the central region which has been called a "cooling flow". The timescale for the jet and the radio activity is naturally shorter than the cluster dynamics and thus is usually assumed to be relatively transient. If this is the case, in any galaxy we are seeing a snapshot of conditions that may either be more or less active than the average. It seems quite possible that the synchrotron emission could die away completely at times.

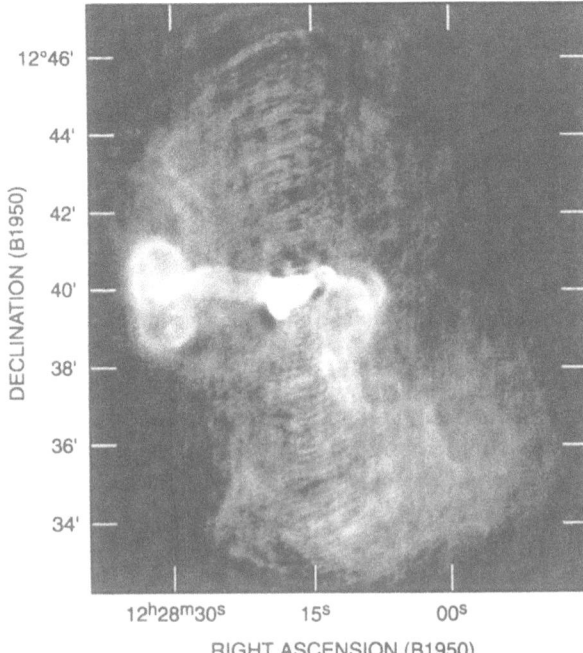

Fig. 2. 330 MHz image of M 87

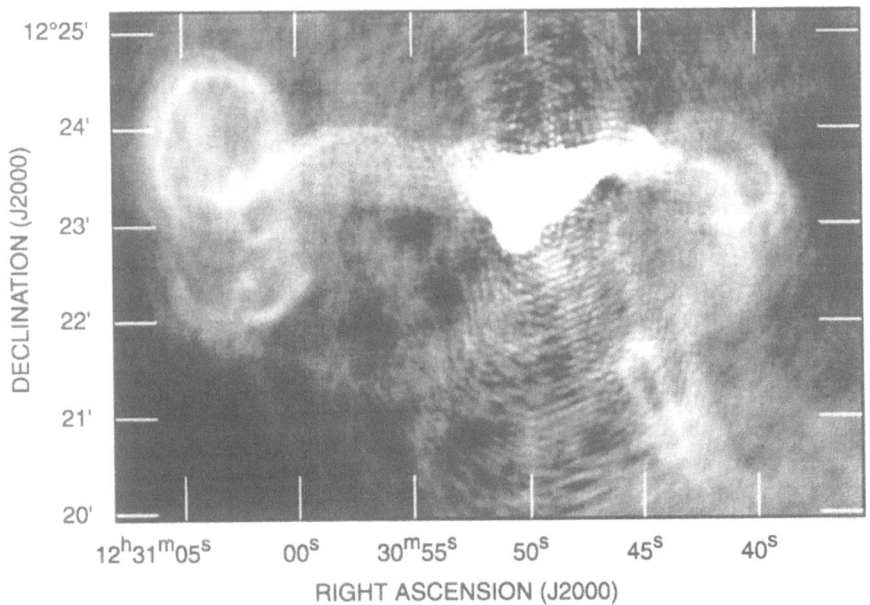

Fig. 3. 1400 MHz blowup of the central region

The energy deposited in the form of heat in the external medium will then be radiated away on much longer timescales. When the central source is inactive the system must slowly lose energy and contract. In an active state the opposite can be true. M 87 seems likely to be in this active state. What is interesting is

- that the inferred kinetic energy in the jet can exceed the X-ray luminosity,
- that most cooling flows have a central radio source ,*e.g.* Baum 1992,
- and that in M 87 we appear to be seeing the heating process occurring on the right scale for such a picture.

As argued by Binney (1999)and Jaffe 1992, a picture like this may apply to all cooling cores instead of the popular "cooling flow" model. Thus perhaps cooling flows are not powered by the cluster gravitational field but by the field of the central black hole and its transient energy output. This could explain why we have had so much trouble finding the mass which must be deposited in the core of the cluster by a slow, inward cooling flow with no central source of energy to counteract this process. The structure of the M 87 "halo" appears to be the "smoking gun".

References

Baade, W. & Minkowski, R. (1954):ApJ, **119**, 215

Baum, S. A. (1992): *in Clusters and Superclusters of Galaxies*, ed A. C. Fabian, Kluwer, Dordrect, p 171.

Bicknell, G. V. (1999): these proceedings

Binney, J.: these proceedings

Böhringer, H., Nulsen, P. E. J., Braun, R. & Fabian. A. C. (1995): MNRAS, **274**, L67-L71

Eilek, J. A., Owen, F. N. & Zhou, F. (1999): this meeting

Fabricant, D. & Gorenstein, P. (1983): ApJ, bf 267, 535

Hines, D. C., Owen, F. N., & Eilek, J. A. (1989): ApJ, **347**, 713

Jaffe, W. (1992): *in Clusters and Superclusters of Galaxies*, ed A. C. Fabian, Kluwer, Dordrect, p 109.

Keel, W. C., Owen, F. N. & Eilek, J. A. (1996): in P. E. Hardee, A. H. Bridle & J. A. Zensus, eds, *Energy Transport in Radio Galaxies and Quasars* (San Francisco: ASP), 209-214

Mills, B. Y. (1952):Nature, **170**, 1063

Stewart, G. C., Canizares, C. R., Fabian, A. C., Nulsen, P. E. J. (1984): ApJ, **278**, 536.

The Outer Radio Halo: Typical of Atypical Radio Sources?

Jean Eilek

New Mexico Tech
Socorro, NM 87801
USA

Abstract. The outer radio halo of M 87 is one of an unusual class of radio halo sources attached to cluster-center galaxies in cooling cores. The M 87 halo appears to be a young source, partially mixed with the ambient thermal plasma, expanding rather like a bubble into that plasma. The radio jet provides a significant energy source to the thermal core plasma. In turn, flows in the X-ray emitting plasma appear to disrupt the radio jet and account for formation of the radio halo.

1 Cluster-Center Radio Sources

Nearly all large-scale radio galaxies fall into one of two morphological types. Type II sources are rare but bright. They contain highly collimated, probably supersonic jets which impact the ambient medium through a set of shocks, leaving the hot jet plasma behind in a visible radio lobe. Type I sources are more common but generally fainter. They also begin with collimated jets, which often spread or decollimate close to the parent galaxy. The jetted plasma continues to propagate into the ambient medium, perhaps subsonically, perhaps buoyantly. These two types separate well in radio power and parent galaxy magnitude: sources with $P_{1.4\,\mathrm{GHz}} \gtrsim CL_{\mathrm{gal}}^{1.65}$ (C is a numerical constant) are Type II, while others are Type I (Owen 1991, Ledlow & Owen 1996).

Despite their different appearances, both Type I and Type II sources are similar in their underlying dynamics. Their linear size is determined, for a given external density and pressure, by the momentum flux in the beam (perhaps modulated by buoyancy and intergalactic winds in the case of Type I sources). Their volume is determined by the energy flux in the flow.

Where does M 87 fit into this scheme? I believe it is neither a Type I nor a Type II. I think it is typical of amorphous radio galaxies which result from the unusual conditions at the centers of clusters with strong cooling cores.

Recently, Owen & Ledlow (1997) have compiled an excellent data base of radio galaxies in clusters. Nearly all of these sources fall in the Type I region of the (radio power, optical magnitude) plane. Approximately 150 of their sources have radio images which are good enough to reveal morphological information. Nearly all of these sources appear to be described by the basic dynamical model described above; but a few are different. Five out of the

150 are diffuse, closer to circular than linear (in projection), and generally amorphous in appearance. They all extend to tens of kpc from their cores. When spectral information is available, these sources have unusually steep radio spectra. None of them shows evidence of a linear, collimated flow farther a few kpc from the core.

It is tempting to view these sources as truly amorphous — bubbles rather than jet-driven flows. As they are all associated with bright, cluster-center galaxies, they are unlikely to be normal Type I sources seen in projection.[1] In Owen & Eilek (1997) we argued from combined radio and X-ray data that another member of this set, 3C 338 in A 2199, shows evidence for a strong interaction between the radio jet and the local X-ray plasma. It appears that the radio jet has been disrupted by local conditions in the X-ray gas. Such an early disruption of the radio jet is likely to lead to an amorphous radio source, which expands like a "bubble", driven by its internal energy rather than by the momentum flux in its jet.

The outer radio halo of M 87 (Klein 1999, Owen (1999)) fits well into this set of amorphous, cluster-center radio sources. It is very similar in linear size, radio power, radio spectrum and location in a cooling core. In this paper I consider the M 87 radio halo in this light.

2   The Radio Halo of M 87

2.1   What Are the Conditions in the Halo?

The outer radio halo of M 87 is probably in pressure balance with its surroundings. The pressure in the ambient gas is $p_x \sim 2 - 4 \times 10^{-11} \mathrm{dyn/cm^2}$ (Nulsen & Böhringer 1995). The low-resolution equipartition pressure in the radio lobe is $p_{eq} \sim 1 - 2 \times 10^{-12} \mathrm{dyn/cm^2}$ (Feigelson et al. 1987). However, this very likely underestimates the true pressure of the radio plasma, due to inhomogeneity, deviations from equipartition, and the possibility of mixing of thermal plasma. Thus, the radio lobes are probably in overall pressure balance with the ambient plasma.

The radio halo is inhomogeneous. Filaments are apparent in new images from Owen. The halo contains large-scale filaments emanating from the core, and smaller ones throughout the lobe (or possibly on its surface). Such filaments are regions of high magnetic field, high pressure, or both.

The radio halo is magnetized. The low-resolution equipartition field is $B_{eq} \sim 4 - 8\,\mu\mathrm{G}$ (Feigelson et al. 1987). As with the pressure, this is probably an underestimate of the field. The field which would balance the ambient

---

[1] All 5 amorphous sources belong to a "cluster-center" subsample of the Owen-Ledlow sources, defined as those identified with a bright galaxy, and located within 100 kpc of the X-ray peak as identified from ROSAT by Ledlow (1997, in preparation). In addition, the 4 of them with cooling-flow information (Peres & Fabian 1997) have strong cooling cores.

pressure is $B_p \sim 25 - 30\,\mu G$. Faraday rotation data find $B_{RM} \sim 20 - 50\,\mu G$ in the inner few kpc, and suggest that the field is inhomogeneous, probably clumped in high-field flux ropes, and maintained by turbulence (Eilek 1999a).

2.2   What's Happening in the Halo?

The new observations give insight into the physical state of the radio halo.

First, how well has the radio-loud plasma in the halo mixed with the ambient X-ray-loud gas? Older models of these amorphous sources discussed microscopic diffusion of relativistic particles through the ambient medium. Others suggested that macroscopic instabilities could lead to full mixing of the radio plasma with the X-ray gas. Both of these seem unlikely in the case of M 87. The radio source has a well defined, sharp outer edge, which is apparent at all frequencies. Polarization in the south lobe (Rottmann *et al.* 1996) shows that magnetic field is parallel to edge of source. Both of these facts suggest a well-defined boundary, separating the radio plasma from the ambient X-ray plasma. As plasma cannot easily cross magnetic surfaces, this suggests the radio plasma and X-ray ambient gas remain separated.

However, the situation may not be quite so simple. There is no sign of a "hole" in X-rays (as has been detected around Cyg A by Carilli, Perley & Harris 1994). Thus it may be hard to invoke a fully isolated radio-loud bubble, despite the arguments above. The answer may lie in the fact that mass-loading across a well-defined MHD surface is possible through local, small-scale reconnection events. This is occurring, for instance, at the earth's magnetopause (*e.g.*, Lui 1985), indicating that some fraction of the incoming mass flux in the solar wind is admitted to the earth's magnetosphere. It may be that similar processes are occurring at the boundary of the M 87 radio halo, admitting some slow mixing of thermal plasma with the relativistic, radio-loud plasma.

A second question is, how does the kpc-scale jet connect to the outer lobes? The new radio images clearly show filaments leading from the inner lobes into and throughout the outer lobes. I find this strong evidence that the outer lobes are still "alive", that is still being supplied with mass and energy, rather than being a relic of previous active epochs.

A third question is, how does the entire system connect to the dynamics of the X-ray gas? In addition, the X-ray isophotes show complex structure in this region. On large scales, $\gtrsim 50\,\mathrm{kpc}$, they are symmetrical and slightly elliptical (*e.g.*, Nulsen & Böhringer 1995), agreeing with approximate PA of stellar galaxy. However, the isophotes rotate and become strongly elongated in the inner region. Residuals from a spherical fit show a ridge-like X-ray excess which approximately coincides with the bright radio filaments (Feigelson *et al.* 1987, Böhringer *et al.* 1995). New ROSAT HRI data (Harris 1999) reveal this excess has the appearance of a bent sheet, or bow shock, centered on the core of the galaxy. Harris notes that the X-ray ridge appears to pass directly through the knot A-C complex in the jet. Thus, it seems clear that the thermal cluster gas has disrupted the jet of M 87. This disruption may well be what has turned this galaxy into an amorphous radio source.

2.3   How Old Is the Radio Halo?

If the outer radio halo is still "alive", then we can apply the simple model of an energy-driven bubble, expanding at constant pressure. This has $V(t) \simeq (P_b/p_x)t$ (Eilek & Shore 1989), if P_b is energy input to the lobe, and $V(t)$ is the volume of the lobe, and p_x is the ambient pressure in the X-ray gas. Clearly this is a naive picture, ignoring gradients in p_x, mass loading from the ambient gas, and projection effects; however, these caveats should amount only to a factor of a few. Various estimates for the jet power in M 87 agree on the range $P_b \sim 10^{43} - 10^{44}$ erg/s (Bicknell, Cameron & Gingold (1990), Reynolds 1999). Taking $P_b = 10^{43}$ erg/s gives an age ~ 170 Myr for the halo; taking the higher power gives an age ~ 80 Myr. The radio halo is still quite young.

What can the radio spectrum tell us? The lobe shows a range of steep spectral indices (Rottmann & Klein 1999), which are usually interpreted as evidence for particle aging. The situation is more complicated, however. The synchrotron age of electrons radiating at 1 GHz is 220 Myr in a 4 μG field, but only 10 Myr in a 20 μG field. This last is much shorter than the dynamical age of the source. Thus, either the particles spend much of their time in high field regions and undergo local re-acceleration in the lobe, or they spend most of their time in low-field (interfilament) regions and radiation rapidly when they hit the filaments (Eilek, Melrose & Walker 1996). In either case, these short lifetimes also argue against the radio halo being a relic source.

3   The Energy Budget of the Virgo Core

The core of the Virgo cluster is an active place. It is anything but a slow, spherically symmetric cooling inflow. The radio data tell us that outflows exist, on a scale $\sim 20 - 40$ kpc, driven by the central galaxy. The X-ray data show us direct evidence of a non-static atmosphere, and (if the X-ray sheet is a bow shock) strongly hint at transonic motion of M 87 relative to the cluster. The radio also tells us that dynamically significant magnetic fields exist in the very inner core, and probably throughout the entire core region. These fields are very likely to be maintained by transonic turbulence (of which we see a trace in the velocities of the emission line clouds in the inner core; SBM).

The total energy thermal energy in the core is $U_{th} \sim 1 \times 10^{59}$ erg (Nulsen & Böhringer 1995); turbulence and magnetic fields probably boost this by a factor of order unity (Eilek 1999a). We can measure the energy losses from this region directly. The total luminosity in the emission line clouds is $P_{EL} \sim 10^{42}$ erg/s (Sparks $et\,al.$). The bolometric radio power is $P_{rad} \sim 10^{42}$ erg/s

(Herbig & Readhead 1992). The X-ray luminosity of this region is $P_X \sim$ 1.5×10^{43} erg/s (scaling the full X-ray excess, from Peres & Fabian 1997, to this inner region). The overall radiative loss time is thus $\sim 300\,$Myr. In Eilek (1997) I argued the turbulent decay time $\sim 20 - 100\,$Myr. There must be an ongoing energy resupply to offset these rapid losses.

The conventional picture is that this energy is supplied by the slow release of gravitational potential energy in a cooling inflow. However, the complex nature of the Virgo core is clearly inconsistent with a simple cooling flow picture. We have two other power sources. One is the radio jet itself, which puts out $P_b \sim 10^{43} - 10^{44}$ erg/s (§2.3). This input alone is comparable to the total radiative losses from the region. Another important driver is large-scale mass motions in the Virgo cluster. I argued above that the inner X-ray ridge (Harris 1999) may be evidence that some of this energy is transmitted to the core. Only a slightly transonic flow, $\mathcal{M} \sim 1.1$, is needed to account for the X-ray excess in this region (using data from Feigelson et al. 1987 and from Harris 1999). Such a flow, driving a bow shock on an scale $R \sim 20\,$kpc powers the core at $P_{flow} \sim \rho v^3 R^2 \sim 3 \times 10^{43}$ erg/s. This is strikingly similar to the energy provided by the radio galaxy, and again supports the picture of the Virgo core being an active and interesting region.

References

Bicknell, G. V. (1999): these proceedings

Böhringer, H., Nulsen, P. E. J., Braun, R. & Fabian. A. C. (1995): MNRAS, **274**, L67-L71

Carilli, C. L., Perley, R. A. & Harris, D. E. (1994): MNRAS, **270**, 173-177

Eilek, J. A. (1999): this meeting

Eilek, J. A. & Shore, S. N. (1989): ApJ, **342**, 187-207

Eilek, J. A., Melrose, D. B. & Walker, M. A. W. (1996): ApJ, **483**, 282-295

Feigelson, E. D., Wood, P. A., Schreier, E. J., Harris, D. E. & Reid, M. J. (1987): ApJ **312**, 101-110

Harris, D. (1999): this meeting

Herbig, G. & Readhead, A. (1992): ApJS, **81**, 83-124

Klein, U. (1999): this meeting

Ledlow, M. J. & Owen, F. N. (1996): AJ, **112**, 9-22

Lui, A. T. Y. (1985): *Magnetotail Physics* (Baltimore: Johns Hopkins University Press)

Nulsen, P. E. J. & Böhringer, H. (1995): MNRAS, **274**, 1093-1106

Owen, F. N. (1991): in H.-J. Röser & K. Meisenheimer, eds, *Jets in Extragalactic Radio Sources* (Berlin: Springer-Verlag), 273-278.

Owen, F. N. (1999), this meeting

Owen, F. N. & Eilek, J. A. (1997), submitted to ApJ.

Owen, F. N. & Ledlow, M. J. (1997), ApJSupp, **108**, 41-98

Peres, C. & Fabian, A. (1997): in preparation

Reynolds, C. (1999): this meeting

Rottmann, H. & Klein, U. (1999): this meeting

Rottmann, H., Mack, K.-H., Klein, U. & Wielebinski, R. (1996): A&A, **309**, L19-L22

Sparks, W. B., Ford, H. C. & Kinney, A. L. (1993): ApJ, **413**, 531-541

The Intracluster Medium in the X-Ray Halo of M 87

Hans Böhringer

Max-Planck-Institut für extraterrestrische Physik
D-85740 Garching
Germany

Abstract. The halo of M 87 is an interesting study object for X-ray and radio astronomy. X-ray observations allow to study the hot gaseous environment of M 87 and provide a means to determine the galaxy halo mass. The X-ray halo is actually an indistinguishable part of the intracluster medium of the Virgo cluster, which features a moderate cooling flow focused onto M 87. There is a clear interaction effect between the relativistic plasma in the radio lobes and the thermal intracluster medium, featuring cooling gas in the regions of the lobes.

1 Introduction

The giant elliptical galaxy M 87 embedded in the cluster environment of the Virgo cluster displays several interesting astrophysical phenomena which can be studied in this place better than in any similar galaxy, because M 87 is the closest interesting central cluster elliptical. The dominating role of this galaxy in the Virgo cluster is nicely demonstrated in X-ray images of the Virgo cluster, as shown in the contribution by Bruno Binggeli in this volume. The X-ray emission which extends over at least about 8 degrees diameter is strongly peaked on M 87. This X-ray emission originates in hot thermal gas that fills most of the entire cluster volume with plasma densities of $3 \cdot 10^{-5}$ electrons cm^{-3} in the outer regions where very faint X-ray emission is still detected ($\sim$ 1.5 to 1.8 Mpc radial distance from M 87) and peak values of about $1 \, \text{cm}^{-3}$ in the center of M 87.

The study of this hot thermal intracluster medium allows us to characterize the cluster environment, to determine the gravitational mass in a volume around M 87, and last not least to study the interesting interacting effects of the relativistic plasma expelled from the active nucleus of M 87 in the form of a jet and radio lobes with the thermal plasma environment. This contribution describes some of the progress in understanding the structure of the intracluster medium around M 87 studied mainly by X-ray observations. For the physical parameters that scale with distance we adopt a value of 20 Mpc for the distance to M 87.

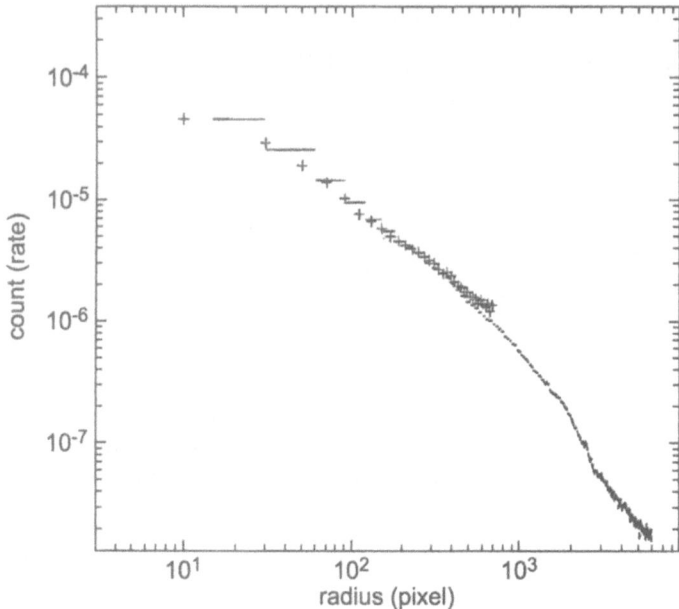

Fig. 1. X-ray surface brightness profile of M 87 as observed with the ROSAT HRI (crosses) and ROSAT PSPC (bars with error bars). The HRI count rate is scaled with the corresponding sensitivity ration of the 2 instruments to be comparable to the PSPC data. The radial scale is given in pixel, with 1 pixel = 0.5″.

2   The X-ray Halo of M 87

The ROSAT All-Sky Survey image of M 87 and the Virgo cluster shows that the galaxy halo blends smoothly over into the cluster halo, and there is no clearly visible boundary between the interstellar and intracluster medium. This is also seen in the X-ray surface brightness profile of the halo shown in Fig. 1. This profile can approximately be characterized by a β-model of the following functional form

$$S(r) = S_0(r) \left(1 + \frac{r^2}{r_c^2}\right)^{-3\beta+1/2} \tag{1}$$

with values for the core radius of about 10″ (1 kpc), which can only be resolved in the ROSAT HRI image, and a slope parameter of about $\beta = 0.47$ (Böhringer *et al.* 1994). The slope is not exactly described by a power law, however, as can be seen in Fig. 1. A deep ROSAT PSPC observation of the halo of M 87 in which about 1.8 million photons were detected in 30 ksec allows a detailed study of the gas density and temperature distribution of the plasma halo. From these data a quite accurate mass profile for the gas mass and the gravitational mass of M 87 and its halo region can be derived

(Nulsen & Böhringer 1995). The gravitational mass inside a radius of 200 kpc for example can be conservatively constraint to the range $1.5 - 3.6 \cdot 10^{13} M_\odot$. One of the best fitting mass models was found to have the following functional form

$$M(r) = \mu r + M_0 \left(\frac{r}{a} - arctan(\frac{r}{a})\right) \tag{2}$$

where the first term on the right hand side of the equation represents the mass of M 87. The best fitting parameters are: $\mu = 3.6 \cdot 10^{10} M_\odot \, \text{kpc}^{-1}$, $a = 56 \, \text{kpc}$, and $M_0/a = 1.24 \cdot 11^{10} M_\odot \, \text{kpc}^{-1}$. The addition of an extra mass for M 87 provided a significant improvement of the model fit (see Nulsen & Böhringer 1995).

In the inner region, out to a radius of at least half a degree, the X-ray emission has a significant ellipticity (Fig. 2). The orientation of the ellipticity is exactly the same as that of the optical galaxy, within the measurement error of less than 10 degrees. The ellipticity value is with $\epsilon = 0.1 - 0.16$ about a factor of 2 to 3 lower than the optical value. But this has to be expected from theoretical arguments (see e.g. Buote & Canizares 1996). The ellipticity of the gravitational potential of the Galaxy originates in an anisotropic velocity dispersion of the dark matter particles and the galaxies, while the hot gas fills the gravitational potential with an isotropic distribution function and therefore with a higher degree of spherical symmetry.

3   The Cooling Flow in M 87

While it is in general difficult to determine the temperature of the hot gas from ROSAT PSPC spectra due to the limited energy resolution and the comparatively soft energy band, the M 87 halo observations offer the unique opportunity for a study of the temperature distribution of the intracluster medium, because the temperature is relatively low with values of 1 to 3 keV and easier to asses with the ROSAT PSPC and because of the very good photon statistics (about 1.8 million photons registered in the 30 ksec observation). Fig. 3 shows for example an integral spectrum of the halo region inside a radius of 5'. An excellent fit of a model spectrum for hot thermal gas is obtained for the bulk plasma parameters $T = 1.8 \, \text{keV}$, metallicity = 0.45 solar, and $N_H = 1.5 \cdot 10^{20} \, \text{cm}^{-2}$ (where the resulting value for the interstellar absorption is somewhat smaller than the value found in 21 cm measurements, see Lieu et al. 1996).

A more detailed analysis in smaller radial regions shows, however, that there is a significant temperature gradient with decreasing radius. The temperature drops from about 3 keV in the outer intracluster medium to about 1.1 keV in the inner region with $r \leq 1'$ (Nulsen & Böhringer, 1995). This temperature structure is consistent with the model of a cooling flow obtained if the pressure and temperature loss due to cooling of the central plasma is

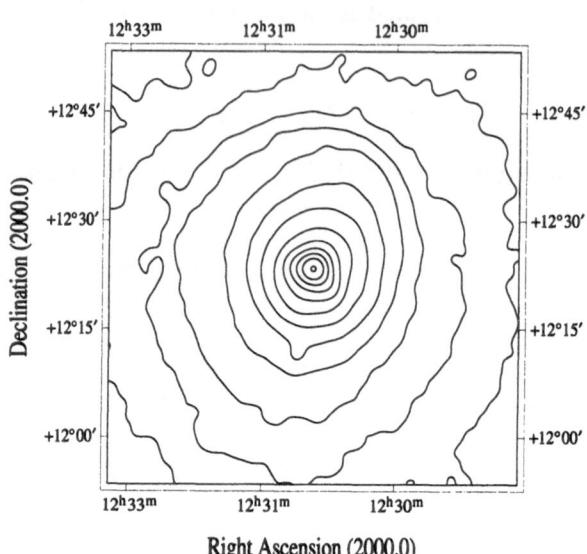

Fig. 2. Contour plot of the ROSAT PSPC image of the X-ray halo of M 87. The image covers a scale of a degree on each side, which corresponds to a physical scale of about 340 kpc at M 87. The surface brightness distribution is elliptical with a position angle of $PA = (158 \pm 10)$ deg and an ellipticity of 0.1 to 0.16 in the radial range from $4'$ to $15'$ (Böhringer *et al.* 1997).

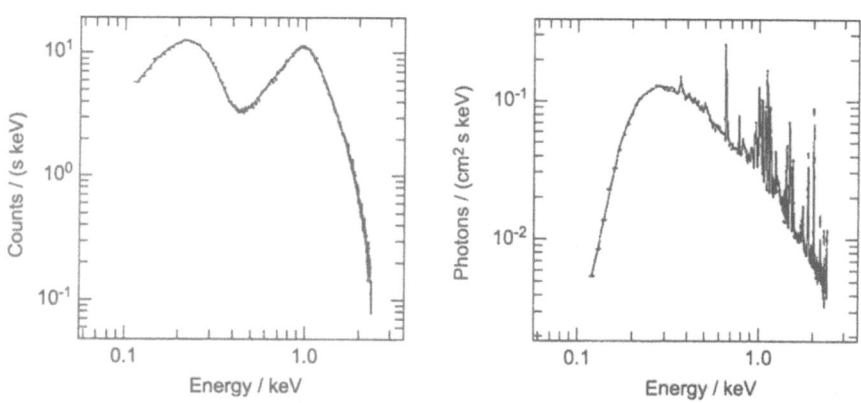

Fig. 3. ROSAT PSPC spectrum of the inner region of M 87 with a radius of about $5'$. The left panel shows the observed spectrum and the model fit while the right spectrum shows the input model with absorption and the correspondingly rescaled data points. For details of the thermal gas model see the text.

compensated by a steady state inflow of hot intracluster plasma from the outer regions (see *e.g.* Stewart *et al.* 1984, Fabian *et al.* 1984). For the standard cooling flow model a decrease of the mean temperature with radius is predicted. The typical mass flow rate determined for the M 87 halo from imaging and spectroscopic data assuming a steady state cooling and inflow is $\sim 10\,M_\odot\,y^{-1}$ (Stewart *et al.* 1984, Canizares *et al.* 1982).

The morphology of the central part of the cooling flow can be inspected in more detail in the ROSAT HRI image taken of M 87 (Fig. 4). The most prominent feature of this image are two central point sources corresponding to the nucleus of M 87 and the first bright knot of its jet. These two point sources have been removed in Fig. 4 to show the underlying cooling flow structure more clearly. At lower surface brightness we can observe the center of the cooling flow region which has the following characteristics: (i) the broad maximum of the cooling flow emission is not exactly centered on the nucleus of M 87 but displaced to the North-West by about $10'' - 15''$, ii) the surface brightness distribution of the X-ray emission after removal of the point sources of the nucleus and jet can roughly be described by a beta model distribution with a core radius of about $10''$ ($\sim 1\,kpc$), iii) the emission in the very center appears lumpy. Thus in the HRI image the core radius of the cooling flow emission is for the first time resolved. The exposure time of the HRI image is with 14.2 ksec still too small to clearly show the significance of the inhomogeneous structure of the central part of the cooling flow. A careful statistical analysis shows that there are at least two significant surface brightness excess regions to the north of the nucleus with a significance of at least 3σ. The two knots have a diameter of about $15''$ to $20''$ ($\sim 1.5 - 2\,kpc$) and an X-ray luminosity of about $2 - 3 \cdot 10^{39}$ erg s^{-1}. The cooling losses would correspond to a steady state mass deposition rate of 0.1 to $0.2\,M_\odot\,y^{-1}$. Thus these features are really small and have smaller mass deposition rates than the rate assumed for example for the galactic fountain of our galaxy. These two knots are very much smaller than the bright X-ray knot observed in the central cooling flow region of NGC 1275, which is also disturbed by the radio lobes of the galaxy (Böhringer *et al.* 1993) with an X-ray luminosity two orders of magnitude higher than in the case of M 87. There seems to be more structure in this central cooling flow region which can be analyzed in detail in the deeper ROSAT HRI observation of M 87 (see contribution by D. Harris in this volume).

4   Interaction of the Radio Lobes with the ICM

The structure that is clearly observed in the cooling flow of M 87 is correlated with the radio lobes. Figure 5 shows a PSPC image of the M 87 halo on a scale of $20'$ ($\sim 120\,kpc$) across. Besides an almost spherically symmetric halo image which is slightly elliptical one notes an arclike structure that extends from the center to about $3'$ to $5'$ in eastward and southwestward direction.

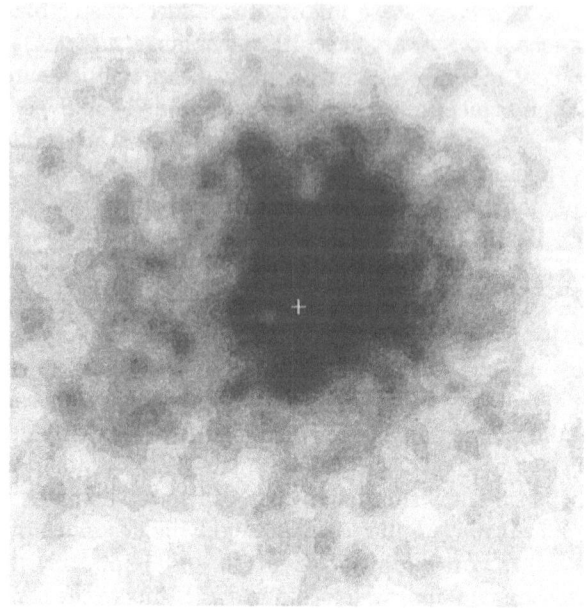

Fig. 4. ROSAT HRI image of the central region of the cooling flow in M 87. The X-ray point sources of the nucleus and jet of M 87 have been removed. The position of the nucleus is marked by a cross. The scale of the image is $2.13' \times 2.13'$.

Subtracting a spherically symmetric model from the image of Fig. 5 yields two very compact residuals, shown in Fig. 6 (Böhringer *et al.* 1995). They coincide very closely with the radio lobe structure in M 87 at a radial scale of 3' to 5' (see Böhringer *et al.* 1995). The excess emission structure was already observed, only less clearly, in an earlier EINSTEIN HRI observation by Feigelson *et al.* (1987). It was then speculated that the excess emission could be due to inverse Compton emission caused by the relativistic electrons which are responsible for the observed radio synchrotron radiation in the radio lobes. Simple modeling of the effect showed, however, that the predicted effect fell short by about a factor of 8 to explain the observed emission. The ROSAT PSPC spectral resolution is just sufficient to analyse the nature of the X-ray emission in more detail. It is found that the emission is thermal and the observation of an iron line feature at 1 keV clearly excludes the possibility of a non-thermal emission mechanism. Figure 7 shows the spectrum of the excess emission observed in the Eastern X-ray emission knot in the radio lobes together with the fitted theoretical spectra for a thermal plasma model and a power law spectrum for the inverse Compton emission.

Further analysis shows that the temperature in the lobe regions is lower than the ambient temperature of the ICM at the same radial distance from the center of M 87. This can be explained phenomenologically by the simple

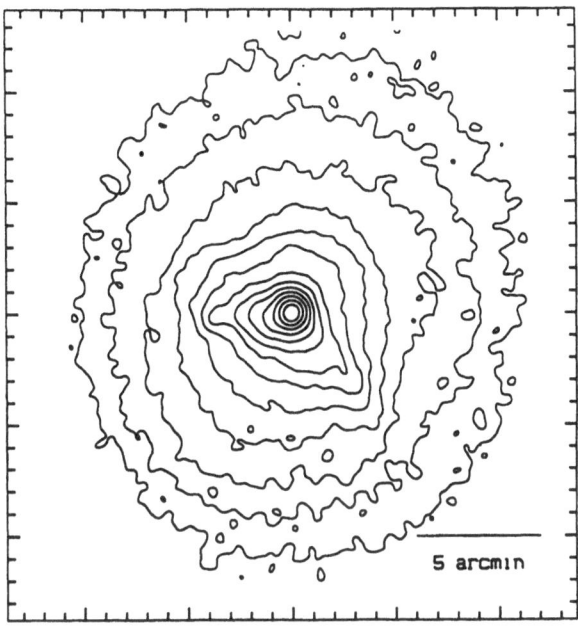

Fig. 5. ROSAT PSPC image of the X-ray halo of M 87. The image was produced from the photons received in the 0.5 to 2.0 keV energy band. The image has been slightly smoothed. The contours are spaced logarithmically, increasing by a factor of 1.4.

fact that the gas is approximately in pressure equilibrium (neglecting the possible extra pressure of the relativistic plasma and the magnetic field in the lobes). Therefore the regions of lower temperature feature a higher gas density giving rise to an enhanced X-ray surface brightness. This does not explain, however, why the ICM is colder in the lobe regions. A simple assumption, that the interaction effects of the relativistic plasma with the thermal ICM would lead to a heating of the gas, leads to the opposite conclusion. Therefore we can only speculate: either the interaction with the radio lobes leads to an enhanced clumping of the ICM in the lobe regions which accelerates the cooling, or magnetic bubbles with lower mass density create a convective flow which lifts colder material from the central region in the lobes to larger radii where it is colder than the surroundings.

Another interesting observation concerning the nature of the excess emission was recently made by Rottmann *et al.* (1996) who found that the lobe structure is not only reflected in the X-ray surface brightness but also in the hardness ratio of the emission. The lobes feature a harder spectrum than the surroundings in spite of the lower temperature. A more detailed inspection of the X-ray colour distribution in the halo region of M 87, is shown in Fig. 8, where the X-ray image in 8 narrow spectral bands within the ROSAT PSPC

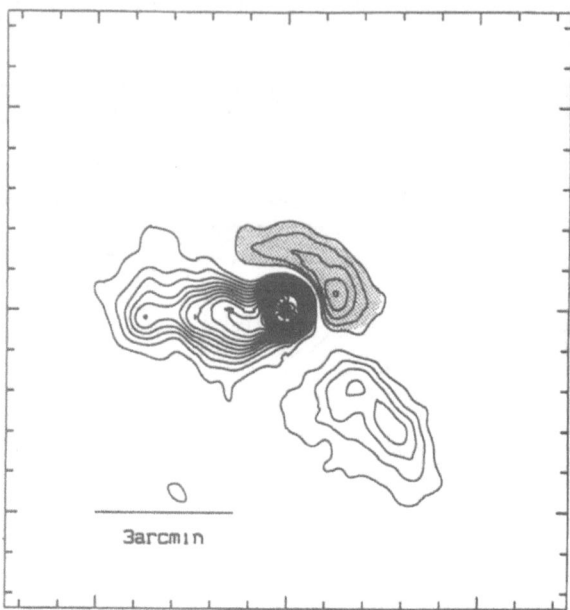

Fig. 6. Residual image obtained after subtraction of a spherically symmetric model image from Fig. 5. The scale is smaller by a factor of 2 than Fig. 5. The contours are linearly spaced and the shaded region is negative.

energy window are displayed. The energy channels used for the bands are indicated in the figure in the upper right corner of each panel (see also Snowden *et al.* 1994; the photon energy roughly corresponds to the channel number in units of 10 eV). We note that the lobes are most prominent in the R5 and R6 bands. Thus the lobe regions appear harder in a comparison of the energy bands R4 (channel 52 - 69) and R5 (channel 70 - 90), while for the two bands R6 (channel 91 - 131) and R7 (channel 132 - 201) the lobes have a softer spectrum. This can well be explained by the fact that for the lower temperatures in the lobes the iron line feature around 1 keV becomes more prominent, such that for a comparison of the intensities in a band around 1 keV with a band at lower (higher) energies the colour of the colder gas becomes harder (softer). A detailed spectral analysis of the lobe region and the surroundings shows that this effect can be accounted for qualitatively with measured lobe temperatures around 1.25 keV compared to ambient temperatures of about 1.6 keV. Thus we obtain a consistent picture of the temperature distribution in the M 87 halo with colder lobes and with a constant value of the metallicity (with a value between 0.4 to 0.5 solar).

It still remains a puzzle how the two plasma media, the thermal intracluster gas and the relativistic plasma in the radio lobes apparently penetrate each other. Alternatively the radio lobes could also consist of a set of bub-

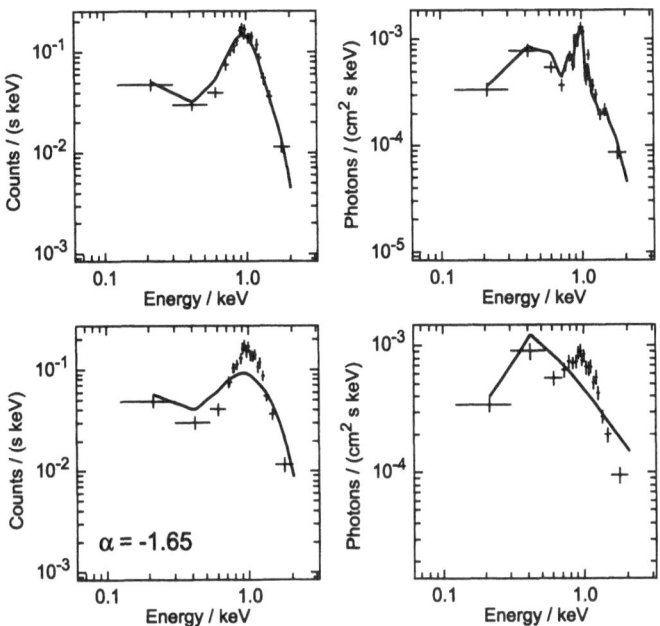

Fig. 7. Spectrum of the excess X-ray emission in the Eastern knot of the X-ray halo of M 87. The two right panels show the observed spectra with different model fits and the two left spectra give the input model spectra with rescaled data. The upper part of the figure shows the fit of a thermal model spectrum to the data which gives a very satisfactory result, while the fitting of power law spectra as expected for inverse Compton emission and as shown in the lower part of the figure, fails to explain the observations.

bles or filaments which are unresolved by the present observations. Future high resolution radio observations and scheduled X-ray observations with AXAF and XMM will provide very exciting new observational clues on these interesting plasma physical phenomena. The AXAF images with subarcsec resolution will definitely be of prime importance.

References

Böhringer, H., Voges, W., Fabian, A.C., Edge, A.C., Neumann, D.M., 1993, MN-RAS, 264, L25

Böhringer, H., Briel, U.G., Schwarz, R.A., Voges, W., Hartner, G., Trümper, J., 1994, Nature, 368, 828

Böhringer, H., Nulsen, P.E.J., Braun, R., Fabian, A.C., 1995, MNRAS, 274, L67

Böhringer, H., Neumann, D.M, Schindler, S., Huchra, J.P.; 1997, ApJ, 485, 439

Buote, D.A., & Canizares, C.R., 1996, ApJ, 457, 565

Canizares, C.R., Clark, G.W., Jernigan, J.G., Markert, T.H., 1982, ApJ, 262, L32

Fabian, A.C., Nulsen, P.E.J., & Canizares, C.R., 1984 Nature, 310, 733

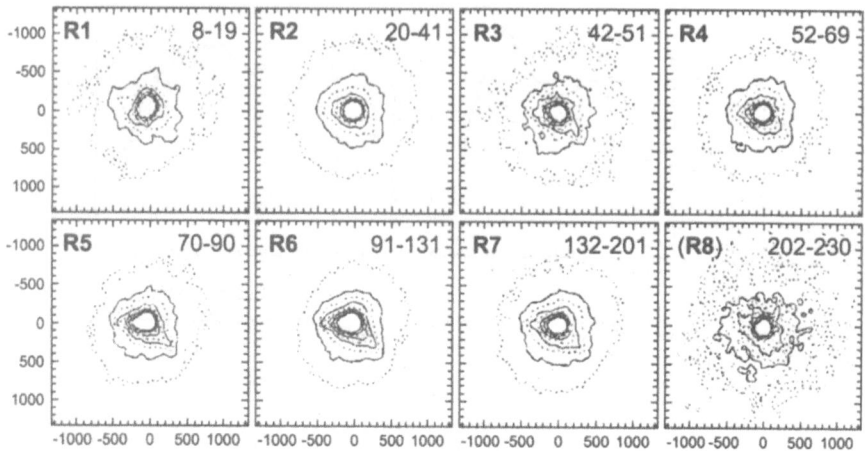

Fig. 8. X-ray image of the X-ray halo of M 87 in 8 different narrow PSPC passbands. The PSPC enerby channels used for the bands are indicated in the upper right corner of the panels. For details on the energy bands see the text. The X-ray features correlated with the radio lobes are most clearly seen in the R5 and R6 energy bands.

Feigelson, E.D., Wood, P.A.D., Schreier, E.J., Harris, D.E., Reid, M.J., 1987, ApJ, 312, 101

Lieu, R., *et al.* 1996, ApJ, 458, L5

Nulsen, P.E.J. & Böhringer, H., 1995, MNRAS, 274, 1093

Rottmann, H., Kerp, J., Mack, K.H., 1996, A&A, submitted

Snowden, S.C., Mc Cammon, D., Burrows, D.N., Mendenhall, J., 1994, ApJ, 424, 714

Stewart, G.C., Canizares, C.R., Fabian, A.C., Nulsen, P.E.J., 1984, ApJ, 278, 536

ASCA Observation of M 87

Hironori Matsumoto

Department of Physics
Faculty of Science
Kyoto University
Sakyo-ku
Kyoto 606-8502
Japan

Abstract. We present the ASCA results of spatially resolved X-ray spectra in the central regions of the Virgo cluster, near to the central elliptical galaxy M 87 and 40' northwest of M 87 (M 87 NW). Since the spectra of the M 87 region is complex, an adequate fit requires at least two thin thermal-plasma components. The temperatures of the hot and cool components are approximately 3.0 and 1.3 keV, respectively, and the temperature of these components are both nearly constant over the central 10' radius from the center of M 87. The spectrum of M 87NW, on the other hand, can be well fitted by a single thin thermal-plasma model of $\sim$ 2.7 keV temperature. The abundances of Fe, Si, and S rise towards the center of M 87, though the ratios of Si/Fe and S/Si remain constant. We also found, contrary to the Einstein results, that the abundance ratio of oxygen to iron (O/Fe) is smaller than the solar value. We found no strong nuclear activity from M 87.

1 Introduction

X-ray observations of the intracluster gas in clusters of galaxies provide detailed information concerning the nature of the dynamical and chemical evolution of clusters and their dark-matter distributions. Virgo, the nearest cluster, is commonly believed to be dynamically young (Binggeli *et al.* (1987)). Its high flux and close location allows for a detailed study of its intracluster gas. The X-ray emission from Virgo cluster is centered on M 87, the dominant galaxy in the cluster.

The ASCA satellite's (Tanaka *et al.* (1994)) imaging capability coupled with its good energy resolution over a wide energy band (0.4 – 10 keV) has not been available with previous satellites. These capabilities allow, for the first time, spatially resolved energy spectra of the Virgo cluster. We present an analysis of the data from a 10' circle centered on M 87, and from the region 40' northwest of M 87 (hereafter M 87 NW). For more detailed information, please see Matsumoto *et al.* (1996). We assume the distance of M 87 to be 15 Mpc.

2 Spectral Analysis and Results

M 87 was observed on 1994 June 7 and 8, pointed at $(\alpha, \delta)_{J2000} = 12^h\,30^m\,32^s$, $12°\,25'\,57''$ (M 87) and $12^h\,28^m\,24^s$, $12°\,38'\,51''$ (M 87 NW). The observation was made with the two SISs and the two GISs at the foci of 4 thin-foil XRT onboard the ASCA satellite.

Since the X-ray image of M 87 obtained with ASCA is roughly circular, we assumed spherical symmetry for simplicity. For a further analysis, we divided the central region of M 87 into 5 concentric annuli, each with a 2' width, and obtained X-ray spectra from each annulus. For the M 87 NW region, the spectra was extracted from the whole field of view.

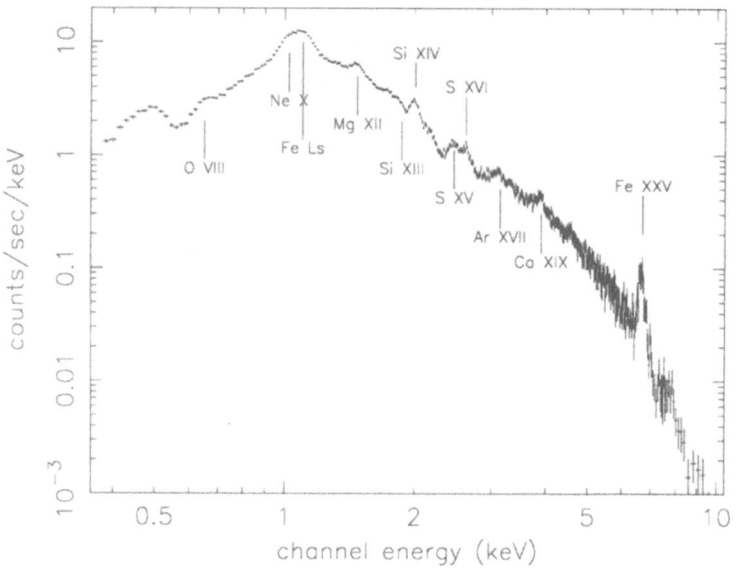

Fig. 1. SIS spectrum of M 87 within 10' radius.

Many emission lines detected with ASCA tells us that the X-ray emission of M 87 is due to optically thin plasma (Figure 1). Therefore, we tried to fit the SIS and GIS spectra simultaneously with a thin thermal plasma model (Raymond and Smith (1977)) modified by interstellar absorption. However, the one-temperature thermal plasma model could not fit the spectra except for the M 87 NW spectra. We then tried a two-temperature model for the central region of M 87, assuming that the metal abundances for the cool component is the same as that of the hot component.

The best-fit temperatures from each region are given in Figure 2. We found that the temperatures of the hot and cool components are almost constant within the M 87 region. The mean value of the best-fit temperature of the

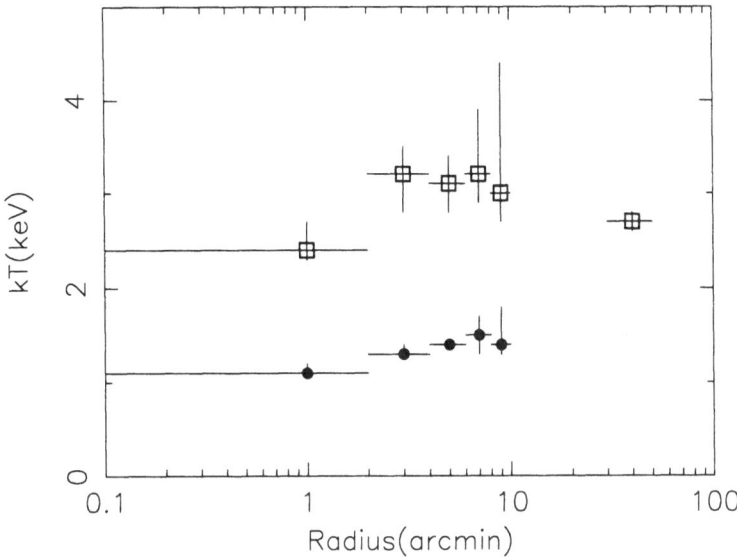

Fig. 2. Temperature distribution of the two-temperature plasma model.

hot and cool component weighted using relative errors is 3.0 and 1.3 keV. The derived kT for the cool component is consistent with that derived by the Einstein SSS (Lea *et al.* (1982)), while the temperature of the hot component is almost the same as that obtained with non-imaging instruments with a wider field of view than ASCA. This suggests that the hot component extends beyond 40' from the center.

In Figure 3, we show the X-ray surface brightness distribution of each component. We can see that the cool component decreases more rapidly than that of the hot component.

We then conclude that the hot component is due to the intracluster plasma which pervades the Virgo cluster with a nearly constant temperature of 3.0 keV [however, we note that the temperature near to the M 49 subcluster has been reported to be higher than that of the M 87 main cluster (Koyama *et al.* (1991); Takano (1990))]. Since the surface brightness of the cool component drops more rapidly with distance from M 87 than that of the hot component, and the temperature of the cool component is also constant within the 10' region centered on M 87, it is tempting to associate the cool component not with the intracluster gas, but rather with the M 87 galaxy.

Plots of the radial distribution of the best-fit abundances for each element are given in Figure 4. We can see that the abundances of Si, S, and Fe have a strong concentration toward the M 87 center. On the other hand, the O abundance is rather constant from place to place. There is no evidence of a gradient of S/Fe or Si/S over the entire image, and the observed Si/Fe ratio is similar to that found in clusters of galaxies (Mushotzky *et al.*

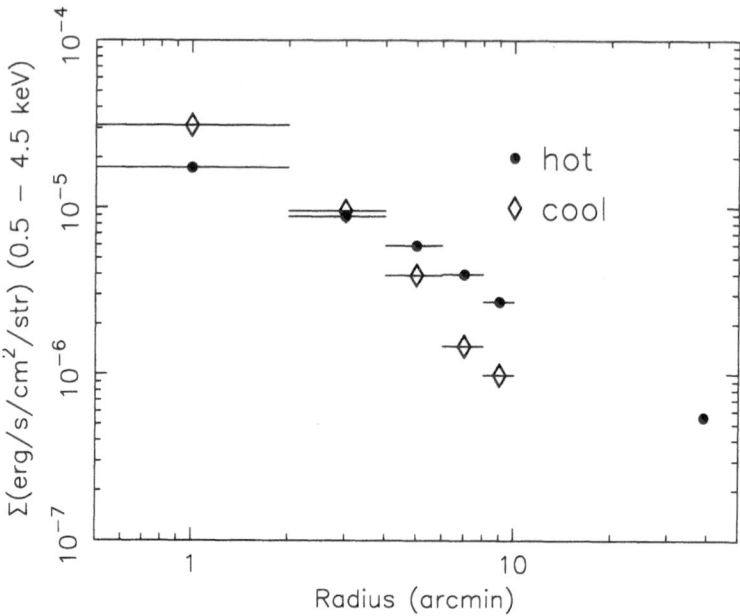

Fig. 3. X-ray surface brightness of each component. Dots show the hot component, and diamonds show the cool component.

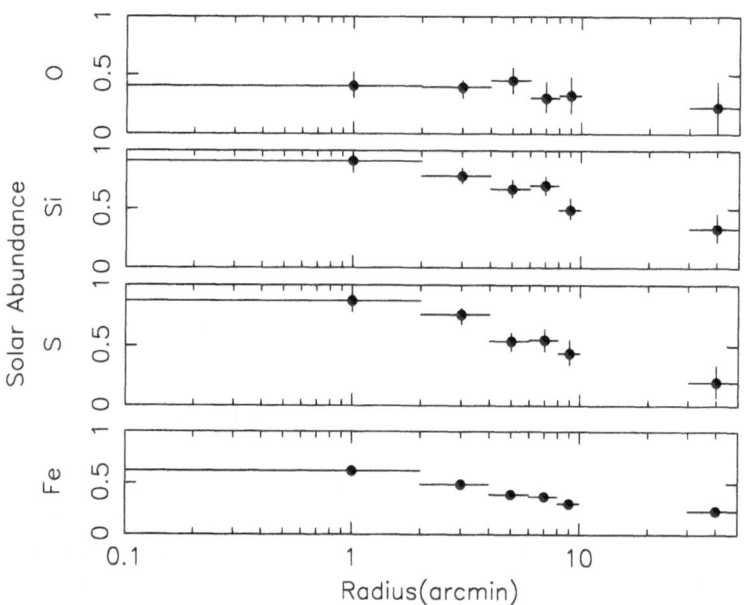

Fig. 4. Abundance distribution of O, Si, S, and Fe.

(1996)). Although a marked gradient in the Fe abundance has been found in some clusters of galaxies (Ohashi *et al.* (1994); Fukazawa *et al.* (1994); Ikebe (1996)), such a gradient has not been previously observed in the Si and S abundances. We have confirmed the Ginga result (Koyama *et al.* (1991); Takano (1990)) that Fe is concentrated toward the M 87 center, and have extended this result to the inner region of M 87. The obvious spectral distinctions between the "cluster gas" and the gas associated with M 87 is the rise in the absolute abundance and the increasing prominence of a cooler spectral component. However, the O abundance shows no significant radial variation.

The abundance ratio of O/Fe is found to be smaller than the solar value, which disagrees with the Einstein FPCS result; the O/Fe ratio is 3 – 5 times the solar value (Canizares *et al.* (1982); Tsai (1994)).

3 X-Ray Emission from AGN of M 87

A power-law component, probably due to an AGN located in the nucleus of M 87, has previously been reported with HEAO-1, Einstein, and Ginga (*e.g.* Lea *et al.* (1981); Schreier *et al.* (1982); Takano (1990); Takano and Koyama (1991); Hatsukade (1989)). We therefore examined for the presence or absence of this power-law component in the ASCA spectra and image.

The power-law component dominates the Ginga spectra above the 4 keV band (Takano (1990); Takano and Koyama (1991); Hatsukade (1989)). Therefore we study the hard band image of M 87. In Figure 5(a), we show the X-ray image of M 87 taken with the GIS in the 7 – 10 keV band. Figure 5(b) shows the radial profile of this image and the point spread function (PSF). Since the radial profile is much broader than the PSF, we concluded that we can see no point-like object at the position of M 87 in the ASCA image. When we fitted the ASCA spectra of the innermost region of M 87 with the two-temperature thermal plasma model, there is no significant residual in the high energy band. Therefore we concluded that we can find no power-law component in the ASCA spectra. The upper limit of the absorption-corrected flux in the 0.5 – 10 keV band of the power-law component is $2.5 \times 10^{-12} erg/cm^2/s$. These suggest the nucleus of M 87 has exhibited a long-term variability.

4 Summary

We summarize the ASCA results as follows:

1. We detected two-temperature plasma components from the region near to M 87. The hot component is likely to be intracluster gas of the Virgo main body (M 87 cluster), and the temperature of about 3.0 keV is nearly constant from place to place. The temperature of the cool component is ~ 1.3 keV, and shows no variation over the field. The flux of the cool

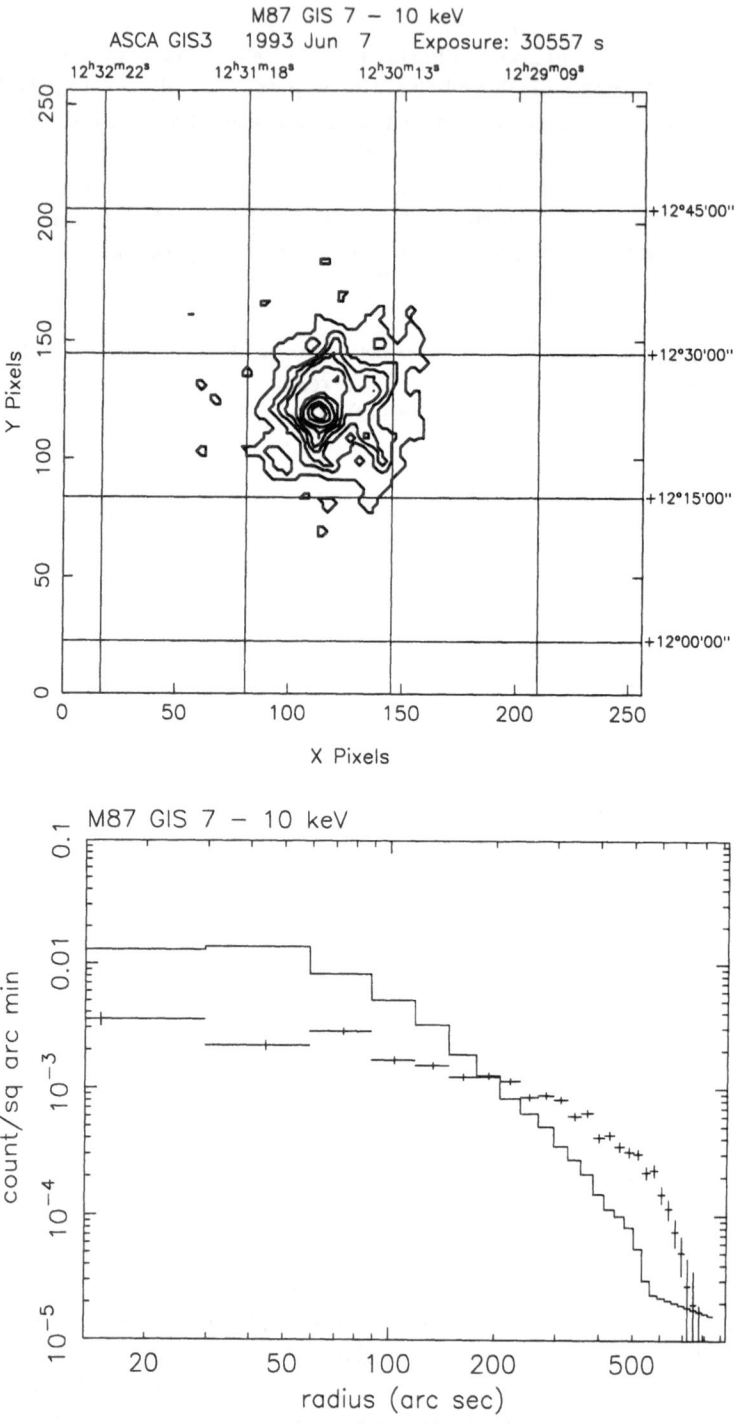

Fig. 5. (a) Hard band (7 – 10 keV) X-ray image of M 87 taken with the GIS; (b) Radial profile of (a). The solid line shows the expected radial profile for a point-like source.

component is more concentrated toward the center of M 87 than that of the hot component.

2. We found a clear abundance gradient of iron, silicon, and sulfur with no significant gradients on the ratios of Si:S:Fe. The Si:Fe ratio is rather similar to that found in rich clusters (Mushotzky *et al.* 1995), while the S:Fe ratio is rather larger. However, there is no significant radial variation in the abundance of O.

3. The abundance ratio of O/Fe is smaller than the solar ratio. This combination of low O:Fe and high Si:Fe is rather difficult to account for in relatively straightforward models of chemical evolution (Loewenstein and Mushotzky (1996)), since both O and Si tend to come from the same massive stars.

4. We found no strong nuclear activity from M 87. The upper limit of the flux of the power-law component is $\sim 2.5 \times 10^{-12}\,\mathrm{erg\ cm^{-2}\ s^{-1}}$ in the 0.5–10 keV band.

References

Binggeli, B., Tammann, G. A., and Sandage, A. 1987, AJ, 94, 251

Böhringer, H., Nulsen, P. E. J., Braun, R. *et al.* 1995, MNRAS, 274, L67

Canizares, C. R., Clark, G. W., Jernigan, J. G., *et al.* 1982, ApJ, 262, 33

Fabricant, D., Lecar, M., and Gorenstein, P. 1980, ApJ, 241, 552

Fabricant, D., and Gorenstein, P. 1983, ApJ, 267, 535

Fukazawa, Y., Ohashi, T., Fabian, A. C. *et al.* 1994, PASJ, 46, L55

Hanson, C. G., Skinner, G. K., Eyles, C. J. *et al.* 1990, MNRAS, 242, 262

Hatsukade, I. 1989, Ph.D. Thesis, Osaka University

Ikebe, Y. 1996, Ph.D. thesis, University of Tokyo

Koyama, K., Takano, S., and Tawara, Y. 1991, Nature, 350, 135

Lea, S. M., Reichert, G., Mushotzky, R. F. *et al.* 1981, ApJ, 246, 369

Lea, S. M., Mushotzky, R., and Holt, S. S. 1982, ApJ, 262, 24

Loewenstein, M., and Mushotzky, R. F. 1996, ApJ, 466, 695

Matsumoto, H., Koyama, K., Awaki, H. *et al.* 1996, PASJ, 48, 201

Mushotzky, R. F., Done, C., and Pounds, K. A. 1993, ARA&A 31, 717

Mushotzky, R. F., Loewenstein, M., Arnaud, A. *et al.* 1996, ApJ, 466, 686

Ohashi, T., Fukazawa, Y., Ikebe, Y. *et al.* 1994, in "New Horizon of X-Ray Astronomy", ed F. Makino, T. Ohashi (Universal Academy Press, Tokyo) p273

Raymond, J. C., and Smith, B. W. 1977, ApJS, 35, 419

Schreier, E. J., Gorenstein, P., and Feigelson, E. D. 1982, ApJ, 261, 42

Stewart, G. C., Canizares, C. R., Fabian, A. C. *et al.* 1984, ApJ, 278, 536

Takano, S. 1990, Ph.D. Thesis, The University of Tokyo

Takano, S., and Koyama, K. 1991, PASJ, 43, 1

Tanaka, Y., Inoue, H., and Holt, S. S. 1994, PASJ, 46, L37

Tsai, J. C. 1994, ApJ, 423, 143

The Large Scale X-Ray Emission from M 87

D. E. Harris[1], J. A. Biretta[2], and W. Junor[3]

[1] Smithsonian Astrophysical Observatory
60 Garden St
Cambridge, MA 02138, USA
[2] Space Telescope Science Institute
3700 San Martin Dr.
Baltimore, MD 21218, USA
[3] University of New Mexico
800 Yale Blvd. NE
Albuquerque, NM 87131, USA

Abstract. We describe asymmetrical features in a long exposure X-ray map of M 87 made with the ROSAT High Resolution Imager (HRI). A bright triangular region is marked by a linear 'spur' along one edge. The structure of this spur suggests an interpretation of a tangential view of a shock front 18 kpc long. None of the brighter features are spatially coincident with radio or optical structures so we concur with earlier investigators that most of the emission arises from thermal processes.

1   Introduction

In addition to the strong X-ray emission (roughly circularly symmetric) from the gas associated with M 87 and the Virgo cluster, asymmetric X-ray features have been known since the Einstein Observatory observations were obtained 15 years ago (Schreier, Gorenstein, and Feigelson (1982), Feigelson *et al.* (1987)). From these data, it was noted that the large scale X-ray emission which is asymmetric was roughly correlated with the intermediate scale radio structure, and both inverse Compton emission and thermal bremsstrahlung emission were considered as possible origins. More recently it has become clear that there is not a tight spatial correlation between radio and X-ray emissions (Böhringer *et al.* (1995)), and thus the inverse Compton process is not the major contributor to the observed X-rays.

In this paper, we present further information on these structures derived from an analysis of a ROSAT HRI map consisting of many monitoring observations. The effective integration time for this map is 202 ksec and we are thus able to delineate weaker features with relatively high resolution. The X-ray emissions from the core and knot A in the radio jet will not be dealt with here (see Harris, Biretta, and Junor, this volume, for the latest variability data on the core and knot A).

2   Construction of the Image

The inherent spatial resolution of the ROSAT HRI is close to 5″. However, the effective resolution can be degraded by two types of aspect (*i.e.* pointing) problems: the star trackers occasionally make gross errors of order 10″ and there is often a residual error associated with the spacecraft wobble. The first type of error results in erroneous locations in celestial coordinates and is seen most often when combining observations taken 6 months apart, although occasionally there will be an aberrant observation interval in the typical observation consisting of 10 or more intervals. The second type of error produces an ellipsoidal point response function (PRF) which can smear the 5″ inherent PRF to something like 7″ by 10″. Since the current objective is to study the larger scale structures, we have not addressed these errors except for the alignment of observations.

Our monitoring program has resulted in the accumulation of 5 observations ranging from 30 to 44 ksec each, at 6 month intervals between 1995 June and 1997 June. To these we added the 13.9 ksec HRI observation obtained in 1992 June, resulting in a total of 202 ksec. Before adding each observation, we measured the position of both the core and knot A, and shifted each observation (up to 6″) so as to align all the data to the reference frame of the 1992 June observation. The resulting events file is represented by the map in Fig. 1, to which we have added labels for some of the features described in section 3.

Since most of the asymmetrical structure occurs to the East and SW of the core, we made a radial profile only for the hemisphere between PA=240° and 60°. We chose the center for the radial profile to be ≈ 7″ north of the core since that position appeared to be the center of the quasi circular 'bulge' of high brightness surrounding the core and knot A. We made no effort to fit this central bulge (radius ≈ 40″) but found that a power law with slope of 1.10 was a reasonable fit to the brightness between 40″ and 900″. We then subtracted this power law $(1+(r/a)^{-1.1})$ from the image, adjusting the normalization until there were few negative regions. The results are shown in Fig. 2.

3   Description of Discrete Features

For the most part, the following sections will describe some of the more obvious features of the X-ray distribution without a quantitative analysis. The latter subject will be addressed elsewhere.

3.1   SW spur

The most striking X-ray feature evident in Figs. 1 and 2 is the spur of emission extending from the central region ≈ 4′ toward the south west in a slightly

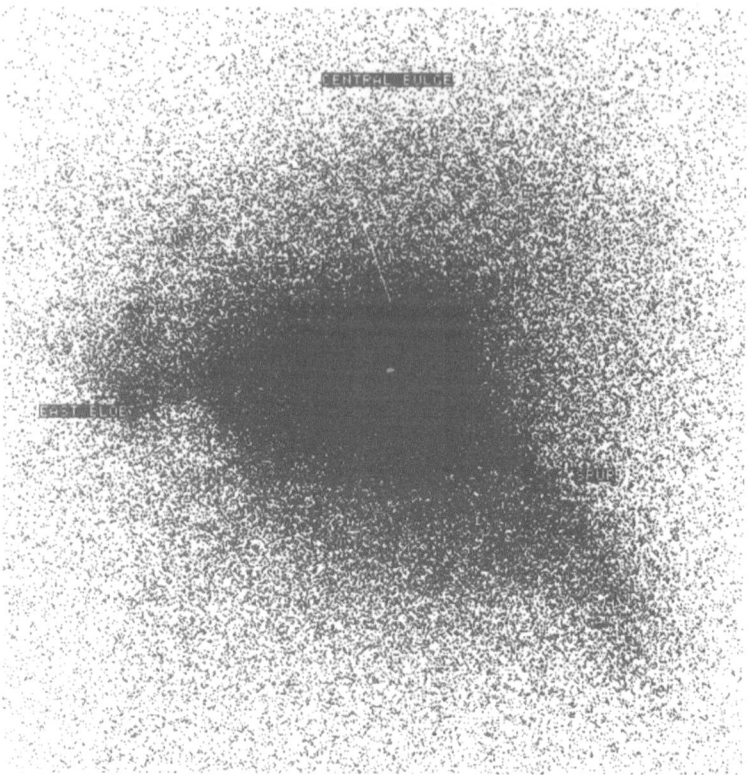

Fig. 1. The stacked image of M 87 from 6 observations. Each side of the figure has a length of 256 arcsec. The core and knot A are within the burned out central region.

curved trajectory. For a distance of 16 Mpc to M 87, 4′ corresponds to 19 kpc. The NW edge is essentially unresolved but the brightness falls off more gradually behind it towards the SE. This is demonstrated in Fig. 3.

There is no optical or radio feature corresponding to the spur, and this agrees with the results of Böhringer *et al.* (1995) who used a spectral analysis of the ROSAT Position Sensitive Proportional Counter observation to conclude that the asymmetric X-ray emission was thermal, but from a cooler gas than the surrounding cluster emission. However, from a preliminary hardness ratio map of our HRI data, there is some indication that the spur is actually harder than the surrounding emission.

As is often the case, it is difficult to reproduce the detail that can be obtained by visual inspection of various versions of an image. For this reason, we have added asterisks at several positions, measured by eye, of the ridge line of the spur (Fig. 2). This aid in tracing the spur closer into the core allows us to conjecture that the spur may be causally related to knot A. A possible

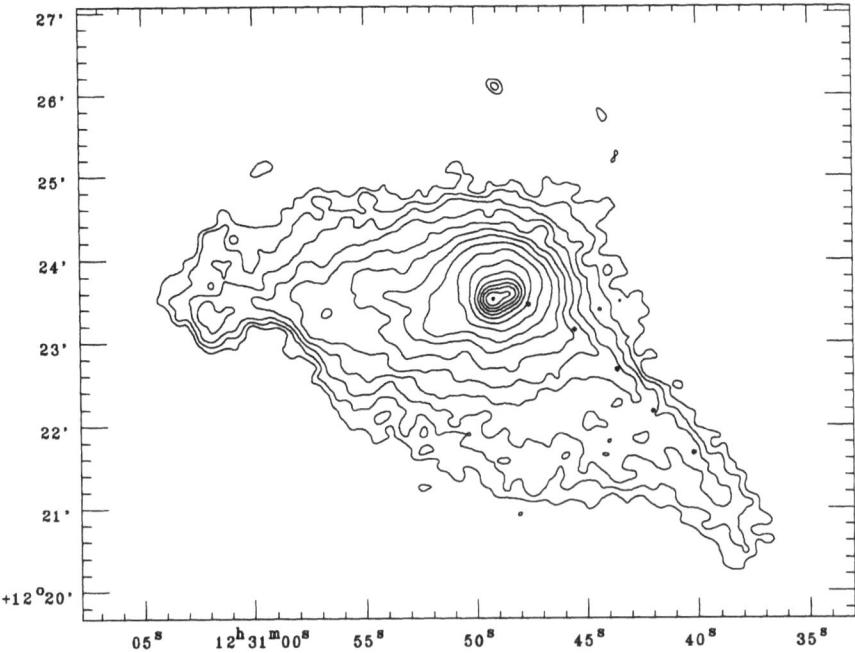

Fig. 2. The residual X-ray map after subtraction of the power law model. Contour intervals are logarithmic, starting at one count per pixel and ending at 155 counts per pixel. The pixel size is one arcsecond. The map has been smoothed with a Gaussian of FWHM = 8″. The asterisks indicate measured positions along the ridge line of the SW spur.

explanation for the spur (and perhaps for much of the Eastern arm) would be that it is a tangential view of a bow shock associated with the radio jet: either knot A itself or just beyond where the jet bends sharply to the South. The excess X-ray emission which we call the spur would then come from a relatively long path length through the higher density gas compressed behind the bow shock.

A quite different scenario, but also involving a shock front, turns the argument around. If the shock is not caused by the radio jet, it would still provide a discontinuity in the ambient density, temperature, and pressure. It is just such conditions which can be the causal agent for the genesis of an internal shock in a radio jet (*e.g.* Hooda and Wiita (1998)); *i.e.* the change in external pressure is the reason that knot A exists at this location. For this scenario, the spur would be the emission from lines of sight which are tangential to the curved bow shock. From the scale of the spur and eastern arm, the most likely cause of such a shock would be the interaction of the galaxy's ISM with the ICM. Although M 87 is not thought to be moving substantially with respect to the Virgo ICM, Binggeli (1998) has argued that

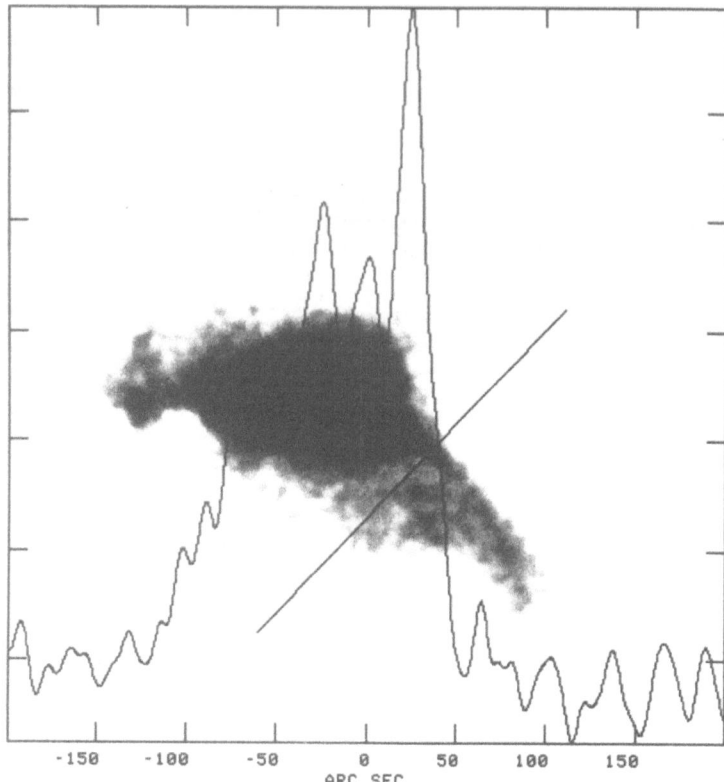

Fig. 3. A profile though the X-ray spur. Note the sharp leading edge on the right side. The X-ray map has had the model power law subtracted and was smoothed with a Gaussian of FWHM = 8″. The straight line shows the location of the measured profile.

a merger between the M 87 subgroup and the M86 subgroup is in progress, and this could provide the required relative velocity between gas distributions to form a weak shock.

The chief criticism of the association between an ICM/ISM shock and knot A is that one would not expect knot A to be seen, at least in projection, near the edge of the shock. This is because the standard interpretation of the M 87 jet is that it is aimed towards us, not much more than 30° to the line of sight.

3.2   Other Features Comprising the Triangular Emission Region

The brightest part of the residual image (Fig. 2) forms a roughly triangular region on the sky with the spur delineating the NW edge. The southern 'base' of the triangle is not straight, nor is it well defined; the emission falls away gradually to the south. The brightest part of the triangle aside from the core and knot A is the central bulge, a quasi circular distribution with a characteristic radius of 40″. Extending towards the East is the 'eastern arm' which has a complex morphology. The easternmost feature in Fig. 2 is a quasi circular emission region, 'the eastern blob'. Note the sharp notch in the base of the triangle near the eastern end of the base. It is this notch which defines the eastern blob, and it is tempting to interpret this notch as the effect of absorption. However, there is no corresponding feature on the hardness ratio map.

3.3 Lower Brightness Extended Emission

Referring now to Fig. 1, we see a lower brightness region to the north of the triangle. This region is remarkable in that it has a well defined northern boundary which is quasi circular, with a radius of 3.5′, centered about 30″ East of the core of M 87. The circular boundary originates in the eastern blob and can be traced for over 90°. At a somewhat lower brightness, one can see extended emission below the 'base' of the triangle; on some displays, it appears to be somewhat striated. Finally, to the NE of the eastern Blob, outside of the circular boundary, there is a 'fan' of lower brightness emission which ends near the left border of Fig. 1.

4 Acknowledgments

The work at SAO was partially supported by NASA contract NAS5-30934 and grant NAG5-2960; that at STScI by NASA grant NAG5-2957.

References

Binggeli, B. (1998): *Ringberg Workshop on M 87* (Springer, Berlin, Heidelberg)

Böhringer, H., Nulsen, P.E.J., Braun, R., and Fabian, A.C. (1995): The interaction of the radio halo of M 87 with the cooling intracluster medium of the Virgo Cluster. MNRAS **274**, L67–L71

Feigelson, E.D., Wood, P.A.D., Schreier, E.J., Harris, D.E., and Reid, M.J. (1987): X-Rays from the Radio Halo of M 87. ApJ **312**, 101–110

Harris, D.E., Biretta, J.A., and Junor, W. (1997): X-ray Variability in M 87. MNRAS **284**, L21–L27

Hooda, J.S. and Wiita, P.J. (1998), ApJ **493**, 81–90

Schreier, E.J., Gorenstein, P., and Feigelson, E.D. (1982), ApJ **261**, 42–50

Cluster Turbulence

Michael L. Norman[1,2] and Greg L. Bryan[3]

[1] Astronomy Department and NCSA
   University of Illinois
   Urbana, IL 61801
   USA
[2] Max-Planck-Institut für Astrophysik
   D-85740 Garching
   Germany
[3] Princeton University Observatory
   Peyton Hall
   Princeton, NJ 08544
   USA

Abstract. We report on results of recent, high resolution hydrodynamic simulations of the formation and evolution of X-ray clusters of galaxies carried out within a cosmological framework. We employ the highly accurate piecewise parabolic method (PPM) on fixed and adaptive meshes which allow us to resolve the flow field in the intracluster gas. The excellent shock capturing and low numerical viscosity of PPM represent a substantial advance over previous studies using SPH. We find that in flat, hierarchical cosmological models, the ICM is in a turbulent state long after turbulence generated by the last major merger should have decayed away. Turbulent velocities are found to vary slowly with cluster radius, being $\sim 25\%$ of σ_{vir} in the core, increasing to $\sim 60\%$ at the virial radius. We argue that more frequent minor mergers maintain the high level of turbulence found in the core where dynamical times are short. Turbulent pressure support is thus significant throughout the cluster, and results in a somewhat cooler cluster ($T/T_{vir} \sim .8$) for its mass. Some implications of cluster turbulence are discussed.

1 Introduction

Our conception of galaxy clusters as being dynamically relaxed systems has undergone substantial revision in recent years. Optical observations reveal substructure in 30–40% of rich clusters (Geller & Beers 1982; Dressler & Shectman 1988). A wealth of new X-ray observations have bolstered these findings, providing evidence of recent mergers in clusters previously thought to be archetypal relaxed clusters (*e.g.* Briel *et al.* 1991). Also eroding the conventional view has been the success of "bottom-up" or hierarchical models of cosmological structure formation in accounting for the formation of galaxies and large scale structure in the universe (*e.g.* Ostriker 1993). Within such models, a cluster sized object is built up through a sequence of mergers of lower-mass systems (galaxies $\rightarrow$ groups $\rightarrow$ clusters). In a flat universe ($\Omega_o = 1$) as predicted by inflation, mergers would be ongoing at the present epoch. In open models ($\Omega_o < 1$), mergers cease at a redshift $z \sim \Omega_o^{-1} - 1$, and clusters become relaxed by today. The amount of substructure observed in X-ray clusters of galaxies at $z \sim 0$ is thus a powerful probe of cosmology. Evrard *et al.* (1993) and Mohr *et al.* (1995) have explored this "morphology-cosmology" connection, and concluded that a high Ω universe is favored. Interestingly, Tsai & Buote (1996) reach the opposite conclusion.

Cluster mergers have been explored numerically by several groups (Schindler & Müller 1993; Roettiger, Loken & Burns 1997; Roettiger, Stone & Mushotzky 1998). In these hydro/N-body simulations, two hydrostatic King models are collided varying the cluster–subcluster mass ratio. It is found that major mergers induce temperature inhomogeneities and bulk motions in the ICM of a substantial fraction of the virial velocity ($> 1000 \, km/s$). Roettiger *et al.* suggest that these bulk motions may be responsible for the observed temperature substructure seen in some X-ray clusters, as well as bending Wide-Angle Tailed radio sources, energizing cluster radio halos, and disrupting cooling flows.

If hierarchical models are correct, the thermal and dynamical state of the ICM could be considerably more complex than the above mentioned simulations indicate. In a flat universe, for example, the ICM would be constantly bombarded by a rain of minor mergers in addition to the occasional major merger. Also omitted in those simulations are a variety of cosmological effects which may be important, including memory of the complex formation history of the merging clusters, infall of matter along filaments, accretion shocks, large scale tides, and cosmic expansion.

In this paper we show results of numerical simulations that take all these effects into account. We find in two flat models investigated, that quite generally the ICM of rich galaxy clusters is in a turbulent state. The turbulent velocities are typically 60% the virial velocity at the virial radius, decreasing inward to roughly 25% within the core. The relatively slow decline in turbulence amplitude with decreasing radius suggests that frequent minor mergers are an important driving mechanism in addition to rare massive mergers. In addition, we find ordered fluid circulation in the core of one well-resolved cluster which is likely the remnant of a slightly off-axis recent merger.

2 Simulations

The simulations are fully cosmological. That is, the formation and evolution of the clusters is simulated by evolving the equations of collisionless dark matter, primordial gas and self-gravity in an expanding FRW universe (see *e.g.* Anninos & Norman 1996). Initial conditions consist of specifying linear density and velocity perturbations in the gas and dark matter in Fourier space with power spectrum $P(k)$ and random phases. We have simulated two cosmological models which differ primarily in their assumed $P(k)$'s: CDM, with power normalized to reproduce the abundance of great clusters at z=0, and CHDM, normalized to the COBE measurement on large scales. We assume the gas is non-radiative, which is a good approximation except in the cores of cooling flow clusters. The statistical properties of X-ray clusters in these models (as well as an open model) are presented in Bryan & Norman (1998a). The internal structure of a smaller sample of X-ray clusters computed at higher resolution are presented in Bryan & Norman (1998b). Here we summarize the key findings from Bryan & Norman (1998b), restricting ourselves to the properties of four clusters drawn from the CDM simulations. Table 1 summarizes the clusters' bulk properties.

Two different numerical gridding techniques have been employed. The first uses a uniform Eulerian grid with 512^3 cells in a comoving volume of 50 Mpc, for a cell resolution of $\sim 100\ kpc$. While unable to resolve the cluster core, this calculation provides good coverage in the cluster halo and beyond. Three clusters, called CDM1-3, are taken from this simulation. The second employs adaptive mesh refinement (AMR; Bryan & Norman 1997a) which automatically adds high resolution subgrids wherever needed to resolve compact structures, such as subclusters forming at high redshift or the cluster core at $z = 0$. We have computed a single rich cluster, called SB, with 15 kpc resolution in the core (Bryan & Norman 1997b). This cluster has been simulated by a dozen groups in the "Santa Barbara cluster comparison project", (Frenk et al. 1998). Together, these simulations allow us to characterize the thermal and dynamical state of the ICM across a wide range of scales. Both simulations use the piecewise parabolic method (PPM) for gas dynamics, modified for cosmology (Bryan et al. 1995), and the particle-mesh method (PM) for the dark matter dynamics. The excellent shock capturing and low numerical viscosity of PPM make it ideal to study cluster turbulence.

Table 1. Cluster parameters

cluster	r_{vir} (Mpc)	M_{vir} ($10^{15} M_\odot$)	T_{vir} (keV)	σ_{vir} (km/s)	Δx
CDM1	2.58	0.890	4.63	861	98
CDM2	2.32	0.647	3.74	774	98
CDM3	2.40	0.716	4.00	801	98
SB	2.70	1.1	4.71	915	15

3   Turbulence in the Halo

In Figure 1, we plot the azimuthally averaged total velocity dispersion, radial component of the velocity dispersion and the radial velocity for both the dark matter and gas components of clusters CDM1-3. They are normalized by the virial values from Table 1 and all velocities are relative to the center-of-mass velocity of the matter within r_{vir}. The dispersion in the radial direction is around the net radial velocity of that shell: $\sigma_r^2 = \langle (v_r - \langle v_r \rangle)^2 \rangle$.

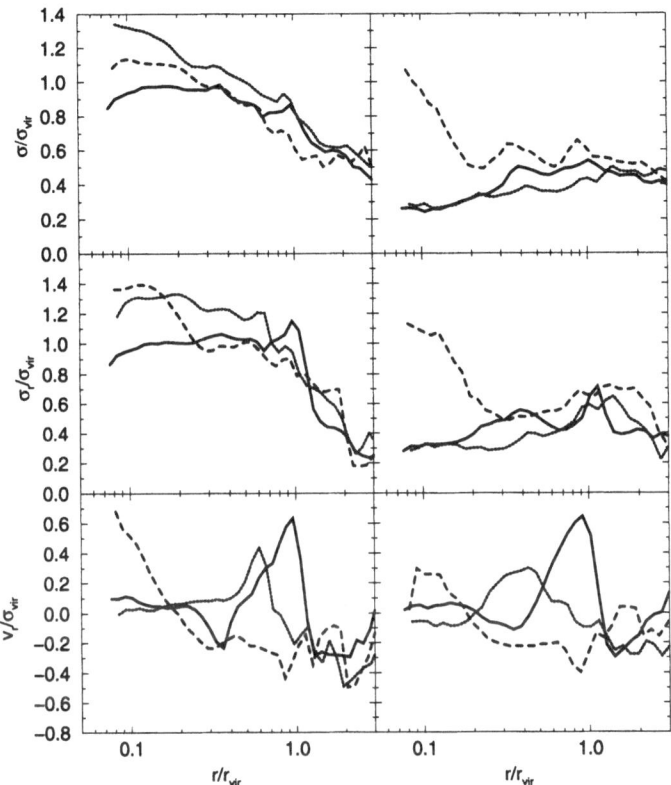

Fig. 1. The velocity dispersion (top panels), radial velocity dispersion (middle panels) and radial velocity in shells for the dark matter (left side) and gas (right side) of the three largest clusters in the CDM512 simulation (solid/dotted/dashed lines correspond to clusters designed as CDM1/CDM2/CDM3). Profiles are normalized by their virial values (see Table 1).

Focusing first on the dark matter, the velocity dispersion profiles are roughly compatible with their virial values within the virial radius, but fall off quickly beyond that point. There is some preference for radial orbits around and slightly beyond r_{vir}, but at low radii, the velocities are isotropic. The radial velocity profile (bottom panel) shows evidence for infall in the $1-4r_{vir}$ range. The third cluster in this sample (dashed line) is undergoing a major merger and shows signs of enhanced bulk motions in the inner 400 kpc, although the velocity-dispersion profiles are not strongly disturbed.

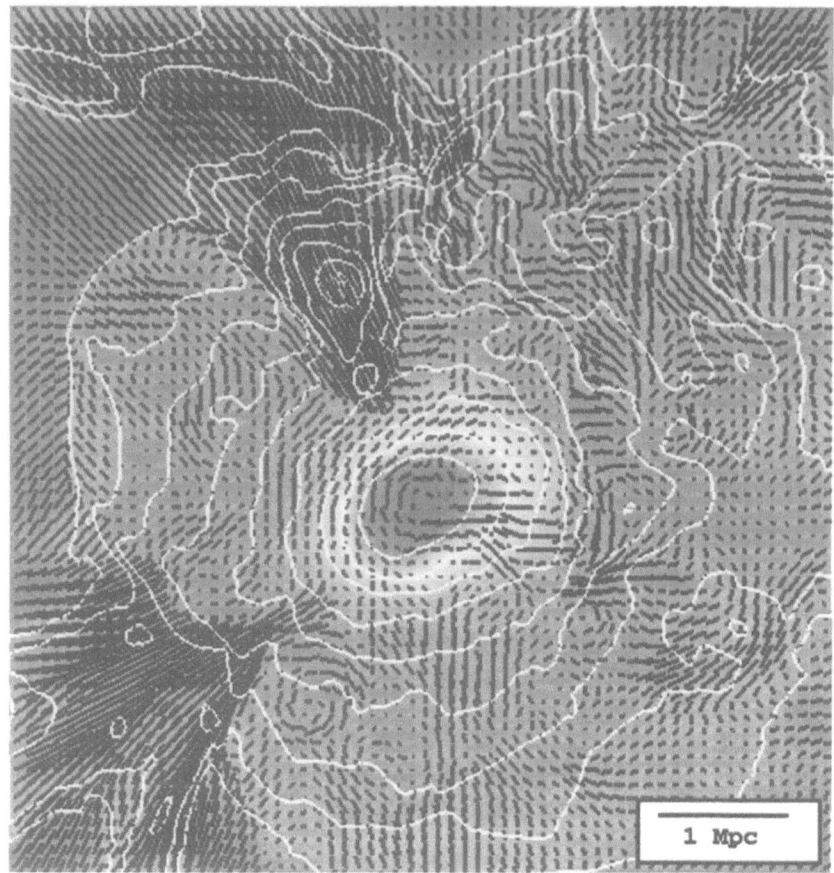

Fig. 2. The large scale velocity field on a thin slice though the center of cluster SB shown overtop the logarithm of gas density (image, contours). The maximum velocity vector is 2090 km/s. The image is 6.4 Mpc on a side.

The gas velocity dispersions range between 0.25 and 0.6 σ_{vir}, considerably below their dark matter counterparts, but are not insignificant. In fact, these motions contribute some additional support beyond that provided by the mean baryonic pressure gradient. We may approximate this by applying Jean's equation to the coherent clumps of gas with velocity dispersion σ and density ρ_c. Ignoring differences between the radial and tangential velocity dispersion this becomes:

$$\frac{1}{\rho_c}\frac{d(\rho_c\sigma^2)}{dr} + \frac{1}{\rho}\frac{dP}{dr} = -\frac{GM(r)}{r^2}. \tag{1}$$

Since $P = \rho kT/\mu m_h$, where μm_h is mean mass per particle, we see that the temperature and σ^2 combine to support the cluster gas against gravitational collapse. We can directly compare T/T_{vir} against σ^2/σ_{vir}^2, so the temperature provides about 80% of the support. This provides an explanation for the observation (Navarro, Frenk & White 1995; Bryan & Norman 1998a) that the mean cluster temperatures were, on average, about 0.8 of its virial value.

Thus we see that the gas has not completely virialized and sizable bulk motions exist. Since the mean entropy profile increases with increasing radius, the halo is globally stable, so this turbulence must be driven by external masses falling into the cluster and damped by viscous heating. The turbulence amplitude in the halo appears to be roughly compatible with this explanation since the driving time scale — approximately the Hubble time — is slightly larger than the damping time scale which is essentially the crossing time. Moreover, σ^2 seems to drop (and T approaches T_{vir}) as $r \to 0$ and the crossing time decreases. We discuss this point further in the last section.

Figure 2 shows the chaotic flow field on a slice through the center of cluster SB. Velocity vectors for the gas are superposed on the log of the gas density. High velocity streams seen at 8 and 11 o'clock are caused by inflow of low entropy material along large scale filaments. This low entropy gas sinks to the center of the cluster. Generally, subclusters fall in along filaments, and their passage through the cluster generates vorticity, seen here as large scale eddies, via the baroclinic mechanism (*e.g.* Stone & Norman 1992). The eddies are $\sim 500\,\mathrm{kpc}$ in diameter and have a velocity of $\sim 1000\,\mathrm{km/s}$. Between the filaments, gas can actually move outwards. In this cluster, a portion of the main accretion shock is visible in the upper left corner. The infalling gas impacts the shock with a range of angles. When the velocity is normal to the shock front, the gas is almost completely virialized, however, oblique impacts generate substantial vorticity in the post-shock gas. This is another source of turbulent motions in the cluster gas.

4 Bulk Motions in the Core

Inside $1\,\mathrm{Mpc}$, we see coherent bulk motions with typical velocities of $\sim 500\,\mathrm{km/s}$ and correlation scales between $100\,\mathrm{kpc}$ and $1\,\mathrm{Mpc}$. The geometry of the flow is complex, changing character on different slices. The slice shown in Fig. 2 shows a large-scale circulation about the cluster core.

Using the high resolution model SB, we can probe the velocity field on scales down to $0.01r_{vir} = 27\,\mathrm{kpc}$. A blow up of the central portion of the cluster shown in Fig. 3. Here, the spacing of the vectors corresponds to our cell size $15\,\mathrm{kpc}$. The clockwise circulation is clearly evident here, and numerically well resolved. Within the central $200\,\mathrm{kpc}$, we can see eddies 4–5 cells in diameter — close to our resolution limit. Thus, turbulence exists even in the cores of X-ray clusters.

Fig. 4 shows the three dimensional velocity field in a $600\,\mathrm{kpc}$ box centered on the core (shaded isosurface). We find that the flow is quite ordered on these scales, with bulk velocities of 300–400 km/s. Here we have rendered

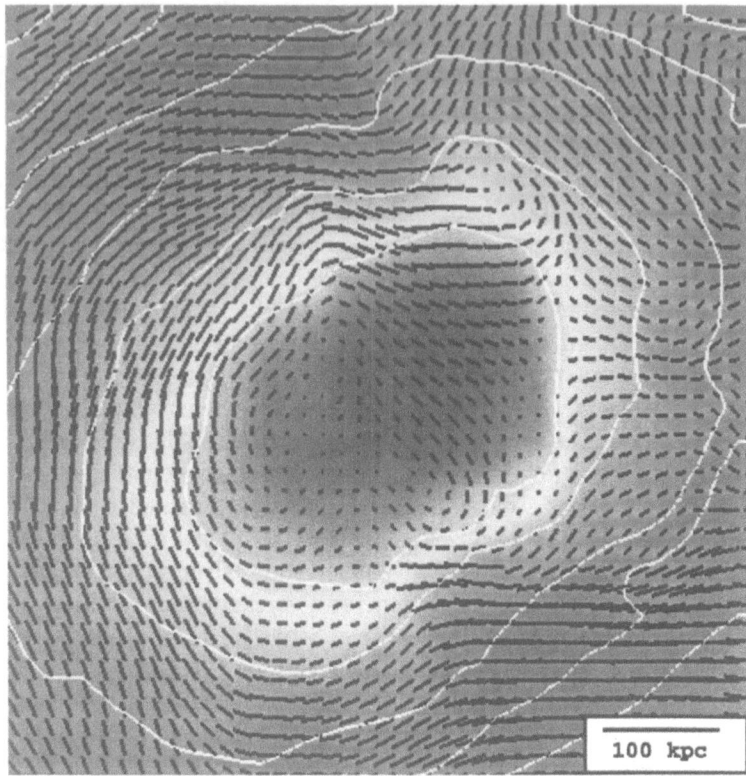

Fig. 3. The velocity field on a thin slice in the inner 600 kpc of cluster SB. The maximum velocity vector is 520 km/s.

fluid "streaklines" passing through the core, which are tangent curves of the instantaneous velocity field. In a steady flow, streaklines and streamlines are identical and trace out the paths that fluid elements follow. In a time-dependent flow, such as we have here, streaklines provide only a sense of the geometry of the velocity field. Close inspection of Fig. 4 as well as 3d rotations on graphics workstation reveal a swirling flow superposed on a linear flow. The linear flow corresponds to the mean peculiar velocity of the cluster core, which points from the origin of the cube to the upper right furthest corner of the cube. The swirling flow can be seen as the bundle of streaklines coming out of the page below the core, passing in front of the core, and going back into the page above the core.

5   Discussion

We have shown using high resolution hydrodynamic simulations that the ICMs in bright X-ray clusters in flat hierarchical models are turbulent through-

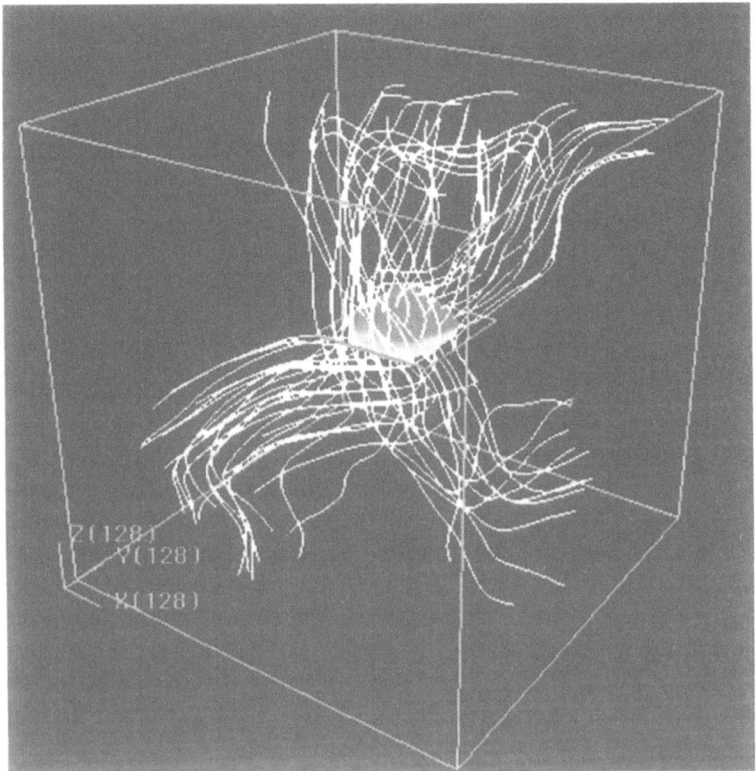

Fig. 4. The 3d velocity field in the inner 600kpc centered on the core.

out. The turbulence in strongest in the outskirts of the cluster and weaker in the core. Due to the declining temperature profile in cluster halos, the turbulence is found to be mildly supersonic ($M \sim 1.6$) near r_{vir}, decreases rapidly to $M \sim 0.5$ at $\sim \frac{1}{3}r_{vir}$, and thereafter declines more slowly to $M \sim .3$ in the core.

Here we argue that infrequent major mergers cannot sustain the observed level of turbulence in the core. It is known from simulations of decaying turbulence in a box that the turbulent kinetic energy decays as $t^{-\eta}$ where t is measured in units of the dynamical time. The exponent η depends weakly on the nature of the turbulence, but is around 1.2 for compressible, adiabatic, hydrodynamic turbulence (Mac Low *et al.* 1998). The time for a sound wave to propagate from the center of the cluster SB to a radius $.01, .1, 1 \times r_{vir}$ is 0.014, 0.173, 3.1 Gyr, respectively. The cluster underwent a major merger at $z = 0.4$, or 5.2 Gyr earlier. Taking the sound crossing time as the dynamical time, we predict that fluid turbulence induced by the major merger at $z = 0.4$ would have decayed to $.006, .017, .56$ of its initial value by $z = 0$.

Several possibilities suggest themselves to account for the high fluid velocity dispersions seen in the core. The first is that energy is somehow pumped into the core by motions in the outer parts of the cluster which relax on longer time scales. However, shock waves generated by supersonic motions in the outskirts would weaken into acoustic disturbances as they propagated into the dense, hotter core. Gravitational accelerations in the core would be dominated by the local dark matter distribution which would relax on a time scale comparable to the turbulence decay time scale. Another pumping mechanism discussed by Roettiger, Burns & Loken (1996) is global oscillations of the cluster potential following a major merger. They find that rms velocities decay to $\sim 200\,\mathrm{km/s}$ by 2 Gyr after core passage, and remain quite constant thereafter. This is substantially less than the velocities we find.

The second possibility, which we consider more likely, is that core turbulence is driven by the more frequent minor mergers. Lacey & Cole (1993) have quantified the merger rates in hierarchical models. They find that the merger rate for CDM scales as $(\Delta M/M_{cl})^{-\frac{1}{2}}$ where ΔM is the subcluster mass. Whereas most of a cluster's final mass is typically accreted in a single major merger, they find that the cluster will typically accrete $\sim 10\%$ of its mass in ten minor mergers of clumps $\sim 1\%$ of its final mass. The most probable formation epoch for a $10^{15}\,\mathrm{M}_\odot$ cluster in the standard CDM model we have simulated is at $.7\,t_{Hubble}$, or 4 Gyr ago. The mean time between minor mergers is thus 0.4 Gyr — comparable to the dynamical time at a tenth the virial radius.

Is there sufficient energy in minor mergers to sustain the turbulence in the core, and if so, how is the energy deposited? The kinetic energy of ten $10^{13}\,\mathrm{M}_\odot$ subclusters is $\sim 10^{63}\,\mathrm{erg}$, as compared to approximately $10^{62}\,\mathrm{erg}$ of turbulent kinetic energy within $0.1r_{vir}$. Thus, a 10% energy conversion efficiency is required for this mechanism to be correct. If the coupling is purely hydrodynamic (*i.e.* shocks), then the energy available is the kinetic energy of the gas in the subcluster, which is down by a factor of Ω_b from the estimate above. Since $\Omega_b = .05$, this energy is insufficient. Thus, it would seem that a substantial gravitational coupling between the ICM and the dark matter in the subclusters is required. This is equivalent to saying that the gas remains bound to the subcluster until it reaches the core. Roettiger *et al.* (1996) found that this is indeed the case.

There are a number of interesting implications to significant levels of turbulence in the cores of X-ray clusters, many of which have already been pointed out by Roettiger *et al.* (1996), including Doppler shifting of X-ray emission lines, bending of Wide-Angle Tailed radio galaxies, and powering cluster radio halos. Our findings strengthen their conclusions. For example, the turbulent amplification of magnetic fields would be expected to be most efficient in cluster cores where dynamical time scales are shortest. Moreover, continuous stirring by minor mergers could modify cooling flows appreciably. Because turbulent pressure "cools" inefficiently compared to atomic processes, turbulent pressure support could become increasingly important in the central parts of a cooling flow. Its effect would be to reduce the mass inflow rate into the cluster center. Secondly, at radii much less than the cooling radius, turbulent motions would concentrate cooling gas into filaments, and possibly account for the observed $H\alpha$ filaments. Finally, we note that ordered circulation in the cores of X-ray clusters such as we have found might account for the S-shaped symmetry of radio tails seen in some sources (*e.g.* M 87; Böhringer *et al.* (1995), Owen, these proceedings.)

Acknowledgements: This work was partially supported by grants NASA NAGW-3152 and NSF ASC-9318185. Simulations were carried out on the Connection Machine-5 and Silicon Graphics Power Challenge Array at the National Center for Supercomputing Applications, University of Illinois.

References

Anninos, P. & Norman, M. L. 1996. ApJ, 459, 12.

Böhringer, H., Nulsen, P., Braun, R. & Fabian, A. 1995. MNRAS, 274, L67.

Briel, U. *et al.* 1991. A&A, 246, L10.

Bryan, G. L., Norman, M. L., Stone, J. M., Cen, R. & Ostriker, J. P. 1995. Comp. Phys. Comm., 89, 149.

Bryan, G. L. & Norman, M. L. 1997. in *Computational Astrophysics*, PASP Conference Series Vol. 123, eds. D. Clarke & M. West, (PASP: San Francisco), 363.

Bryan, G. L. & Norman, M. L. 1998. to appear in *Structured Adaptive Mesh Refinement Grid Methods*, ed. N. Chrisichoides, IMA Conference Series, in press (astro-ph/9710187).

Bryan, G. L. & Norman, M. L. 1998a. ApJ, in press (astro-ph/9710107).

Bryan, G. L. & Norman, M. L. 1998b. NewA, submitted.

Dressler, A. & Schectman, S. 1988. AJ, 95, 985.

Evrard, A., Mohr, J., Fabricant, D. & Geller, M. 1993. ApJ, 419, L9.

Frenk, C. S. *et al.* 1998. ApJ, in press.

Geller, M. & Beers, 1982. PASP, 94, 421.

Lacey, C. & Cole, S. 1993. MNRAS, 262, 627.

Mac Low, M.-M., Klessen, R., Burkert, A. & Smith, M. D. 1998. Phys. Rev. Lett., submitted (astro-ph/9712013).

Mohr, J., Evrard, A., Fabricant, D. & Geller, M. 1995. ApJ, 447, 8.

Navarro, J., Frenk, C. & White, S. 1995. MNRAS, 275, 720.

Ostriker, J. P. 1993. ARAA, 31, 689.

Roettiger, K., Burns, J. O. & Loken, C. 1996. ApJ, 473, 651.

Roettiger, K., Stone, J. M. & Mushotsky, R. 1998. ApJ, 493, 62.

Schindler, S. & Müller, E. 1993.

Stone, J. M. & Norman, M. L. 1992. ApJ, 389, L17.

Tsai, J. C. & Buote, D. A. 1996. MNRAS, 282, 77.

M 87 and Cooling Flows

James Binney

Theoretical Physics
Oxford University
Keble Road
Oxford OX1 3NP
UK

Abstract. The hot gas in M 87 is dense enough to have a cooling time that is significantly shorter than the Hubble time out to a radius $r_{cool} \sim 70$ kpc. For historical reasons this gas has often been assumed to be in a steady state in which gas sinks inwards until at some radius it condenses out into undetected compact objects. A considerable body of observational evidence now suggests that the gas, far from being in a steady state, is interacting violently with the AGN which it feeds. There are good observational and theoretical grounds for believing that the nuclear activity is episodic and that the enveloping plasma is in a highly dynamic state completely unlike that assumed in the classical cooling-flow picture.

1 Introduction

M 87 is the best-studied "cooling-flow" galaxy, and as such an object of great importance for extragalactic astronomy as a whole. To understand the literature of cooling flows a knowledge of its historical context is essential. Therefore I now summarize the history of the subject as a whole before turning to the particular case of M 87 in some detail. Since it is the work that I know best, I shall give more prominence to my own work than it merits.

2 Historical background

Soon after it was discovered that many clusters of galaxies are filled with X-ray emitting gas, it was realized that, near the centres of clusters, the cooling time of the gas was frequently less than the Hubble time. The early models of this phenomenon (Cowie & Binney 1977, Fabian & Nulsen 1977) assumed that cooling at the centre led to a steady inflow of material. It was tentatively assumed that the flux through this inflow was simply dumped at $r = 0$, without much consideration of what became of the deposited material. When radial surface brightness profiles were furnished by the *EINSTEIN* satellite, it was found that these could not be fitted by the models; the model profiles were too centrally concentrated.

At this point it would have been natural to suspect that the problem with the models was that they assumed that the gas was in a steady state.

Indeed, the arguments originally offered in support of the assumption of a steady state were weak, largely because the assumption was motivated more by computational convenience than by physical insight. However, instead of suspecting the steady-state assumption, the mainstream of research assumed that the problem lay with insistence upon mass conservation.

The reason mass conservation was considered suspect was the following. In a classic study Field (1965) had shown that hot radiatively cooling gas is thermally unstable. In physical language, the Field instability works as follows. A parcel of gas that is slightly more dense than its surroundings cools faster than the surroundings. In order to remain in pressure equilibrium with its surroundings, it then contracts, which enhances the difference between its cooling rate and that of its surroundings. A classical exponential increase in the parcel's cooling rate clearly follows.

In the original papers Cowie & Binney (1977) and Fabian & Nulsen (1977) suggested that this thermal instability might cause mass to "drop out" of a cooling flow and lead to the formation of stars at $r > 0$. When *EINSTEIN* showed that models in which there was no mass drop-out could not fit the data, it was natural to fit the data with models that included mass drop-out (Mathews & Bregman 1978). In fact, direct evidence for drop-out came from two observations. One was the detection of filaments of $H\alpha$ emission in several clusters. This emission indicated that there is cool gas at $r > 0$ just as the drop-out picture predicts. The low-energy X-ray spectra of clusters with suspected cooling flows also indicated the presence of gas which is substantially cooler than the bulk of the material. Unfortunately, these spectra had insufficient spatial resolution to say where the cool gas was.

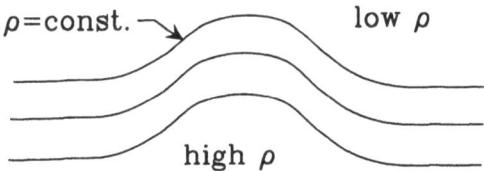

Fig. 1. In a gravitationally stratified fluid, a region of enhanced density is associated with an upward distortion of isodensity contours. When released from this configuration, the isodensity surfaces will oscillate up and down like waves on the surface of the sea.

In 1987 Malagoli, Rosner & Bodo showed that the Field instability does not apply to cooling flows. The relevant mathematics is complex and was only satisfactorily cleaned up by Balbus & Soker (1989) and Tribble (1989) but the physical point is simple enough — see Figure 1. A parcel of cooling-flow gas that is denser than its environment will sink in the prevailing gravitational field on a timescale that is much shorter than the differential cooling time. As it sinks, it moves through gas of ever lower specific entropy. Eventually it encounters gas that is of even lower specific entropy than itself, and is then *less* dense than its environment. As it is carried still lower by its downward momentum, our parcel experiences an upwards buoyancy force. Hence its downwards motion eventually ceases and the parcel starts to rise. Soon it rises to near the height from which it started. Then the cycle repeats. During this cycle the parcel is sometimes over-dense and sometimes under-dense. On average it cools at about the same rate as material of the same specific entropy, and there is no runaway instability.

Strangely, the discovery that cooling flows are not subject to the Field instability has by no means led to the abandonment of steady-state cooling flow models with mass drop-out. Two arguments are given for continuing to include mass drop-out in models. First, it is argued that cooling-flow gas at large r is seeded with inhomogeneities of non-linear amplitude, so linear analysis is irrelevant (Nulsen 1988). Second, it is argued that magnetic fields prevent parcels of over-dense gas from oscillating vertically as in the linear analysis; with over-dense blobs pinned in place by magnetic field lines, the Field instability proceeds classically (Loewenstein 1990).

Assuming that blobs are pinned by magnetic field lines (i) requires that a typical thermally unstable parcel has radius $\lesssim 20$ pc (Hattori *et al.* 1995), and (ii) implies that field lines pass from the hot substrate into the over-dense parcel. Electrons will efficiently conduct heat along these field lines and will eliminate the temperature differential unless the parcel's radius exceeds (McKee & Cowie 1977)

$$R = 7 \left(\frac{n}{2 \times 10^3 \, \text{m}^{-3}} \right)^{-1} \left(\frac{T}{10^7 \, \text{K}} \right)^2 f^{1/2} \, \text{kpc}, \tag{1}$$

where $f \leq 1$ is the factor by which magnetic fields lower the coefficient of thermal conductivity below Spitzer's (1962) value. At a distance $r \sim 10$ kpc from the centre of a typical large galaxy, the brackets in equation (1) evaluate to unity. Tao (1995) argues that $f \gtrsim 10^{-2}$, while Chandran & Cowley (1998) argue that $10^{-2} \gtrsim f \gtrsim 10^{-3}$. So small-scale ($R < 20$ pc), potentially unstable inhomogeneities are certain to be eliminated by thermal conductivity. We may be sure that many field lines will pass out from a cool blob into the hotter substrate because if, initially, it happened that there were no such interpenetrating field lines, they would soon form by reconnection in the turbulent boundary layer between the blob and the ambient medium.

During galaxy formation, the ISM will be thoroughly stirred. As our every-day experience with cream stirred into coffee demonstrates, in the high Reynolds number regime, stirring mixes fluids most effectively. Against this point of view it is sometimes argued that the ISM of the Galaxy is extremely inhomogeneous. This is true, but the lesson we should draw from this observation is not the one drawn by the advocates of steady-state cooling flows.

Near the Sun cooling causes cold gas to collect in a very thin and cold layer in the plane. Massive stars form in this layer and blast holes in it, pushing filaments of cool gas high above the plane. The observed local ISM represents a complex balance between the shredding effects of massive star formation and the propensity of the gas to become stably stratified about the mid-plane — it is, in fact, the *response* of a cylindrical system to the cooling catastrophe that leads to the formation of stars and the release of energy.

The situation in an elliptical galaxy is different. Since the gas in ellipticals is not loaded with angular momentum, it tends to stratify not on planes, but on spherical surfaces. Thus, in the absence of star formation, dense, cool gas would accumulate at the centre of the galaxy, rather than in an extensive plane. Massive star formation within this dense core would push filaments of cold gas out into the overlying hot gas, but, as in the solar neighbourhood, the scale-height of the distribution of such filaments would be less than a kiloparsec, rather than the tens of kiloparsecs required by models of distributed mass drop-out. And, of course, if the ISM were made multiphase by massive young stars, these stars would be observationally conspicuous, which they are not.

Hypothesizing that the gas around cooling-flow galaxies is seeded with non-linear inhomogeneities is unattractive from the point of view of the scientific method because it deprives the resulting theory of all predictive power: it implies that the radial mass-deposition profile, and thus the X-ray brightness profile, is determined by the initial spectrum of mass inhomogeneities (Nulsen 1988). Since we have no way of predicting the initial fluctuation spectrum, in this theory we have no way of predicting the X-ray brightness profile — it must simply be taken from the observations. What the theory *does* predict is the distribution of young stars. Spectroscopy in the near infrared has now imposed a very low limit on the upper mass of the stars that can form: Prestwich *et al.* (1997) show for the case of NGC 1275 that the upper mass cannot exceed $0.1 \, M_\odot$.

In reality the addition of distributed mass drop-out to models of cooling flows was motivated by the desire to build successful *steady-state* models. Let us examine the conceptual basis of the assumption of a steady state. Cowie & Binney (1977) argued that "the steady-state model should ... give an approximate description, provided the cooling time is much shorter than the Hubble time". This argument is *wrong* as Murray & Balbus (1992) showed by computing a time-dependent solution to the cooling-flow equations without mass drop-out. It turns out that in these equations the partial derivatives with respect to time *increase* in importance rather than deceasing. The reason is that their order of magnitude is given not by the value of the differentiated variable divided by the age of the cooling flow, but by the value of the differentiated variable divided by the time that must elapse until that fluid element cools right down. Thus the partial derivatives with respect to time increase in importance everywhere, but most strongly at small radii. A cooling-flow is simply not one of those frequently encountered systems in which a time dependent flow asymptotes to a time-independent flow.

From all these arguments we should acknowledge that steady-state cooling-flow models were from the outset a mistake, and that it is pointless to keep them alive by adjusting the cooling-flow equations in a physically doubtful way.

3   Response to the cooling catastrophe

The large masses of heavy elements, especially O and Si, detected in intra-cluster gas (Lowenstein & Mushotzky, 1996) imply that the interstellar media of cluster ellipticals were early on strongly heated and ejected by large numbers of type II supernovae. Subsequently some of this gas will have cooled and fallen back in the cluster potential (Loewenstein & Mathews, 1987; Ciotti *et al.*, 1991). Eventually a cooling catastrophe will have occurred at the cluster centre. In the case of the Virgo cluster, this catastrophe must have occurred in the core of M 87.

What response did this catastrophic cooling evoke? Two points are clear: (i) the response will come from the centre; (ii) by le Chatelier's principle "the effect will tend to diminish the cause". Since the cause is a drop in the central temperature T and an increase in the central density ρ, the effect will be a release of energy that will heat the central gas, increasing T and decreasing ρ.

There are two obvious central energy sources to consider: (i) massive star formation; (ii) an active galactic nucleus (AGN). Although galaxies such as M 87 have formed stars in the last few billion years, they do not host type II supernovae and their spectra indicate that they do not contain significant numbers of O and B stars. Moreover, searches with ever increasing sensitivity for the cool gas that is invariably associated with young stars have detected none in any cooling-flow galaxy (O'Dea *et al.*, 1995; Braine *et al.*, 1997). Hence, nuclear activity powered by accretion onto a central massive black hole is the most likely source of restorative energy for the ISM.

There is now clear evidence that massive black holes reside in the nuclei of giant galaxies. Fabian & Canizares (1988) pointed out the inevitability of a central black hole accreting at a significant rate when it is immersed in a dense ISM. As the cooling catastrophe develops and the central ISM becomes more dense, the rate of accretion onto the black hole must increase. Our task is to understand how the accretion energy acts by le Chatelier's principle to reduce the central density of the ISM.

We again have to choose between two mechanisms: (i) through Compton scattering by the ISM of hard photons radiated by the accretion source; (ii) through the action of radio jets on the ISM. Ciotti & Ostriker (1997) have explored models in which the ISM is heated by Compton scattering. The basic problem such models encounter is that each scattering of a photon of frequency ν transfers to the gas only a fraction $\sim \nu/\nu_{comp}$ of the photon's energy, where $\nu_{comp} = m_e c^2/h$ is the Compton frequency. There are two important consequences of this fact: (i) the heating of the gas is dominated by the most energetic photons, and (ii) for observed quasar spectra, in which most energy is concentrated in photons with $\nu \ll \nu_{comp}$, little energy is transferred to the gas unless the optical depth to Compton scattering is very high. From observation we know that any large optical depth must be confined to the immediate vicinity of the AGN, with the result that an outburst becomes, for all intents and purposes, the response of the ISM to a conventional hydrodynamical explosion at the location of the nucleus.

While Compton-driven explosions may be important in some, radio-quiet systems, the observations of M 87 make it clear that this system is being driven not by a bomb but by the highly collimated flows that are character- istic of radio sources. From observations of the synchrotron radiation from Virgo A, Reynolds *et al.* (1996) infer that the jet of M 87 radiates $\sim 3 \times 10^{35}$ W and that the jet is supplying kinetic-energy to the ISM at a rate $\sim 10^{36}$ W. By contrast, Fabricant & Gorenstein (1983) find that the total X-ray lu- minosity of M 87 within $R \sim 220$ kpc, three times the cooling radius, is a mere 1.6×10^{35} W. Thus there can be no doubt that at the present time the AGN is pumping in significantly more energy than the entire 'cooling flow' is radiating. In short, just now M 87 has a heating flow!

4   Accretion energy and jets

On general grounds it is no surprise to find that jets mediate the response of an AGN to a cooling catastrophe. Indeed, jets are observed in systems that vary in scale from single stars to megaparsec-sized radio sources. Jets are also known with speeds that vary from tens of km s^{-1} (e.g., bipolar flows in star-forming regions) to virtually the speed of light (high-γ radio jets). From this ubiquity and diversity it follows that the physics of jet formation must be simple electrodynamics, even if we have not yet completely understood it. Several likely pieces of the puzzle are already on the table, however. From the work of Balbus & Hawley (1991) it is now widely accepted that magnetic stresses are responsible for making the Shakura-Sunyaev viscosity parameter $\alpha \simeq 1$ in most accretion disks. Moreover, Blandford & Payne (1982) already explained how magnetic stresses can lead to the formation of a jet, which carries away most of the angular momentum that is released as material sinks through an accretion disk. Finally, it should be noted that the process outlined by Blandford & Payne is unlikely to give rise to highly relativistic jets (Pelletier & Pudritz 1992); high-γ jets must form in low-density regions above the poles of the accreting black hole, drawing their enegy either from the hole's spin or from strong Alven turbulence (Henri & Pelletier 1991).

This last point — that the accretion flow that is the ultimate source of all AGN luminosity is likely not responsible for *any* of the energy associated with high-γ jets — has an important corollary: the accretion energy must emerge in some other form, and the likelihood is that it comes in a subrelativistic, Blandford-Payne type, jet. Any high-γ jet is likely to run along the axis of this subrelativistic jet.

If Advection-Dominated Accretion (ADA) is a reality, as has recently been persuasively argued (Narayan & Yi, 1995; Fabian & Rees, 1995), thoroughly subrelativistic jets must be very important. Indeed, when ADA occurs, the central accretion disk is so rarefied that the black hole swallows most of the energy that is released close to the black hole. Consequently, in ADA *all* the energy released by the disk must emerge in the form of a significantly subrelativistic jet. The observation of triple-peaked line profiles from the accretion disk of M 87 that Ford reports in this volume, is most straightforwardly interpreted as direct observational evidence for this proposition.

When seeking observational confirmation of these considerations, we need to bear in mind that we observe jets through the synchrotron radiation. Consequently, their kinetic-energy becomes observable only in so far as it is converted to highly relativistic particles. The efficiency $\epsilon(v)$ with which kinetic energy is channelled into highly relativistic particles surely decreases with decreasing jet velocity v. We do not know how to calculate $\epsilon(v)$ but we do know its value for $v \simeq c/500$: the energy of a supernova emerges in a flow at $v \simeq c/500$. At the rate of a supernova per 30 years, these flows power the ISM of an an L^* galaxy such as the Milky Way at a rate $\sim 10^{35}$ W. Yet the synchrotron luminosity of such a galaxy is a paltry $L_{\text{radio}} \simeq 10^{30}$ W (Condon, 1992). So we should not assume that a subrelativistic flow will be conspicuous, even if it carries considerable power; most of its energy may go straight into heating the ISM.

5   A simple model of back-reaction from a jet

A simple model of the interaction between an AGN and a cooling flow will help us to understand what is happening in galaxies such as M 87, even though the model is crude and aimed at lower-luminosity cooling flows, such as that in NGC 4472.

Details of this model can be found in Binney & Tabor (1995). In brief, a second-order-accurate spherical hydrodynamic scheme was used to integrate the basic cooling-flow equations from the early years of the interstellar medium through the cooling catastrophe. The models had up to 160 logarithmically spaced radial grid points running from $r = 50$ pc to $r = 500$ kpc. On this grid of order 10^6 Courant-limited time-steps are required to reach the cooling catastrophe. The equations include mass injection and supernova heating along the lines advocated by Ciotti *et al.* (1991), but exclude mass drop-out. A diffusive term was included to suppress the Field instability which is otherwise plagues calculations such as these which are restricted to spherical symmetry. The luminosity profile of the galaxy followed Hernquist's (1990) law, while the mass profile was such that the circular speed was constant from $r \sim 5\,\mathrm{pc}$ to $r = 50\,\mathrm{kpc}$.

The effect of an AGN on the cooling flow was simulated as follows. The central boundary condition on the velocity was

$$v(50\,\mathrm{pc}) = \begin{cases} 0 & \text{if } \overline{M}_{20} > -0.05, \\ -100\,\mathrm{km\,s^{-1}} & \text{otherwise,} \end{cases}$$

where $\overline{M}_{20}$ is the mean Mach number at the innermost 20 grid points. When $v(50\,\mathrm{pc}) = -100\,\mathrm{km\,s^{-1}}$ the nucleus is deemed to be ejecting accretion luminosity at a rate $L_j = 0.01\dot{m}_0 c^2$, where $\dot{m}_0$ is the rate of mass inflow at $r = 50\,\mathrm{pc}$. This luminosity is injected into the volume within $r = 1\,\mathrm{kpc}$ at a fixed rate per unit volume. Radial momentum is also injected under the assumption that the energy is carried by jets of velocity $v_j = 0.15c$ — for the rationale behind these numbers, see Binney & Tabor (1995). When applying the model to M 87 one should bear in mind that the important scale $r = 1$ kpcwas motivated by consideration of the parameters of NGC 4472 — a radio map of this galaxy that is shown in Binney (1996). The appropriate scale for the case of M 87 is at least a factor 2 larger — see below.

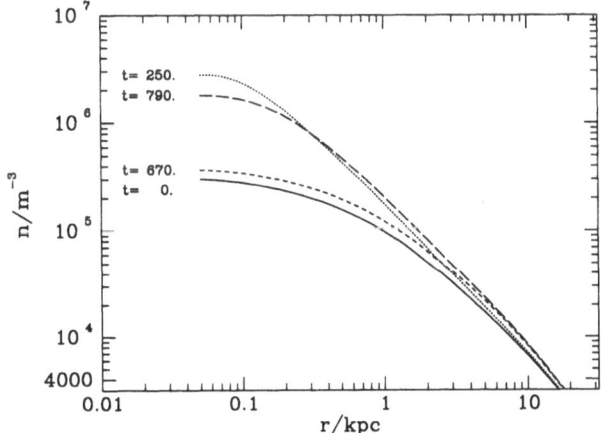

Fig. 2. An accretion-induced outburst nearly puts the clock back for the ISM. The curves are labelled in Myr from an arbitrarily set zero-point of time. The model suffers a cooling catastrophe just after $t = 250$ Myr. The resulting nuclear outburst increases the core radius of the ISM until at $t = 670$ Myr the density profile differs from that at $t = 0$ only by a shift upwards by a factor ~ 1.3. A second cooling catastrophe occurs after $t = 790$ Myr.

Figure 2 shows the response of a typical model to accretion-induced nuclear activity when the model's global X-ray luminosity is 5.2×10^{34} W, one third of the X-ray luminosity of M 87. The curve labelled $t = 0$ shows the particle density after 8.6 Gyr, while the other curves show the density 250, 670, and 790 Myr later. It can be seen that between 250 and 670 Myr, nuclear activity has almost set the clock back to zero; the profile at $t = 670$ Myr is identical to that at $t = 0$ except for a slight upward shift. This shift translates to an increase in the X-ray surface brightness by a factor ~ 1.3.

The AGN diminishes the central X-ray luminosity by an order of magnitude by blasting the ISM in half a dozen very short bursts of power $\gtrsim 10^{34}$ W. The total energy expended by the AGN in the process is negligible compared to the energy radiated in X-rays: the time-averaged nuclear input is 2.6×10^{32} W while the mean X-ray luminosity of the central kiloparsec alone was 4.4×10^{33} W. The nucleus is able to reverse the cooling catastrophe at very modest cost to itself for two reasons. (i) When the cooling catastrophe occurs, there is approximate balance within the central kiloparsec between power input by stars (3.3×10^{33} W) and X-ray luminosity. A small input by the nucleus tips this balance in favour of power input. (ii) When active, the nucleus hits the central kiloparsec of the ISM very hard and promptly staunches the radiative loss of energy by refashioning the central density profile on a dynamical time-scale. The average nuclear power output of the model over the interval between the curves labelled $t = 250$ and $t = 670$ in Figure 2 was 2.6×10^{32} W, about a hundred times smaller than the peak power and more than a thousand times smaller than the minimum current power of the nucleus of M 87. Even when we bear in mind that our model is designed for systems less luminous than M 87, this finding suggests that we are privileged to observe M 87 during a short burst of activity rather than at a typical time.

5.1   Pile-up and mass drop-out

Figure 2 indicates that although a nuclear outburst can restore the core radius of the ISM to its pre-outburst value, in 670 Myr the density of the ISM increases by a factor ~ 1.3 throughout the optical galaxy. This density increase is the inevitable consequence of the fact that the nuclear source heats only the central kiloparsec, while radiative cooling takes place everywhere. Hence cooling continues to feed gas into the galactic centre throughout the outburst. The nucleus accretes only a negligible fraction of this gas, so most of it has to be accommodated in the core by everywhere increasing the ISM density. This increased density shortens the time required for a second cooling catastrophe to take place. Evidently, if mass drop-out does not occur, nuclear outbursts must become more and more frequent.

One expects the increasing density normalization at each outburst to make successive outbursts more violent. This increased violence will have an effect which lies between two extreme cases. At one extreme the volume heated by the jets at each outburst steadily grows, with the result that the jets eventually push the ISM back out of the optical galaxy. At the other extreme the specific entropy to which the jets heat the gas in an essentially fixed volume increases until the interface between the heated gas and overlying gas becomes convectively unstable. Then, and only then, does the gas become thermally unstable (Balbus & Soker 1989) and is mass drop-out to be expected.

If mass drop-out does occur, its radial profile will not coincide with that derived from a steady-state cooling-flow model. It remains to be seen whether mass drop-out in time-dependent models is concentrated near the Hα filaments which are observed in M 87 and other cooling-flows. Steady-state models predict that most mass drop-out occurs at larger radii than the observed filaments. Moreover, it is hard to understand the observed strength of the Hα emission in steady-state models. By contrast, excitation by shocks and plasma turbulence seems capable of producing Hα emission at about the right level and accounting for the observed line rations (Dopita & Sutherland 1995).

5.2   Non-spherical phenomena

Once a cooling catastrophe has developed, the ISM of an elliptical galaxy is an inherently non-spherical system. If jets play a key role, the non-spherical nature of the problem is manifest. If the ISM is heated by some other means, for example by Compton scattering, the system still becomes non-spherical because it will inevitably become unstable to non-spherical perturbations: in spherical symmetry, a bubble of very hot gas at the centre will act as a piston and quickly push the entire ISM out of the galaxy. In reality either convective instabilities or the kind of dynamical non-spherical instabilities that break up supernova explosions will mix the hot bubble in with overlying cooler gas.

Notwithstanding the inherent asphericity of the phenomenon, Binney & Tabor (1995) constructed spherical models of nuclear reheating of the ISM by assuming a characteristic scale of a kiloparsec for the radius within which the nucleus heats the ISM. As stated above, if one were modelling M 87 one would chose this scale to be at least 2 kpc, and then the characteristic times and luminosities of the model would increase. An important task for the future is to model the effects of nuclear jets in cylindrical symmetry. Such models should give insight into both the scale of the jet-heated region, and the predicted elongation of the X-ray distribution on the plane of the sky. This is a challenging task computationally because very small time-steps must be taken if the system is to be modelled with adequate spatial resolution. Fortunately, the initial conditions of cylindrically-symmetric calculations could be taken to be the configuration reached by a spherically-symmetric model at the onset of the first cooling catastrophe. This procedure would allow observations to be matched by models that had been evolved in cylindrical symmetry for only a few hundred Myr rather than ~ 10 Gyr.

6 Implications of observations of M 87

The phenomenology of M 87 is amply covered elsewhere in this volume, so I will mention only some salient points. As has already been mentioned, the jet is currently pumping substantially more energy into the ISM than is being radiated by the *entire cooling flow*.

M 87 is detectable in synchrotron radiation out to at least 8 arcmin ($\sim$ 40 kpc). Since synchrotron-emitting plasma cannot be injected into thermal plasma without doing significant $P \, dV$ work, this fact implies that energy input by the AGN has been important for the ISM to a large fraction of the cooling radius.

The region $r \lesssim 3$ arcmin has clearly been the beneficiary of substantial heating by the nucleus. It looks as if a new jet is currently blasting its way through the region disturbed by an outburst that occurred at least 20 Myr ago (Klein, this volume). Older structures seem to be rotated in a clockwise sense relative to younger structures. This rotation could be the effect of the jet axis precessing, but radio maps suggest that the rotation is caused by rotation within the ISM convecting synchrotron-emitting material as the ISM swirls around the nucleus.

At radii ~ 3 arcmin, the X-ray maps (Böhringer *et al.* 1995 and this volume) clearly show the effects of dynamical disturbance of the thermal X-ray emitting plasma by the jet. Two features of the X-ray maps are notable. First, the distortion of the X-ray contours is not large (perhaps a change in surface brightness by a factor ~ 1.4) even though we know that the relevant region is now being *very* strongly heated by the jet. Second, the X-ray emission is *enhanced* around the jet, rather than diminished.

The first of these points is important because a failure to observe a correlation between X-ray and radio maps of less well studied sources has often been used as an argument against nuclear heating. The case of M 87 demonstrates that one needs extremely high-quality data if one is to see the radio–X-ray connection. The second point is significant because it demonstrates the importance of momentum transfer from the jet to the thermal plasma; the enhancement of the X-ray surface brightness around the jet must reflect upwelling of gas of relatively low specific entropy and therefore high density. By contrast, shocks driven into the thermal plasma by the jet will *lower* the X-ray surface brightness by increasing the specific entropy of shocked gas.

Two remarks are pertinent here. The first concerns the mechanical stability of a cooling flow: in an undisturbed cooling flow, specific entropy increases strongly with radius, so substantial energy and/or momentum must be put into low-lying gas to shift it appreciably upwards. However gas is displaced, it is liable to overshoot its equilibrium location and plunge down again. In the language of stars, jets must be a effective exciters of g-modes. Just as convection in stars gives rise to strong magnetic fields and cosmic-ray production, so it is likely that the g-modes of cooling flows will damp through the production of cosmic rays. Probably very little energy will go into the production of electrons energetic enough to produce observable synchrotron radiation, but, as in the local ISM, the energy density in cosmic rays may nevertheless be dynamically important.

The second remark one might usefully make is that even the *sign* of the X-ray–radio connection that has now been observed cannot be predicted without high-resolution, multidimensional, magnetohydrodynamical simulations that have yet to be made. A sober assessment of the physical complexity of the cooling-flow problem can only be obtained by studying our still incomplete knowledge of the physics of the local ISM.

7   Conclusions

The interstellar media of galaxies like M 87 are inherently time-dependent systems, which cannot be correctly understood in terms of steady-state models.

The first models of these systems assumed a steady state fundamentally for reasons of computational convenience. When the simplest steady-state models were found to conflict with observations, the governing equations were changed to bring theory and observation into agreement. The change in the equations seemed plausible until it was realized that straightforward cooling flows are not thermally unstable. By the time the thermal stability of simple cooling flows had been demonstrated, steady-state models had gained too much intellectual momentum to be stopped dead in their tracks, and further epicycles, in the form of magnetic fields, were added to the equations to keep steady-state models alive. I have argued that these epicycles are internally inconsistent: if a field can pin an inhomogeneity, it will permit thermal conductivity to erase it. Consequently, steady-state models make no sense, and quantities, such as mass drop-out profiles, that are inferred from these models are not credible.

The time-dependence of the ISM in an elliptical galaxy is first driven by radiative cooling and the inherently time-dependent rates at which an evolving stellar population pumps energy and mass into the ISM. After a cooling catastrophe has developed at the centre, the time-dependence is driven by the irregularity with which the nucleus pumps energy into the ISM.

A simple model of this process demonstrates that the nucleus can profoundly affect configuration of the ISM with a remarkably small output of energy. This is the case because (i) the cooling catastrophe is a highly localized phenomenon, and (ii) it arises as a consequence of a *small* imbalance between radiative cooling and heating by stars. In the case of M 87, both X-ray and radio observations provide clear evidence that the nucleus *has* refashioned the ISM out to at least $r \sim 8\,\text{arcmin}$, well over half the cooling radius.

Nuclear heating leads inevitably to violation of spherical symmetry, both because it is likely to be effected by jets, and because one expects convection to play a key role whenever gas is heated from below. Very little work has so far been done on non-spherical models of the interstellar media of elliptical galaxies and one can only speculate as to what picture will finally emerge from this numerically challenging task.

Comparison with the case of the local ISM is valuable for what we see all around us is the response of the Galactic ISM to a cooling catastrophe. The geometry of this catastrophe is differs from that in M 87 because it is planar rather than spherical, and the regulating heat source is different because it is massive star formation rather than nuclear activity. Some points that the two systems probably do have in common include: (i) repeated explosions that keep the ISM dynamical and out of equilibrium; (ii) a dynamically important but observationally inconspicuous magnetic-field/cosmic-ray component; (iii) a multiphase structure in which both hot and cool plasma can be found at any level. In both the local ISM and a cooling flow one expects the cool plasma to be, on the average, descending at any given level, and the fraction of cool plasma will tend to increase as one descends.

This point of view is unattractive in that it suggests that the observed phenomena in galaxies such as M 87 will be hard to model accurately. But the hardness and narrowness of the road to true understanding is no justification for turning down the first broad smooth path one encounters. I have presented powerful arguments for believing that cooling catastrophes do occur in elliptical galaxies, and that the ISMs of these galaxies are profoundly affected by the nuclear outbursts that they provoke. In the case of M 87 we now clearly see an outburst remodelling the surrounding ISM. The crucial evidence has come only recently from observations of the highest quality and similar data are not to hand for more distant cooling flows. The implications for other cooling flows are nonetheless clear.

References

Balbus, S.A., Soker, N., 1989, ApJ, **341**, 611

Balbus, S.A., Hawley, J.F., 1991, ApJ, **376**, 214

Binney, J.J., 1996, in "Gravitational Dynamics," eds Lahav, O., Terlevich, E., Terlevich, R.J., Cambridge University Press, Cambridge, p. 98

Binney, J.J., Tabor, G.R., 1995, MNRAS, **276**, 663

Blandford, R.D., Payne, D.G., 1982, MNRAS, **199**, 883

Böhringer, H., Nulsen, P.E.J., Braun, R., Fabian, A.C., 1995, MNRAS, **274**, L67

Braine, J., Henkel, C., Wiklind, T., 1997, A&A, **321**, 765

Chandran, B.D.G., Cowley, S.C., 1998, Phys. Rev. Lett. sdubmitted

Ciotti, L., D'Ercole, A., Pellegrini, S., Renzini, A., 1991, ApJ, **376**, 380

Ciotti, L., Ostriker, J.P., 1997, ApJ, **487**, L105

Condon, J.J., 1992, ARA&A, **30**, 575

Cowie, L.L., Binney, J.J., 1977, ApJ, **215**, 723

Dopita, M.A., Sutherland, R.S., 1995, ApJ, **455**, 468

Fabian, A.C., Canizares, C., 1988, Nature, 333, 829

Fabian, A.C., Nulsen, P.E.J., 1977, MNRAS, **180**, 479

Fabian, A.C., Rees, M.J., 1995, MNRAS, **277**, 55P

Fabricant, D., Gorenstein, M.V., 1983, ApJ, **267**, 535

Field, G.B., 1965, ApJ, **142**, 531

Hattori, M., Yoshida, T., Habe, A., 1995 MNRAS, **275**, 1195

Henri, G., Pelletier, G., 1991, ApJ, **383**, L7

Hernquist, L., 1990, ApJ, **356**, 359

Loewenstein M., 1990, ApJ, **349**, 471

Loewenstein M., Mathews W.G., 1987, ApJ, **319**, 614

Loewenstein M., Mushotzky, R.F., 1996, ApJ, **466**, 695

Malagoli A., Rosner, R., Bodo, G., 1987, ApJ, **319**, 632

Mathews, W.G., Bregman, J.N., 1978 ApJ, **224**, 308

McKee, C.F., Cowie, L.L., 1977, ApJ, **215**, 213

Murray, S.D., Balbus, S.A., 1992, ApJ, **395**, 99

Narayan, R., Yi, I., 1995, ApJ, **452**, 710

Nulsen, P.E.J., 1988, in *Cooling Flows in Clusters and Galaxies*, ed A.C. Fabian, Kluwer, Dordrect, p. 175.

O'Dea, C.P., Gallimore, J.F., Bau, S.A., 1995, AJ, **109**, 26

Pelletier, G., Pudritz, R.E., 1992, ApJ, **394**, 117

Prestwich, A.H., Joy, .J, Luginbuhl, C.B., Sulkanen, M., Newberry, M., 1997, ApJ, **477**, 144

Reynolds, C.S., Fabian, A.C., Celotti, A., Rees, M.J., 1996, MNRAS, **283**, 873

Spitzer, L., 1962. *Physics of Fully Ionized Gases*, New York, Wiley

Tao, L., 1995, MNRAS, **275**, 965

Tribble, P.C., 1989, MNRAS, **238**, 1247

The Inner Lobes of M 87

Frazer Owen

NRAO
Socorro, NM 87801
USA

Abstract. The radio imaging of M 87, combined with supporting data from other wavelength bands, suggest that the inner 2 kpc of M 87 are a region of transonic turbulence, dynamically important magnetic fields, and synchrotron emitting emitting filaments in a 10^7 K thermal plasma. The energy source for this activity is likely to be the black-hole driven jet which can be seen bending and mixing with the other radio emission from the region.

1 Definition

For the purpose of this review, the "Inner Lobes" are defined as the diffuse, filamented radio emission with about the same linear extent as the jet, about 2 kpc (in projection) from the nucleus. This emission is distinct because it has a much higher surface brightness than the rest of the radio source which extends as much as 40 kpc from the center of M 87. In Figure 1, I show a 4.6 GHz image of this region from the observations of Hines, Owen, & Eilek 1989.

2 History

In their early discussion of the radio source in M 87, Baade & Minkowski 1954 assumed that the synchrotron optical jet is directly related to the radio source, Virgo A. However, the early radio observations of M 87 saw only the 15' scale source (Mills 1952). Macdonald, Kenderdine & Neville 1968 were able to show that there was an inner source of about one arcminute extent. Hogg *et al.* 1969 seem to have been the first to image the inner lobes. Graham 1970 and Turland 1975 produced the first astrophysical discussions of the Inner Lobes as well as the best pre-VLA images. Although they do not clearly show the radio jet, these observations prove that there is a an extended radio source on the same scale.

3 Total Intensity Radio Structure

The VLA has provided us with many detailed images of M 87 which clearly show the great similarity of the radio and optical jet as well as its connection

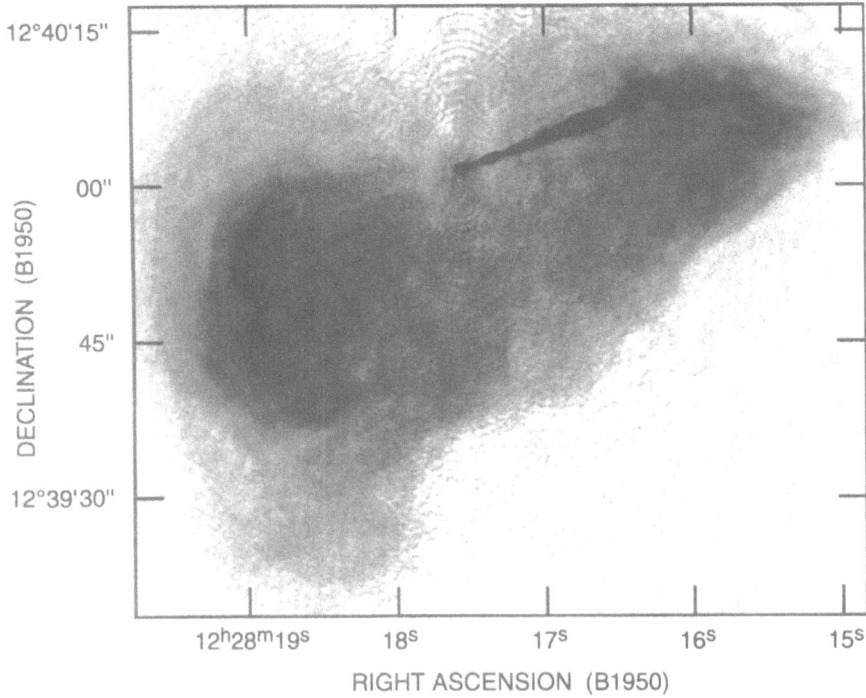

Fig. 1. 4.6 GHz image of the region defined as the inner lobes of M 87

to the Inner Lobes. In Figure 1, one can see the strong asymmetry of the jet, only visible on the west side of the source, combined with the fairly symmetric lobes. This has led many to suggest that there must be a jet on eastern side as well which is beamed away from us by relativistic motion. if this is the case, the lobes also make it hard to make out a low brightness jet against the confusion of the more diffuse structure.

In Figures 2–5, I show 14.4 GHz images of the detailed structure in the lobes. In Figure 2, one can see the jet apparently bending around and merging with the more diffuse structure. In Figure 3, we show this region at lower contrast so that one can see the more complex details of the jet-lobe merger. In Figure 4, a high contrast image of the eastern lobe is shown of the more complex structure in this part of the source. In Figure 5, a the higher brightness structure in the eastern lobe is shown at $0\overset{''}{.}15$ resolution.

This inner lobe is a complex, turbulent region. The brightest finescale structures in Figure 5 also are coincident with optical/IR emission which is discussed elsewhere in these proceedings. The bright structures also are correlated with the highest degrees of polarization as well as the field direction. Against this polarized background Faraday rotation measures up to 8000 rad m^{-2} can also been seen as discussed by Eilek, Owen, & Zhou 1999.

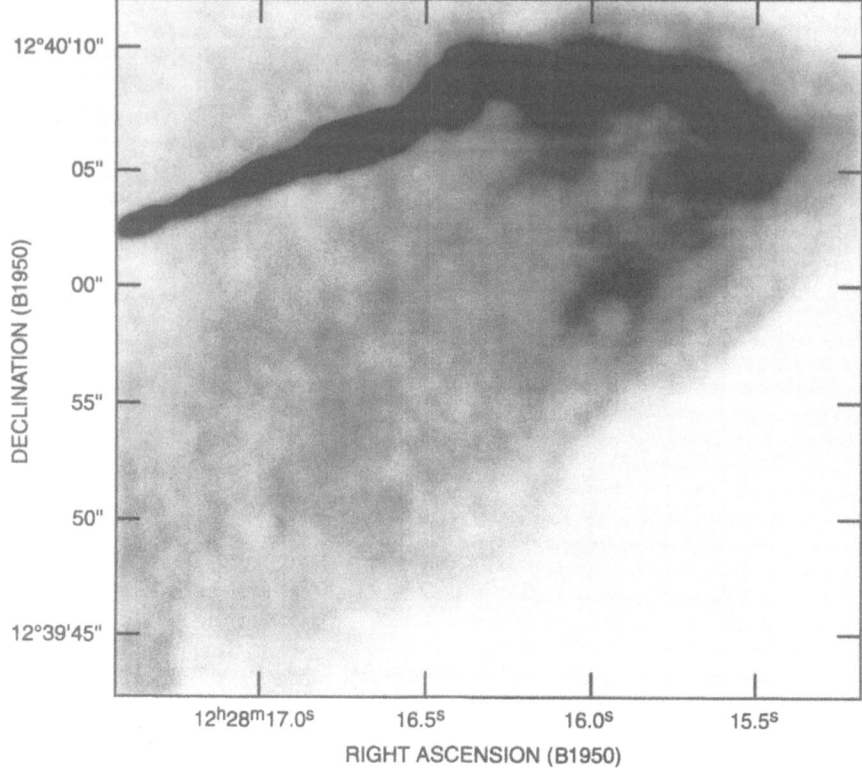

Fig. 2. 14.4 GHz image of west lobe

In Figure 1, one can clearly see that the jet and the entire inner lobe region appears to be deflected to the south of the core. The Hα filaments which surround the region of M 87, mostly avoid the inner lobes (in projection) and lie primarily to the north of the radio emission. The origin of this asymmetry is one of the mysteries about M 87. Supposedly M 87 is almost at rest relative to the cluster and this emission takes place deep in the galaxy's potential anyway. Furthermore, the most diffuse radio emission from M 87 seen on much larger scales does not show the same asymmetry. Thus whatever is causing the bending must now be isolated to the inner part of M 87.

4   Physical Conditions in the Inner Lobes

Hines, Owen, & Eilek 1989 show that the minimum pressures in the bright features in the inner lobes range from about 10^{-9} to 2×10^{-10} dyne cm^{-2}. These pressures are comparable to the inferred X-ray pressures in the same region, perhaps a factor of two higher. The diffuse emission has a minimum

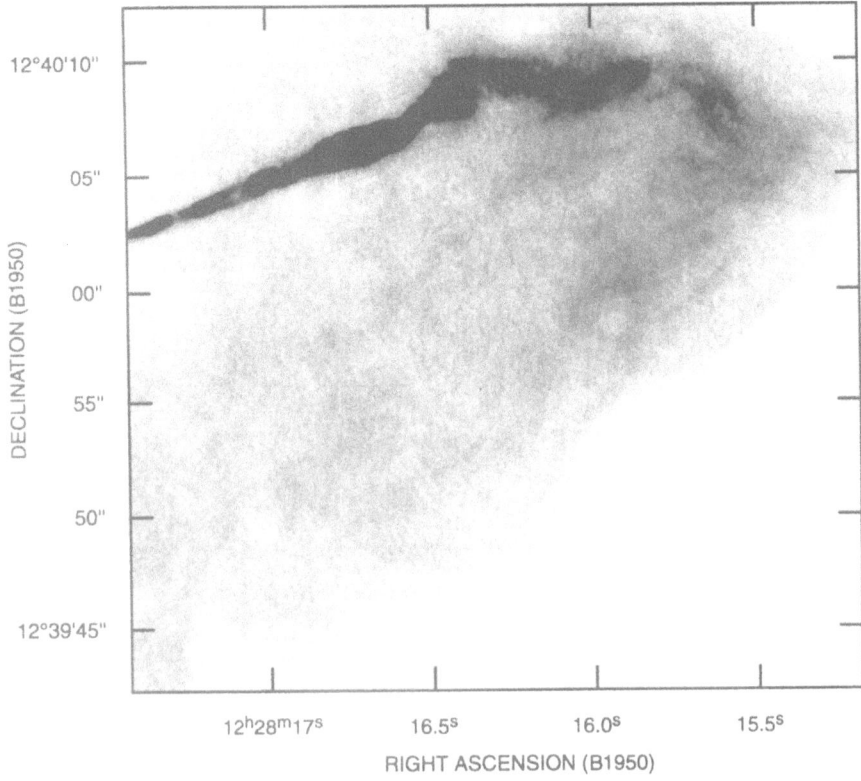

Fig. 3. 14.4 GHz image of the high brightness features in west lobe

pressure of about 4×10^{-11} dyne cm^{-2} assuming it is distributed uniformly and is not just a mass of filaments seen in projection. Thus the data are consistent with the brightest filaments being in pressure balance or slightly above the surrounding regions dominated by the X-ray emitting, thermal gas. The appearance of the emission, the velocities in the Hα emitting clouds (Keel, Owen & Eilek 1996) and the external Faraday rotation all are consistent with a transonic, turbulent environment driven by the kinetic energy flow from the jet into this region.

The similarity of the X-ray and radio minimum pressures also suggests that the system is not very far from equipartition, although one can imagine a conspiracy of physical variables which would mimic this case. The 5 GHz synchrotron lifetimes in the bright features, assuming the minimum pressure magnetic fields (50–100 μG), range from 6×10^5 to 2×10^6 years. These lifetimes are a factor of 10 to 100 shorter than than the time for gas to move across the source at the thermal sound speed (~ 300 km s^{-1}). Thus it seems likely that either there has been *in situ* particle acceleration or the particles have spent most of there lives in much lower magnetic fields. Given

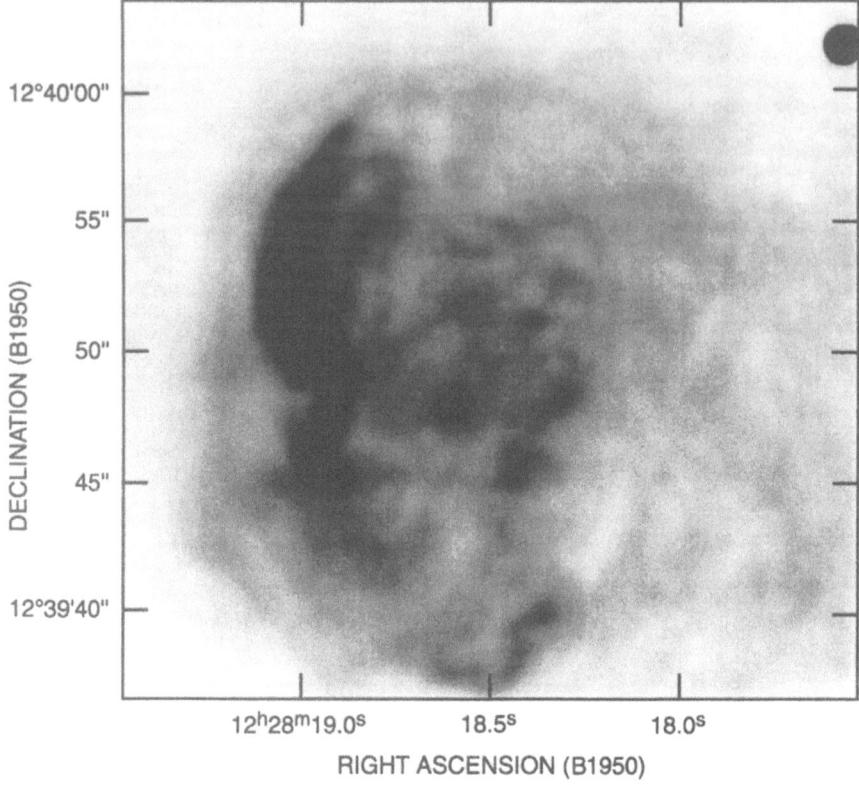

Fig. 4. 14.4 GHz image of the east lobe

the complex structure of the source the latter possibility seems quite possible.

Hines, Owen, & Eilek 1989 show that the filaments are unlikely to result from synchrotron or thermal instabilities. Tearing mode instabilities also take too long. Thus the mostly orgin of the filaments is the transonic turbulence which is consistent with the rest of the picture.

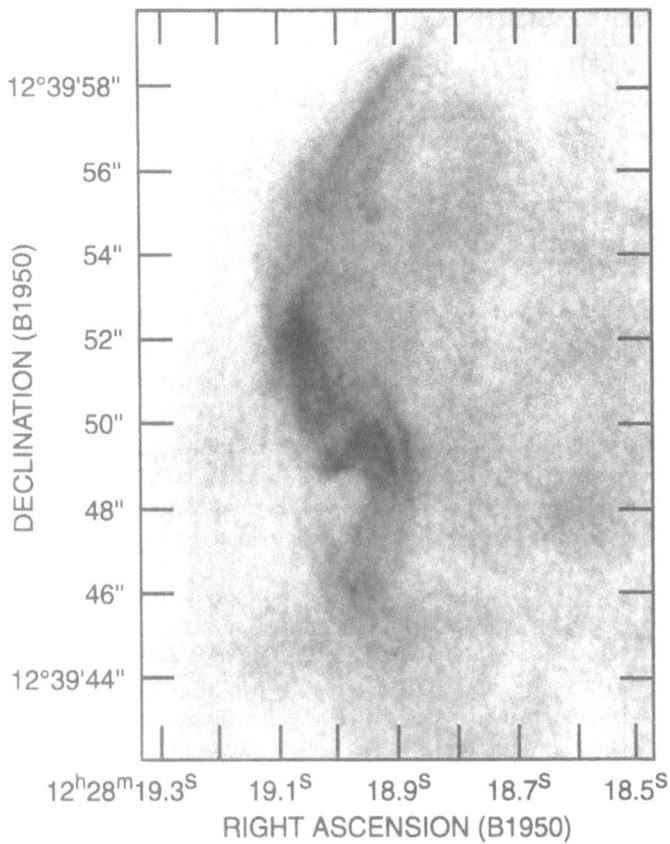

Fig. 5. 14.4 GHz image of the high brightness feature in the east lobe

References

Baade, W. & Minkowski, R. (1954):ApJ, **119**, 215

Eilek, J. A., Owen, F. N. & Zhou, F. (1999): this meeting

Graham, I. (1970): MNRAS, **149**, 319

Hines, D. C., Owen, F. N., & Eilek, J. A. (1989): ApJ, **347**, 713

Hogg, D. E., Macdonald, G. H., Conway, R. G., & Wade, C. M. (1969): AJ, **74**, 1206

Keel, W. C., Owen, F. N. & Eilek, J. A. (1996): in P. E. Hardee, A. H. Bridle & J. A. Zensus, eds, *Energy Transport in Radio Galaxies and Quasars* (San Francisco: ASP), 209-214

Macdonald, G. H., Kenderdine, S., & Neville, A. C. (1968): MNRAS, **138**, 259

Mills, B. Y. (1952):Nature, **170**, 1063

Turland, B. D. (1975): MNRAS, **170**, 281

Magnetic Fields and Turbulence in the Center of M 87

Jean Eilek[1], Frazer Owen[2] and Fang Zhou[1,2]

[1] New Mexico Tech
Socorro, NM 87801
USA
[2] National Radio Astronomy Observatory
Socorro, NM 87801
USA

Abstract. Faraday rotation data reveal that the inner core of M 87 has a strong, ordered, magnetic field. This field is located in front of the inner radio lobes, and has a pressure at least as large as that of the thermal X-ray plasma. This field must be supported by fluid turbulence in the core region: the question is, what drives the turbulence?

1 Radio Polarization Imaging of the Inner Lobes

Radio polarization provides a unique way to determine the magnetic field in the gas surrounding a synchrotron source. We measure the rotation angle of linearly polarized rotation, which is related to the plasma density and magnetic field along the line of sight through

$$\chi = (\text{RM})\lambda^2 \; ; \qquad \text{RM} = \int n\mathbf{B} \cdot d\mathbf{l} \simeq 810n\langle B_{\parallel}\rangle L \; \text{rad/m}^2 \qquad (1)$$

Here, B in μG, L in kpc, n in cm^{-3} and RM refers to Rotation Measure. It is important to note that we do not measure the full magnetic field, but only the algebraic (signed) mean of the line of sight component of that field.

Elsewhere in this meeting Owen (1999) has described the radio structure of the inner lobes of M 87. In Fig. 1 we show our new VLA image of the rotation measure (which improves on our previous work, Owen, Eilek & Keel (1989), by extending the frequency baseline). Several features are worth noting.

- The rotation arises from foreground plasma, rather than from within the radio source. If this were not the case, the simple dependence of χ on λ^2 would be violated.
- The RM is ordered, not random, across the source, and is generally positive. It has RM $\sim 750 - 1000$ rad/m^2 and projected transverse scale $\sim 0.5 - 1$ kpc.
- There is a high-RM filament in the east lobe of the source, with RM ~ 8000 rad/m^2 and projected width $\lesssim 0.1$ kpc.

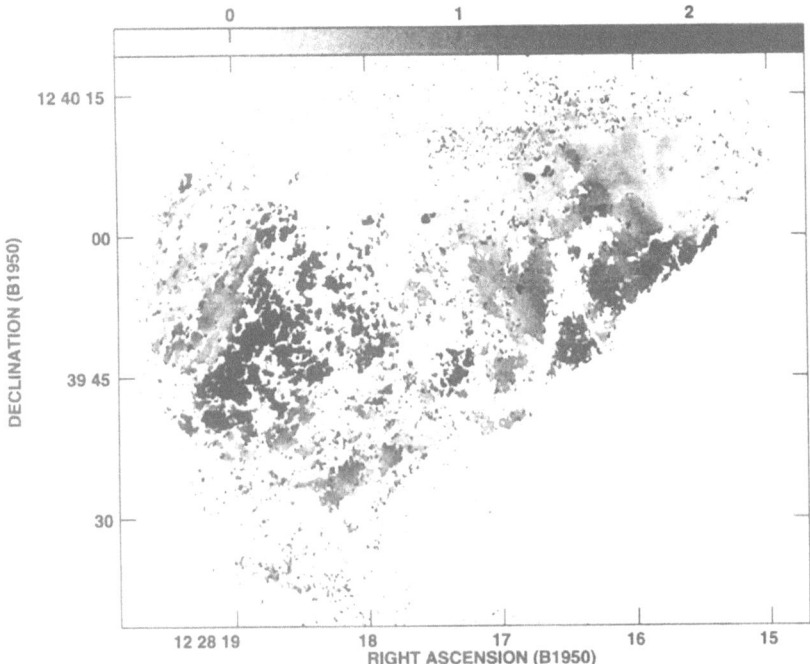

Fig. 1. Rotation measure image of the inner lobes of M 87, based on 4 frequencies around 5 GHz and 4 frequencies around 8 GHz. The grey scale runs from -500 rad/m² to +3000 rad/m². The narrow, high-RM filament in the eastern lobe is in the blackest part of this figure.

- There is a general trend for the RM to increase away from the core. In particular, the RM is much lower, $|\text{RM}| \lesssim 100\,\text{rad/m}^2$, and negative, in front of inner part of the jet.

2 What Is the Nature of the Magnetic Field?

2.1 Evidence from Faraday Rotation

The RM image tells us that there is an ordered magnetic field in the inner core. We start with the simplest geometry in order to estimate the field strength. That is, we assume the ordering scale along the line of sight is similar to the transverse scale, $\sim 0.5 - 1\,\text{kpc}$, and that there is no significant algebraic cancellation. This does not require a particular model of the magnetic field, but only that the covering factor of the field structures is of order unity.

For the gas density, we estimate $n \sim 0.08\,\text{cm}^{-3}$ from Nulsen & Böhringer (1995). We assume the RM arises from magnetic fields in the thermal X-ray loud plasma.[1] Taking a typical RM$\sim 750\,\mathrm{rad/m^2}$, and adding a factor $\sqrt{3}$ to estimate projection effects, we find $B \sim 20 - 50\,\mu G$. Comparing this to the "pressure equipartition" field, $B_p \sim 25 - 30\,\mu G$ (Eilek 1999a), we find that the *typical* field in the inner core is dynamically significant.

Similar calculations for the high-RM filament give $B \sim 2\,\mathrm{mG}$ here, if the internal density is assumed to be the ambient density of the X-ray plasma. Owen, Eilek & Keel (1989) searched for Hα emission and X-ray excesses at the location of this filament, and found no evidence for density higher than the ambient value. Thus, this filament is clearly strongly overpressured relative to its surroundings.

What of the low-RM patch in front of jet? We have two options: either the source sits in a central "hole" in the magnetic field, or the jet lies in front of much of the inner radio lobe, and that the RM is closely associated with this radio lobe.

2.2   Evidence from Depolarization

The polarization data also tell us that, in addition to large-scale order in the magnetic field, the field has a small-scale structure, on scales $\lesssim 10\,\mathrm{pc}$.[2] In particular, we find the source is depolarized: the fractional polarization is smaller at longer wavelengths.

Low polarization at a given frequency comes about because of gradients in the direction of the linear polarization of the synchrotron radiation, on a scale smaller than the radio beam ($\sim 8\,\mathrm{pc}$). The dependence on wavelength tells us that this is caused by gradients in the rotation measure, rather than intrinsic emissivity gradients in the radio loud plasma. Images of M 87 in polarized emission (shown in Zhou 1997) show short, unpolarized *filaments*, of width $\sim$ beam size, at both 5 and 8 GHz. These dark filaments are more numerous at 5 GHz, so that the source is depolarized. The RM gradients must occur within these filaments. These filaments can be unresolved magnetic flux tubes with helical magnetic fields. Alternatively, the filaments may be thin interfaces between regions of uniform field. Either one will depolarize if the field gradient scale is smaller than a beam size.

2.3   Possible Models of the Magnetic Field

We can now compare the data to existing models of magnetized plasma. Because of the large-scale order, we can rule out the old, simple picture of a random field, consisting of many turbulent "cells" (Burn 1966). We can rule out the simple cooling-flow model which predicts many radially stretched loops (Soker & Sarazin 1989), for the same reason.

---

[1] The RM screen is likely to sit inside of the extended radio lobes, Eilek 1999a. If the lobe density is lower than this value, then our derived fields are *lower limits*.

[2] The detection of power on both small and large scales is, of course, characteristic of turbulence.

Bicknell, Cameron & Gingold (1990) have suggested that the RM in Cyg A arises from magnetic fields in the radio-loud plasma, which have mixed with the ambient thermal gas, in a skin of the radio source. This picture has problems explaining trends seen in all cluster-center radio sources. It does not easily account for RM asymmetry or the lack of limb brightening, and it may have trouble with the slow rate of cross-field mixing and the need for strongly overpressured thin skins (Eilek, Owen & Wang 1997). On the other hand, in the case of M 87, the low RM in front of the jet may suggest that the inner lobes sit behind the jet, in which case the RM would have come from a region (skin?) between the inner lobes and the jet. Thus, we cannot rule out this model for this galaxy.

The picture we favour, however, is that the magnetic field is maintained, and shaped by, turbulence in the core region. We have direct evidence of transonic velocities very close to the core. The emission line clouds move at ~ 200 km/s (SBM), which gives a Mach number $\mathcal{M} \sim 0.4$ in a plasma at $T \sim 1.5$ keV. The clouds are very likely to represent the velocity field in the surrounding X-ray gas. In addition, observations of the outer radio halo strongly suggest that the entire Virgo core is disordered and turbulent (Eilek 1999a). Magnetic fields supported by turbulence can reach a level,

$$\frac{B^2}{8\pi} \lesssim \frac{1}{2}\rho v_t^2. \tag{2}$$

For the conditions in the core of M 87, this dynamo-supported field can be estimated as $B \lesssim 25\mathcal{M}$ μG. Thus, Mach 1 turbulence, if it exists, can produce fields comparable to those observed. We think it likely this is the origin of the fields, and that they exist throughout the inner Virgo core.

The geometry of a magnetic field in a turbulent plasma depends on the nature of the turbulence. Isotropic turbulence leads to intermittent magnetic fields: the strong field, satisfying (2), tends to be localized in filaments or flux ropes (*e.g.*, Ruzmiakin, Sokoloff & Shukurov 1989, Miller *et al.* 1996). If the turbulence is helical (meaning that it has a particular small-scale anisotropy), it also supports magnetic fields ordered on large scales. These fields may well be in a force-free state (Taylor 1986). Their geometry can be described as nested toroids, which lead to a central "hole" in the RM image (Eilek 1997b). The observations in hand do not allow us to discriminate between helical or symmetric turbulence. However, the depolarization data seem to us to support the picture of small-scale intermittent fields.

3   The Virgo Core Compared to Other Clusters

Whatever the details of the dynamo, the data tell us that the inner core of M 87 is magnetized, at a level comparable to the thermal gas pressure. This striking result is not unusual for cluster-center galaxies. Ten such galaxies have been imaged in rotation measure. (The situation is described in Eilek,

Owen & Wang 1997). The radio sources range in size from a few kpc to ~ 150 kpc. Their cooling cores range in strength from $\sim 600\,M_\odot/\mathrm{yr}$, to $\lesssim 1\,M_\odot/\mathrm{yr}$. All ten sources show ordered RM (projected scales $\sim$ few -20 kpc), and *all show large fields* (ratio of magnetic pressure to gas pressure $\sim 0.1-10$). Thus, the X-ray gas in M 87 is not unusual: it seems very typical in having a dynamically significant magnetic field embedded in its core.

4   Turbulence in the Virgo Core

We have shown that the very center of M 87 contains a strongly magnetized, thermal plasma; we have argued that transonic turbulence must be what supports this field. How extensive is this active region? We have no direct evidence for magnetic fields in the X-ray gas on scales larger than the inner lobe. However, we have strong indirect evidence from the outer radio halo and the X-ray image that the entire Virgo cluster core, on a scale $\lesssim 40$-50 kpc, is disordered and stirred up. Thus, it does not seem unjustified to assume that this entire core region is turbulent, possibly at the transonic level. What does this require of the energy budget?

Turbulence will decay if it is not driven. Scaling arguments show that the decay time $\tau_t \sim f\lambda_t/v_t$, where λ_t, v_t are the characteristic (outer) scale and velocity of the turbulence, and f is a fudge factor on the order of a few. Estimating $\lambda_t \sim 1-10$ kpc, we find $\tau_t \sim 25 - 100$ Myr. This is a short time; the turbulence must be driven.

What driving power is needed? The X-ray data (Nulsen & Böhringer 1995) tell us that the total thermal energy in this region is $U_{th} \simeq 1 \times 10^{59}$ erg. The turbulent energy is then $U_t \simeq \mathcal{M}^2 U_{th}$. Dividing the turbulent energy by its decay time, we find the necessary driving power: $P_t \simeq 3 \times 10^{44} \mathcal{M}^3/\lambda_{t,kpc}$ erg/s. This is very similar to current estimates of the power deposited by the radio galaxy, and also to the power deposited in the core by large-scale motions in the cluster (as summarized in Eilek 1999a). This reinforces our picture of the Virgo core being an active, turbulent region.

References

Bicknell, G. V., Cameron, R. A. & Gingold, R. A. (1990): ApJ, **357**, 373-387
Burn, B. J. (1996): MNRAS, **133**, 67-83
Eilek, J. A. (1999a): this meeting
Eilek, J. A. (1997b): preprint
Eilek, J. A., Owen, F. N., & Wang, Q. (1997): preprint
Miller, R. S., Mashqyek, F., Adumitroaie, V. & Givi, P. (1996): Phys. Plasmas, **3**, 3304-3317
Nulsen, P. E. J. & Böhringer, H. (1995): MNRAS, **274**, 1093-1106
Owen, F. N. (1999): this meeting
Owen, F. N., Eilek, J. A. & Keel, W. C.: (1990), ApJ, **362**, 449-454

Ruzmiakin, A., Sokoloff D. & Shukurov, A. (1989): MNRAS **241**, 1-14
Soker, N. & Sarazin, C. L. (1989): ApJ, **348**, 73-84
Sparks, W. B., Ford, H. C. & Kinney, A. L. (1993): ApJ, **413**, 531-541
Taylor, J. B. (1986): Rev. Mod. Phys., **58**, 741-763
Zhou, F. (1997): Ph.D. thesis, New Mexico Tech

Extended Emission-Line Gas in M 87

William B. Sparks

Space Telescope Science Institute
3700 San Martin Drive
Baltimore, MD 21218
USA.

Abstract. The extended optical ("warm") emission filaments in M 87 may connect the inner nucleus to the outer coronal halo of the Virgo cluster of galaxies. The historical development of observations of the filament system is reviewed, and placed in context both with the warm gas of elliptical galaxies in general, and within the context of different theoretical scenarios. A detailed morphological picture has emerged, particularly with recent *Hubble Space Telescope* (HST) images in which complex strands and loops of emitting material may be seen extending over many kiloparsec. The emitting gas is dusty at a level consistent with Galactic gas-to-dust ratios. There may also be a component of neutral gas present. Kinematically, the gas system is in outflow within the central few kiloparsec of the nucleus, whereas at greater distances, the gas is infalling.

1 Introduction

The extended optical emission filament system of M 87 provides an interesting arena in which to study the physics of the interstellar material within and around M 87, from regions close to the inner black hole out to the hot corona of the Virgo Cluster of galaxies. The detail in which this relatively nearby system may be studied offers the potential for insight into other more distant cases of extensive optical line emission. The spectral character of this line emitting gas is remarkably consistent moving across these very diverse regimes. Physically, we are observing gas at temperatures of order 10^4 K emitting low excitation optical lines, with forbidden lines of [NII] typically substantially stronger than $H\alpha$. The emission lines are kinematically narrow, although they become broader towards the center of the galaxy, Heckman *et al.* (1989). Within the central few tenths arcseconds, the extended emission filaments appear to connect directly onto the roughly flat disk surrounding the supermassive black hole, Ford *et al.* (1994), Marconi *et al.* (1997). At very large radii, the surface brightness diminishes and the filaments cannot be observed beyond a radius of about 2 arcminutes. A more detailed description of their properties follows, as the historical development of observations and their interpretation is presented. Firstly, though we consider the various contexts in which the study of the M 87 emission filaments may be placed.

2   Context I: The M 87 Local Environment

The optical emission filaments of M 87 occupy a very interesting, perhaps unique, role in linking most of the interesting astrophysical phenomena manifested in M 87. As in other elliptical galaxies, the interstellar medium of M 87 is dominated by a hot X-ray emitting coronal plasma. As in other such galaxies, that hot gas appears to be associated with a system of low excitation optical emission filaments. The two may be related through a cooling-flow, *e.g.* Fabian (1994), or possibly in some other way such as through a merger, Sparks (1997).

Also, the filaments appear to be involved intimately in the powering of the active nucleus at the center of M 87. Ford and co-workers, and Macchetto *et al.* (see elsewhere these proceedings) have shown that there is an inner emission line disk surrounding the central black-hole, and plausibly connected directly to it. Accretion of infalling cooler material is often invoked as the AGN fuel source, and in M 87 we see directly the filaments connecting onto the more regular inner disk as would be expected in that scenario. A complication, however, is that the velocity field indicates more outflow than infall near the center, however this may be a manifestation of angular momentum transport outwards from the center enabling the gas closer to the nucleus to fall onto the nucleus.

Sparks, Ford and Kinney (1993) showed that the emission filaments and the inner radio lobes also appear to be 'aware' of one another. The optical filaments wrap closely around the periphery of the inner lobes, which are presumably powered via the famous jet, and this is of interest to theories in which the warm gas arises as condensation triggered by the pressure of the expanding radio plasma.

Finally, as a galaxy, M 87 hosts an old, evolved stellar population that must be losing mass into the ISM at a significant rate through processes associated with ordinary stellar evolution. Such ongoing stellar mass loss has been invoked as a possible origin for cooler material within elliptical galaxies, and if this transpires to be the explanation for filaments in M 87, we may ask what this teaches us about the late stages of stellar evolution in such galaxies.

3   Context II: Extended Emission Line Gas in Elliptical Galaxies

The ISM of elliptical galaxies has now been observed across most of the electromagnetic spectrum. In the X-ray region of the spectrum, hot coronal gas is observed, with large mass per galaxy. Of order $10^{12} M_\odot$ might be a typical coronal mass for a large, giant elliptical galaxy. Optically, warm gas around 10^4 K is seen, the subject of this review, with masses in the range 10^5–$10^7 M_\odot$ typically. At cooler temperatures and longer wavelengths we see or place good limits on dust, neutral gas and molecular gas. Typically, the masses involved are $\sim 10^5 \, M_\odot$ for dust, and $< 10^9 \, M_\odot$ for the neutral and molecular gas components.

Historically, the existence of an ionized gaseous phase in elliptical galaxies has been known for some time from spectroscopic work, Mayall (1936), Humason, Mayall and Sandage (1956), Osterbrock (1960). Humason, Mayall and Sandage (1956), for example, found that approximately 18% of the 82 elliptical galaxies they observed showed optical line emission. More recently, surveys with imaging detectors using line-isolating narrow-band filters have shown that low excitation gas is a relatively common phenomenon in elliptical galaxies, and that of order 50% of elliptical galaxies now can be shown to host optical line emission, e.g. Goudfrooij (1994), Macchetto (1996). The typical characteristics of the emission are low excitation "LINER" like spectra, with strong $[NII]$ relative to $H\alpha$, Heckman et al. (1989).

While the amounts of matter in the warm phase are rather trivial, the amount of energy required to sustain a typical filament system can be large, and typically the warm gas may be considered as *very important energetically* in elliptical galaxies. Locally, the emission from optical and associated emission lines can easily exceed the emission from X-rays of the hot coronal phase, even with the much smaller mass of gas involved.

Two important correlations have been identified relating to low excitation emission lines. The first is that the presence of $H\alpha$ emission is related to the presence of a thermal X-ray excess, indicative of a hot coronal halo. The second is that the filaments are always associated with dust.

There is genuine scatter in the relation between $H\alpha$ and X-rays, however the correlation does appear to be well established. From early work by Hu (1988), through to the recent study of Macchetto (1996) based on an ESO Key Program, it has been shown that X-rays and optical emission lines are statistically related. Figure 1 illustrates the correlation from Macchetto et al.

In a similar vein, a number of studies have sought direct links between the two components of the ISM through high resolution X-ray morphological studies. Sarazin, O'Connell and McNamara (1992) obtained ROSAT HRI images of the cooling-flow cluster 2A 0335+096 and found that the peak X-ray emission was not centered on the dominant cD, but *was* centered on the peak $H\alpha$ emission. There were other similarities between the distribution of optical line emission and of the X-rays, although there was not a one-to-one correspondence. Similarly, Sparks, Jedrzejewski & Macchetto (1994) and Sparks (1997) presented high resolution X-ray images of NGC 4696 at the center of the Centaurus cluster of galaxies. The morphology of the X-ray emission is very similar in appearance and scale to the optical line emission. Again, the peaks coincide, but offset from the nucleus, and a one-armed spiral morphology is evident in both components.

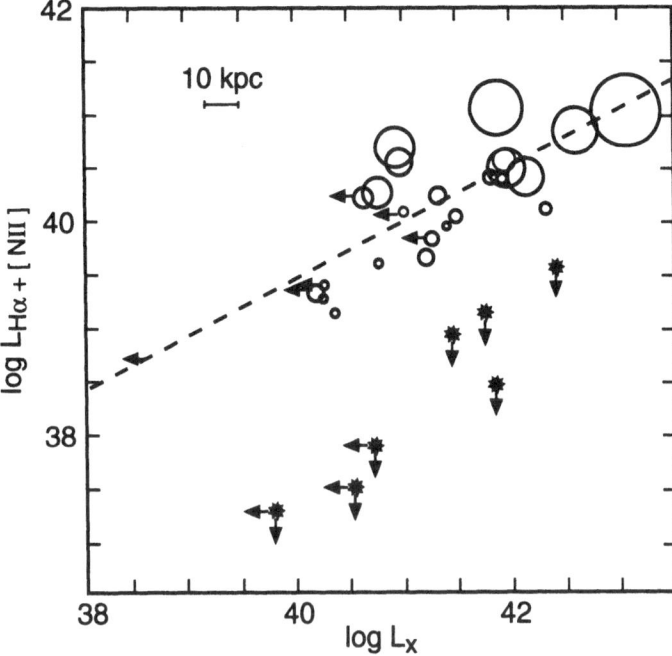

Fig. 1. From Macchetto *et al.* (1996) showing a statistical relationship between optical emission lines and X-ray luminosity, but with substantial scatter.

Hence, given these two lines of approach to the problem, we may conclude that there is indeed a statistical, and probably direct, relationship between the optical emission line gas and the X-ray emission.

Table 1. Dust correlation with $H\alpha$ emission in elliptical galaxies

	Dust	No dust	
line emission	21	10	31
no line emission	1	22	23
Totals:	22	32	54

Concerning the presence of dust in optical emission filaments, it has become clear through many lines of investigation that there is essentially a one-to-one correspondence between optical filaments and dust. Sparks, Macchetto and Golombek (1989) showed that for NGC 4696, the spatial distribution of dust and gas is exact to the limits of the data. They also showed that the wavelength dependence of extinction is indistinguishable from a Galactic extinction law. Table 1 shows a summary of the detection statistics from Goudfrooij *et al.* (1994a).

Goudfrooij *et al.* (1994b) investigated the dust extinction law and showed that a Galactic, or 'normal', extinction law is found in the more chaotic systems, whereas the regular smooth dust lanes have significant departures from normal. They interpret this as evidence of recent infall of normal dust in the chaotic systems, and subsequent processing by the hot gas of the dust once it has settled.

4   Context III: Theory

The primary theoretical question relating to optical emission filaments in elliptical galaxies, and clusters of galaxies, is "what is the origin of the filaments?". Of course we also wish to understand the excitation mechanism of the gas, the ultimate fate of the warm material, and how it interacts with its environment.

An important possibility for explaining the intermediate temperature gas is that it arises from a "cooling-flow", in which the hot coronal X-ray emitting gas cools and, through thermal instability, condenses quickly to lower temperatures in the central high pressure regions. This theory arose from observations of a strong central X-ray peak, inference of a short X-ray cooling time for the coronal gas, and has received support recently from observations of temperature gradients in galaxy clusters, *e.g.* Fukazawa *et al.* (1994). The theory has quantitative problems when confronted with the characteristics of the optical filaments, however, if they arise as a direct consequence of the cooling process. The filaments are overluminous, leading to a need to ionize gas multiple times, "H_{rec}". The spatial and hence surface brightness distribution is problematic, with filaments seen only within a small fraction of the inferred cooling region from X-ray observations. Coupled with the previously mentioned high luminosity, this means that locally, the filaments are very much brighter than expected from straightforward versions of the cooling-flow theory. Also, the presence of dust in the filaments is unexpected if they have condensed from a hot coronal phase in which dust is effectively destroyed. And finally of course, despite serious searches, the reservoir of cold gas that would arise in this picture has never been identified.

An alternative explanation was put forward by Sparks, Macchetto and Golombek (1989) and by de Jong *et al.* (1990). They proposed that capture of a gas-rich, dusty companion through a minor merger, or tidal stripping process, could explain the presence of the cold gas in an otherwise hostile environment. This gives a natural understanding of the presence and characteristics of dust within the filament system, and also yields excellent quantitative agreement energetically. Both globally, the total flux and locally, the surface brightness, give good agreement with predictions. Often, there are other indications of mergers, and hence this appears to remain a viable interpretation of the observations. The X-ray characteristics of central peak, temperature gradient cooler to the center, and isothermal in the outer regions are all consistent with the consequences of allowing heat transfer (through electron conduction) from the hot coronal gas into the cold accreted gas.

For example, Fig. 2 shows an X-ray bright elliptical galaxy with dust filaments apparently in the process of being accreted from a nearby spiral galaxy.

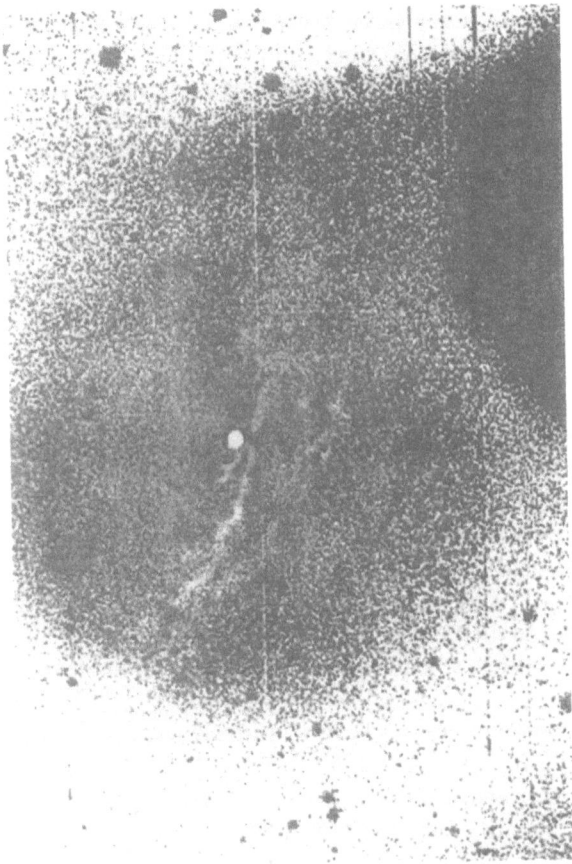

Fig. 2. Image of the dust, shown as white filaments, in NGC 4105 with nearby tidally distorted spiral — a likely donor for the dust.

New theoretical options have gained ground recently, and in particular the possibility that they are triggered by an interaction between the outgoing radio jet and lobes with the ambient hot plasma is intriguing. In the case of M 87, shown below, we see a close spatial relationship between the radio plasma and the $H\alpha$ filaments, Sparks, Ford and Kinney (1993). McNamara *et al.* (1996) and Pinkney *et al.* (1996) show a similar situation in A 1795, with the radio jets and complex interstellar medium strongly interacting, apparently.

Finally, there is the possibility that stellar mass loss within the host elliptical galaxy has given rise to the presence of cooler material within the galaxy. If the process of interaction between the stellar mass loss and ambient hot gas is inefficient, the residue may result in the material now seen. This has a number of problems, including the coherent filamentary structure over many kiloparsec. In many cases, there is a relatively high kinematic rotation associated with the gas and dust, unlike the host galaxy, and the primary axis of the gas and dust often bears little relationship to that of the host, in contrast to what might be expected if the gas arose from the stars of the galaxy. Stellar mass loss therefore seems to be an unlikely explanation for the origin of the filaments.

5   Optical Emission Lines in M 87

5.1   Early Historical Studies

The presence of optical emission lines at the nucleus of M 87 was revealed by spectroscopic observations Baade & Minkowski (1954), Osterbrock (1960) who found strong $[OII]$ and a "peculiar profile" for the lines.

Arp (1967) obtained photographic images using an $H\alpha$ filter with the Palomar 200″ telescope. He found extended features at roughly the same outer radius as the well-known optical jet, but on the diametrically opposite side of the nucleus. The approximately linear feature was termed a "counterjet". Walker (1968) used a long slit spectrograph oriented at a variety of position angles to map out regions where optical emission lines were present. He called the region so-delineated as the self-explanatory "orthogonal fan jet".

In a very detailed study, Ford & Butcher (1979) presented a thorough investigation of the system within approximately 30″ of the nucleus of M 87. They used spectroscopy and electronic imaging to clarify the spatial distribution and to quantify the energetics and spectral characteristics of the gas. Their study showed that the emission filament system was a single complex of filaments, roughly aligned with the radio core but offset primarily to the North. The total luminosity in $H\beta$ was $\approx 10^{39}$ erg/s, and the mass of ionized gas was $\approx 10^4\,M_\odot$. Ford & Butcher (1979) found a high electron density in the central region, with a temperature $T_e < 1.1 \times 10^4$ K. An important observation, potentially, was that nitrogen and sulphur appear to be overabundant relative to solar abundance shock excited planetary nebulae. Their favoured interpretation of the filament origin was the cooling-flow scenario.

5.2   Geometry and Kinematics

Sparks, Ford and Kinney (1993) presented deep ground-based CCD imaging
of the emission filaments, Fig. 3, and tied the imaging properties to kinematic
observations using long-slit spectroscopy to determine the geometrical configuration of the gas with respect to the host galaxy and with respect to the
inner radio lobes and jet.

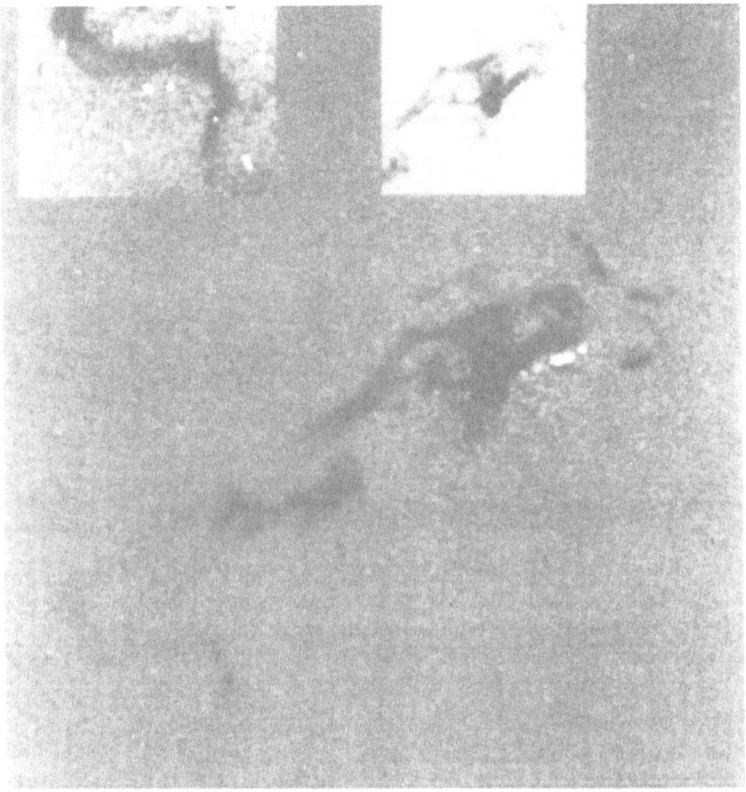

Fig. 3. Ground based images showing the extent of the emission filament system
in M 87. The insets show the nuclear region on a lower contrast display, and a high
S/N detail in the outermost filaments, from Sparks, Ford & Kinney (1993).

Using deep CCD images, the surface brightness distribution of gas was
quantified and it was shown that dust is present within the warm filaments.
The wavelength dependence of dust extinction was shown to be consistent
with the Galactic value, and the inferred mass of gas was $\sim 10^6 \, M_\odot$. The
total $H\alpha + [NII]$ luminosity was $\approx 10^{40}$ erg/sec.

By utilizing the dust properties (extinction and wavelength dependence of extinction) it is reasonable to infer which regions of the filaments lie in the foreground, and can consequently give rise to absorption of the background galaxy light. Primarily, this is the NW quadrant of the filament complex.

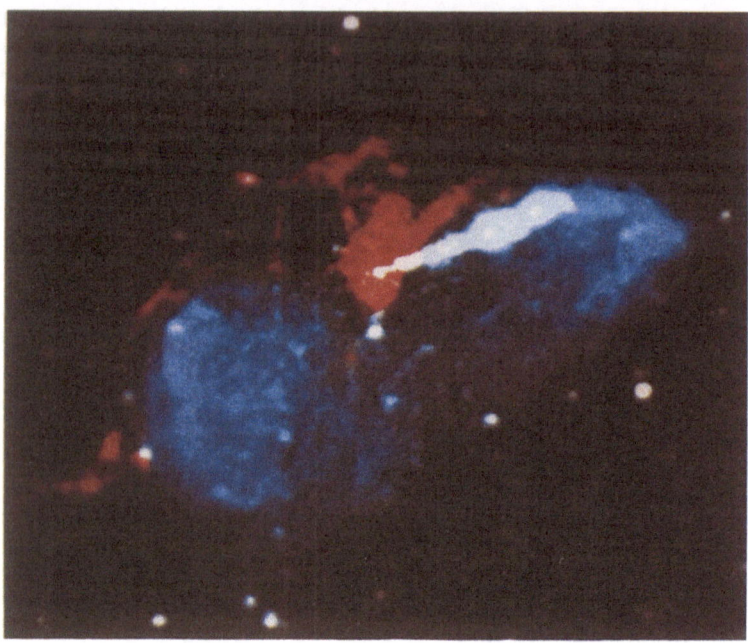

Fig. 4. Composite image showing optical $H\alpha + [NII]$ line emission (red), optical continuum (white) omitting the diffuse underlying elliptical galaxy light, and non-thermal radio emission (blue), illustrating the spatial relationships between thermal gas and synchrotron plasma.

Figure 4 shows a superposition of radio emission (blue), optical line emission (red) and optical continuum after subtracting the elliptical galaxy component (white). Also, in the radio, Faraday rotation observations of the inner lobes indicate that the SE lobe is the more distant of the two. The mere presence of the jet itself is consistent with this geometry since it is generally thought that relativistic beaming of a fluid flow towards us contributes to the high visibility of synchrotron emission from the jet. That requires the NW pointing jet to be in the foreground, relative to the nucleus. Hence, if we also appeal to the close spatial and morphological relationship between the filaments and the inner radio lobes, Fig. 4, and deduce that the filaments are close in three dimensional space to the inner radio lobes, we arrive at the same configuration indicated by the dust, with the NW quadrant in the foreground, and the SE quadrant in the background.

Knowing the three dimensional geometry in broad terms, it is interesting to turn to measurements of the kinematic behaviour. Perhaps unexpectedly, as shown in Fig. 5, from Sparks, Ford and Kinney (1993), the NW filaments primarily show *blueshifts*. Given that we have just determined that they lie in the foreground, this implies that they are *flowing out from the nucleus*. Hence, in the immediate vicinity of the nucleus, but external to the gas disk around the black-hole, we see outflow.

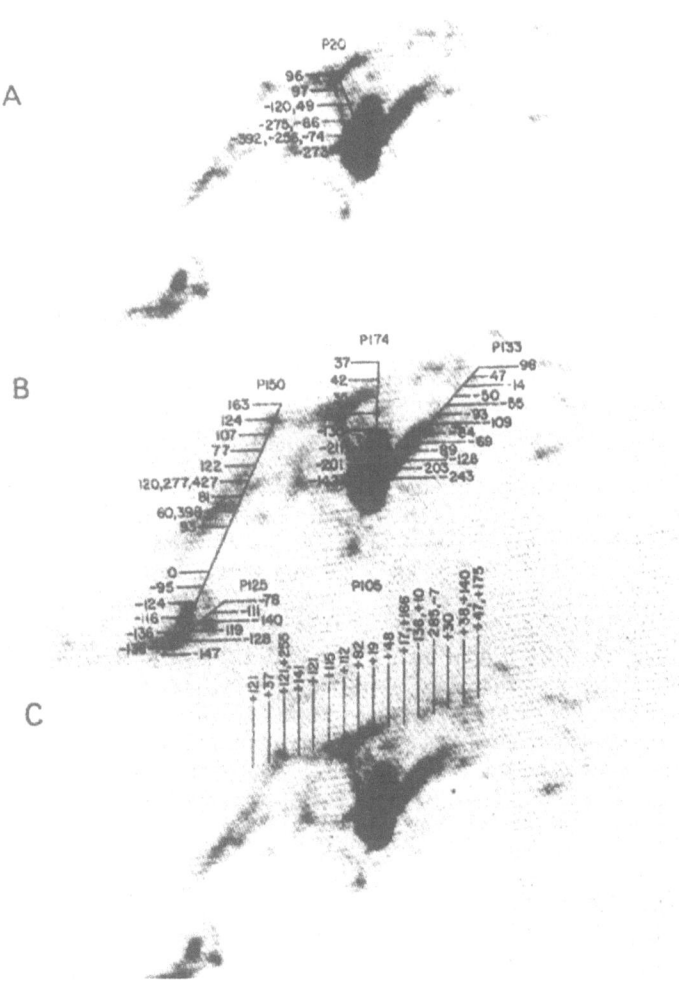

Fig. 5. Velocity field of emission line gas in M 87 relative to systemic, from Sparks, Ford & Kinney (1993). Negative implies blueshifted.

By contrast, the filaments in the SE quadrant that lie *beyond* the radio lobe are also blueshifted, implying since they are in the background that *at large radii the filaments are falling inwards.* This is a potentially important conclusion, as it does suggest consistency with the most likely theoretical scenarios, which require that the gas is infalling. It also suggests that locally, the activity at the nucleus (including the ejection of the radio lobes) is dominating the kinematic behaviour.

5.3   Neutral Gas and X-Ray Observations

Carter, Johnstone & Fabian (1997) showed that there is also blueshifted sodium absorption seen against the bright stellar nucleus. Given the similarity to the $H\alpha$ velocities, they concluded that the neutral gas component is present within the filaments. In a study of NGC 4696 at the center of the Centaurus cluster of galaxies, Sparks, Carollo & Macchetto (1997) showed that interstellar sodium absorption indicative of the presence of neutral gas is present throughout the extended line emission filament system in that object.

Böhringer *et al.* (1995) presented an image of the radio structure on larger scales in M 87, and also compared the radio distribution to the X-ray emission. They found correlations between the presence of radio emission, but also the regions where excess X-ray emission is found follow quite well the areas where optical filaments are located. Future high resolution X-ray observations with AXAF will be interesting to see whether there are close spatial correlations between these components.

6   The Present: Hubble Space Telescope Observations

The next major step forward in studies of the M 87 emission filament system came with imaging observations using the refurbished HST. Ford *et al.* (1994) and Harms *et al.* (1994) presented images taken with *Wide Field and Planetary Camera 2* (WFPC-2) of the $H\alpha + [NII]$ features and spectroscopy in a variety of locations using the *Faint Object Spectrograph* (FOS). They found evidence for an inner gas disk whose velocity field supported the notion of a central supermassive black-hole. Marconi *et al.* (1997) also found evidence for an inner disk, and mapped its velocity field using the *Faint Object Camera* (FOC) long-slit spectrograph. Much more on these topics is to be found elsewhere in these proceedings!

The initial HST imaging data were suggestive, and Ford *et al.* (in preparation) have now obtained new, much deeper, images. Figure 6 shows a high S/N, high resolution image taken with the WFPC-2. This image shows that the extended filament system appears to connect directly onto the inner gas disk. Knowing the kinematics and geometry above, this suggests we may be

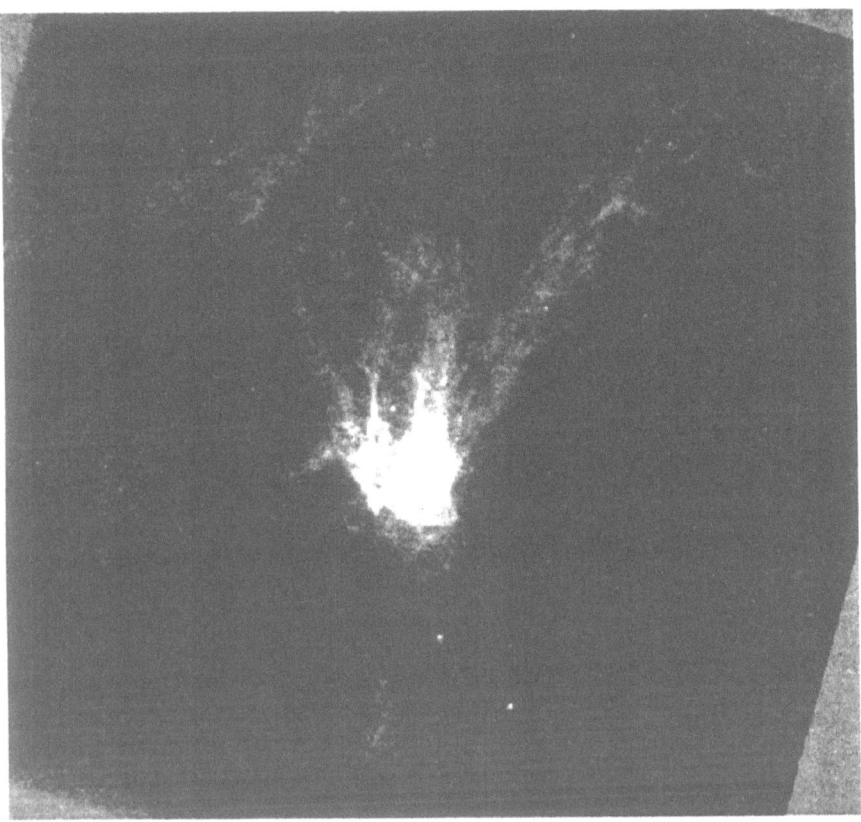

Fig. 6. High resolution image taken with the WFPC-2 on-board HST of $H\alpha+[NII]$ line emission in M 87.

witnessing angular momentum transfer out from the disk to the filaments in the vicinity of the jet and inner radio lobes.

The new HST images also show an intriguing double stranded, twisted morphology over much of the system. In the innermost northern region, there are loop structures reminiscent of solar coronal loops.

As part of a monitoring program on the jet, we have obtained deep new continuum HST images of M 87. These may be processed to reveal the associated dust distribution at very much higher spatial resolution than is feasible from the ground, see Fig. 7. Despite the increase in resolution, the basic conclusion remains the same: the dust and line emission gas are spatially coincident within the limitations of the data. New details may be seen in the NW filament that roughly parallels the jet, and the inner loops referred to above may also be seen to show absorption at least partially, indicating their orientation also. To the SE, we still see no dust absorption consistent with our view that these areas lie behind the galaxy.

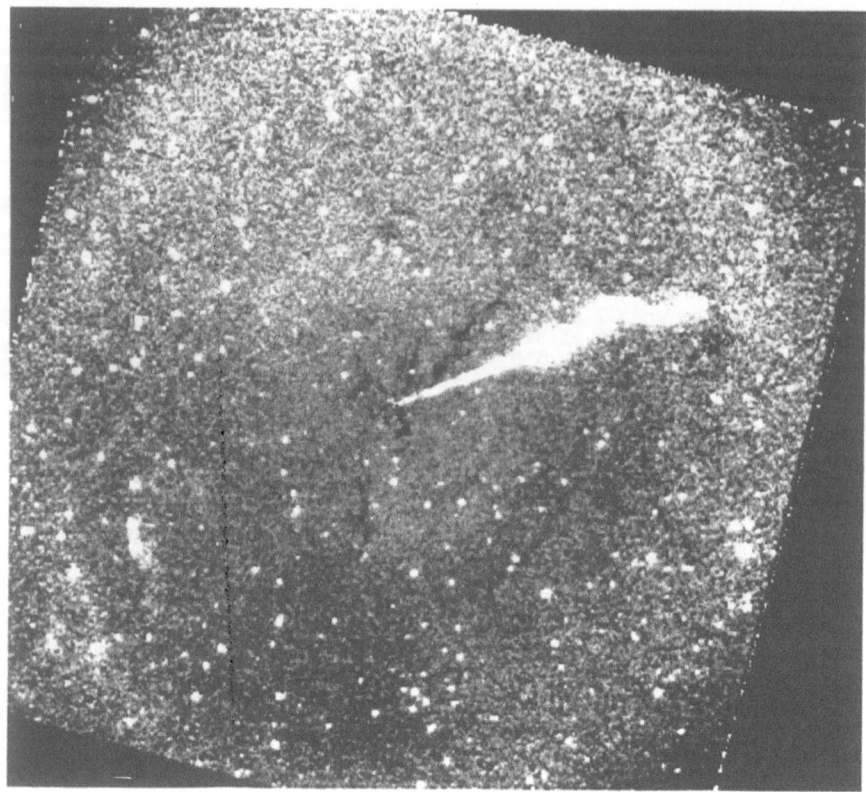

Fig. 7. High S/N image of M 87 taken with the WFPC-2 in V continuum light, divided by a smooth, elliptical model of the galaxy in order to show dust filaments at high spatial resolution in M 87.

There is one very narrow, irregularly shaped dust feature to the West and south of the end of the jet. The extraordinary length/width of this may present interesting constraints on pressures and theoretical options in that structure. There is weak line emission associated with its densest condensations.

7 Theoretical Considerations and Discussion

7.1 Kinematics

In the outer regions, beyond the inner radio lobes, we observe infalling gas, consistent with theoretical expectations in that region.

Closer to the nucleus, the relatively low velocities of the filaments imply a dissipative system. Their small scale length deduced from the high resolution imaging and dust absorption characteristics suggests a low Reynolds number situation with small dense blobs or filaments immersed in a tenuous hot coronal plasma. In such circumstances, the motion of the filaments acts as a tracer of the motions of the hot gas in which they are embedded. Sparks, Ford and Kinney (1993) consequently proposed that a small scale galactic wind, or breeze, is operative in the inner 20″ or so of the nucleus. The energy requirement is modest, with $\sim 10^{38} - 10^{40}$ erg/sec needed to power such an outflow. This is a relatively small percentage of the total nuclear energetic output, and therefore could be powered by the AGN.

Alternatives to direct powering of a galactic wind by the AGN are jet entrainment of its surrounding material and lateral expansion of the surrounding material about the jet. In the first case, as the jet moves rapidly out, it drags the associated material in its vicinity with it, consistent with the velocity field and geometry we see. In the second version, as the jet propagates outwards, it causes an overpressured region around the jet to move laterally away from the jet. There are several unknowns in that picture, however it remains a possibility.

Additional interesting possibilities include the disk being magnetically coupled to the filaments and hence transporting angular momentum from the disk, thereby inducing the motions of the filaments by torquing of the magnetic flux ropes by the disk.

In any event, it appears clear that the nucleus and its associated activity is dominating the energetics of the inner region of M 87.

7.2   Ionization and Origin of the Filaments

There have been numerous proposals for the ionization of the filaments, including shocks, AGN photoionization, hot stars, electron conduction, relativistic electrons, X-ray ionization and recombination and reionization in a cooling-flow. Cooling flow models in particular have been discussed at length elsewhere, *e.g.* Fabian (1994). I will not discuss all these possibilities, but will look at the electron conduction and merger picture. Sparks & Macchetto (1990) plotted the surface brightness of filaments as a function of radius in M 87, shown in Fig. 8.

Taking a model of the hot coronal gas, then there are essentially no free parameters in deducing the energy flux available through electron conduction. If we assume 1% of the energy available appears as $H\alpha + [NII]$, as in the case of NGC 4696 Sparks, Macchetto and Golombek (1989), using the simple model of Schreier *et al.* (1982) for the coronal gas, then the predicted optical surface brightness as shown in Sparks & Macchetto (1990) is in excellent agreement with the data. Recall that *no reference was made to the optical properties* in making the prediction. This agreement shows consistency with the idea that merging infalling gas is powered through electron conduction from the hot corona.

Is there any other evidence of a merger in M 87? Weil, Bland-Hawthorn & Malin (1998) show in a very deep image of the outermost regions of M 87 that

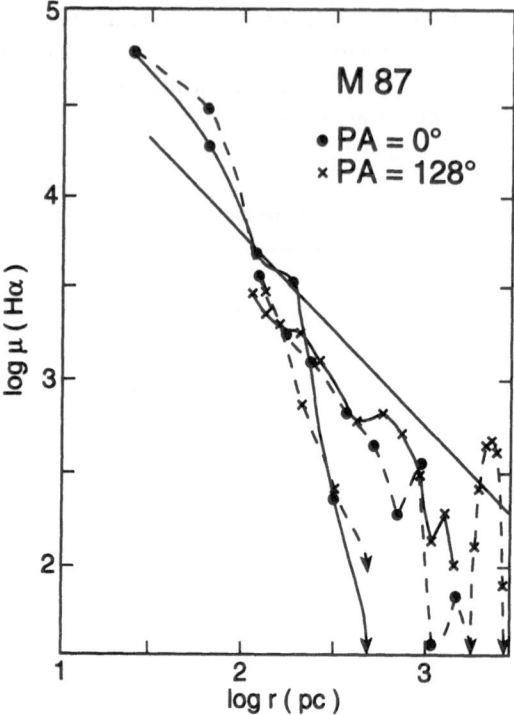

Fig. 8. Comparison of $H\alpha$ surface brightness profile with prediction for electron conduction heating model (straight line).

there is a significant asymmetry to the SE of the galaxy. This is in the direction of the outermost filaments. The proposed explanation for this extended asymmetry is that there has been a merger of a dwarf galaxy with M 87 in the recent past, consistent with the view that mergers may be causing many of the interesting phenomena we observe in elliptical galaxies and clusters of galaxies, including those of M 87.

8   Conclusions

Continually improving technology has enabled us to view the extended emission filaments in M 87 in ever greater detail. From the early work of Arp using photographic plates, to the latest images from the *Hubble Space Telescope*, as sensitivity and resolution have improved, a wealth of morphological detail and interesting interactions with other components of the M 87 system have been revealed The emission line gas is dusty which must provide information on physical processes responsible for the presence of the gas. There is an interesting morphological relationship with the radio emitting plasma suggestive of mutual interaction, and there are hints of a correlation between optical and X-ray emission in spatial terms.

The geometry appears to be quite well established, which, when used together with the kinematics, tells us that near the nucleus we have outflow but further away, infall. The filaments seem to connect to the inner disk around the black hole, and the black hole and other manifestations of nuclear activity dominate the innermost few kiloparsec.

There remain unanswered questions, which revolve around the primary problem of the origin of the material. Subsidiary questions include the nature of the excitation mechanism and the heavy element abundances in the different gas phases. The generality of the characteristics of the M 87 emission filament system is important, and it seems reasonable to assume that the system is no different in its fundamental nature to other emission filament systems in elliptical galaxies, particularly those situated at the centers of relatively rich X-ray emitting clusters of galaxies.

Future AXAF observations and complete mapping of the kinematic velocity field, along with precise abundance determinations in several gas components may succeed in resolving some of the uncertainties associated with this system. Searches for very low surface brightness emission within M 87 and at larger radii may also prove instructive.

References

Arp, H. (1967): ApLett, **1**, 1

Baade, W., Minkowski, R. (1954): ApJ, **119**, 215.

Baum, S.A., Heckman, T., Bridle, A., van Breugel, W. and Miley, G. (1988), ApJ-Supp, **68**, 643.

Böhringer, H., Nulsen, P.E.J., Braun, R., Fabian, A.C., (1995): MNRAS, **274**, L67.

Carter, D.C., Johnstone, R., Fabian, A.C., (1997): MNRAS, **285**, L20.

de Jong, T., Nørgaard-Nielsen, H.U., Jørgensen, H.E. and Hansen, L, (1990): A& A, **232**, 317.

Fabian, A,C., (1994): Ann. Rev. Ast. Ap., 32, 277.

Ford, H.C., Butcher, H.R., (1979): ApJSupp, **41**, 147.

Ford, H.C. *et al.* (1994): ApJ, **435**, L27.

Fukazawa, Y., Ohashi, T., Fabian, A.C., Canizares, C.R., Ikebe, Y., Makishima, K., Mushotzky, R., Yamashita, K., (1994): PASJ, **46**, L55.

Goudfrooij, P., (1994): thesis, University of Amsterdam.

Goudfrooij, P., Hansen, L., Jorgensen, H.E., Norgaard-Nielsen, H.U., (1994): A& ASupp, **105**, 341.

Goudfrooij, P., de Jong, T., Hansen, L., Norgaard-Nielsen, H.U., (1994): MNRAS, **271**, 833.

Harms, R.J. *et al.* (1994): ApJ, **435**, L35.

Humason, M.L., Mayall, N.U. and Sandage, A., (1956): A.J., **61**, 97.

Heckman, T.M., Baum, S., van Breugel, W.J.M., McCarthy, P. (1989): ApJ, **338**, 48.

Hu, E., (1988) in *"Cooling Flows in Clusters and Galaxies,"* ed. A.C. Fabian, Kluwer.

Macchetto, F., Pastoriza, M., Caon, N., Sparks, W.B., Giavalisco, M., Bender, R., Capaccioli, M., (1996): A& ASupp, **120**, 463.

Macchetto, F., Marconi, A., Axon, D.J., Capetti, A., Sparks, W. Crane, P., (1997): ApJ, **489**, 579.

Mayall, N.U., (1936): PASP, **48**, 14.

McNamara, B.R., Wise, M., Sarazin, C.L., Jannuzi, B.T., Elston, R., (1996): ApJ, **466**, L9.

Osterbrock, D., (1960): ApJ, **132**, 325.

Pinkney, J., *et al.* (1996): ApJ, **468**, L13.

Sarazin, C.L., O'Connell, R.W., McNamara, B.R.D., (1992): ApJ, **397**, L31.

Schreier, E.J., Gorenstein, P., Feilgelson, E.D., (1982): ApJ, **261**, 45.

Sparks, W.B., Macchetto, F., Golombek, D., (1989): ApJ, **345**, 153.

Sparks, W.B., Macchetto, F., (1990) in *"Paired and Interacting galaxies"*, IAU Colloquium 124, eds. Sulentic, Keel, Telesco, p403.

Sparks, W.B., Ford, H,C., Kinney, A.L., (1993): ApJ, **413**, 531.

Sparks, W.B., Jedrzejewski, R.I., Macchetto, F. (1994): in *The Soft X-ray Cosmos"*, eds. Schlegel & Petre, AIP Conference procs 313, p. 389.

Sparks, W.B., Carollo, Macchetto, F., (1997): ApJ, **486**, 253.

Sparks, W.B., (1997): in *Galactic and Cluster Cooling Flows"*, ed. N. Soker, A.S.P. Conference Series, vol. 115, p. 192.

Walker, M.F., (1968): ApLett, **2**, 65.

Weil, M.L., Bland-Hawthorne, J., Malin, D.F., in press.

Radio Observations of the M 87 Jet

J. A. Biretta

Space Telescope Science Institute
Baltimore, MD 21218
USA

Abstract. We review observations of the M 87 jet made at radio frequencies and summarize some of the insights gained therefrom.

1 Introduction

Virgo A was among the first discrete sources discovered by early radio astronomy, and its association with the galaxy M 87 (Bolton, Stanley, and Slee 1949) and its "curious" feature (Baade and Minkowski 1954) were quickly recognized. Eventually radio interferometer images resolved the bright double radio lobes and the counter-part of optical jet itself (Turland 1975). Similarly, the jet would be detected at X-ray frequencies, once the resolution was adequate to separate it from the bright thermal emission (Schreier, Gorenstein, and Feigelson 1982). While several hundred jets have now been discovered (Bridle and Perley 1984; Keel 1988; Fraix-Burnet *et al.* 1990; Liu and Xie 1992), M 87 remains one of the nearest examples, making it an ideal target for study of the extragalactic jet phenomenon. Only one other extragalactic jet is substantially closer, Centaurus A (NGC 5128), but it is at a very low declination and therefore is difficult to study with northern radio interferometers.

The radio source associated with M 87 (Virgo A, 3C 274, 1228+127) is classified as an FR-I source based on its low luminosity ($P_{178\,MHz} \sim 1 \times 10^{25}$ W Hz^{-1}) and edge-darkened morphology (Fanaroff and Riley 1974). In spite of this classification, its one-sided jet (Fig. 1) is more typical of FR-II sources, and perhaps this is related to its situation near the FR-I/II division at $P_{178\,MHz} \sim 3 \times 10^{25}$ W Hz^{-1}. Bridle (1984) has also noted that it has an unusually weak core for a one-sided jet. Its distance has been measured to be 15.9 Mpc using a redshift independent technique (Tonry 1991; $z = 0.0043$), and we will assume this distance herein. This distance gives a linear scale of 78 pc per arcsecond; a proper motion of 1 milliarcsecond (mas) per year corresponds to $0.254c$, where c is the velocity of light.

In the following sections I will review radio observations of the M 87 jet. I will begin at the nucleus on sub-parsec scales, and work outwards to include the kpc-scale jet, and finally the jet-like plumes with scales of 10's of kpc. Other papers at this conference discuss the largest-scale radio structures.

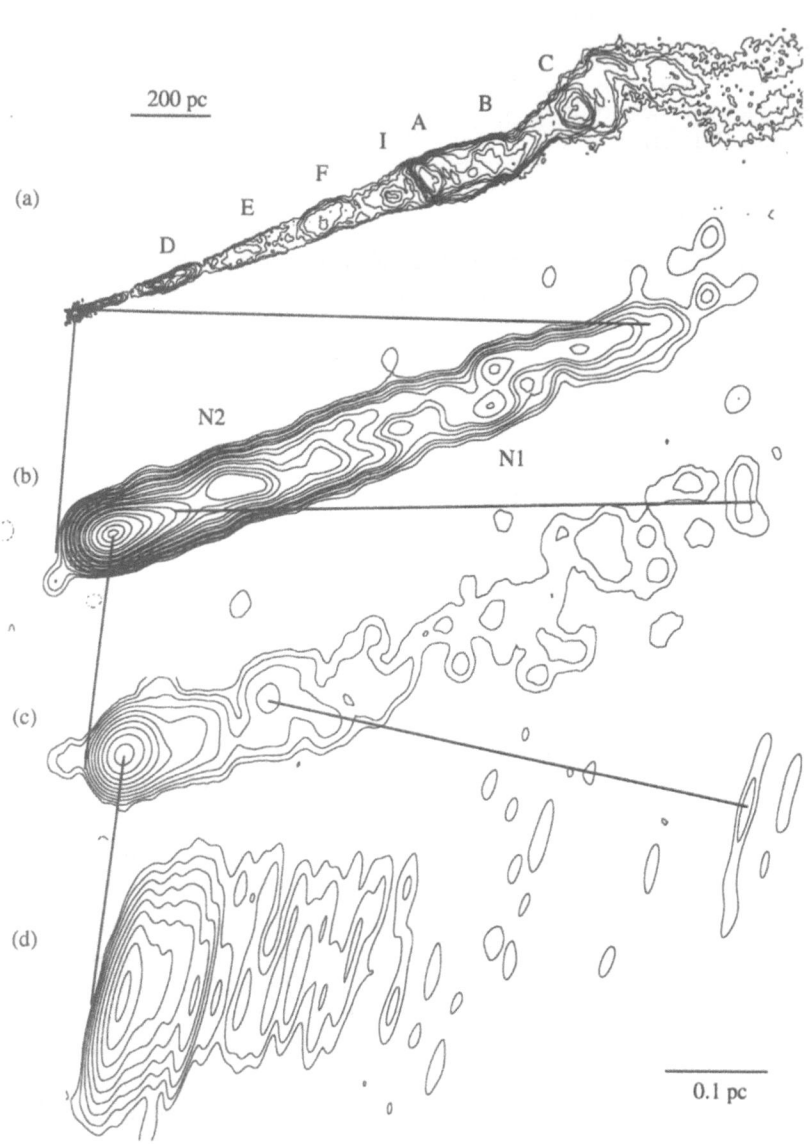

Fig. 1. Radio observations of the M 87 jet on scales between 0.1 pc and 2 kpc. (a.) VLA image at $\lambda = 2$ cm with 150 mas resolution. (b.) $\lambda = 18$ VLBI image at 4 mas resolution (Reid, *et al.* 1989). (c.) Tapered VLBA+EVN at $\lambda = 1$ cm with 1 mas resolution. (d.) VLBA+EVN at $\lambda = 1$ cm with 1×0.14 mas resolution from Junor and Biretta (1995). North is up; all contours have logarithmic spacing.

2   The Parsec-Scale Jet

The nucleus of M 87 was among the first sources to be studied with VLBI techniques (Cohen *et al.* 1969; Kellermann *et al.* 1973). Early 18 cm VLBI images (Reid *et al.* 1982) revealed a typical one-sided core-jet morphology which was well aligned with the kpc-scale jet. Later observations with more antennæ and better resolution (Fig. 2, Reid *et al.* 1989) revealed a 15 pc-long jet having a wealth of complex structure. The jet's brightness falls off fairly smoothly with distance from the core, although three prominent features (N2, N1, and L) stand out above the smooth emission.

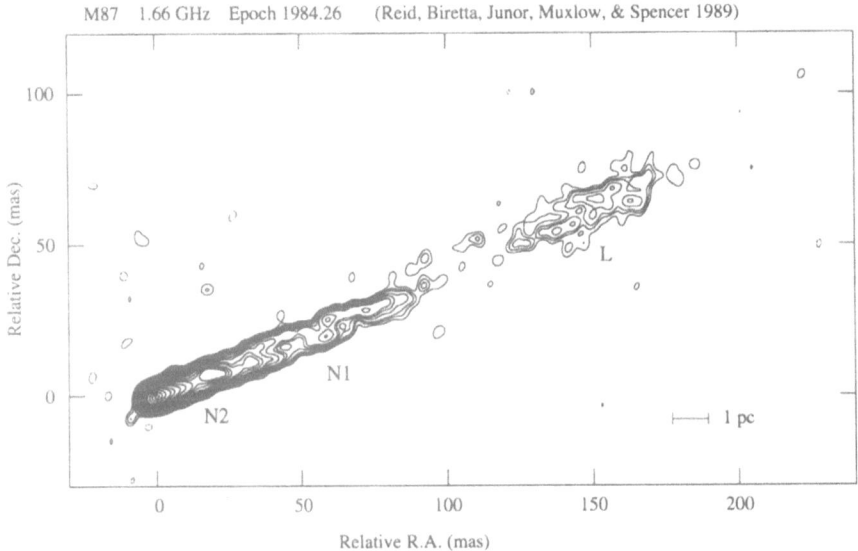

Fig. 2. Global VLBI image of M 87 pc-scale jet at frequency 1.66 GHz from epoch 1984.26. The restoring beam is a 4 mas FWHM circular Gaussian function. The lowest contour is at 2 mJy/beam. North is up. (Reid *et al.* 1989.)

"Filamentary" or narrow elongated structures can be traced along much of the jet in Fig. 2. For example, a linear feature can be traced along the northern limb between core distances $r \sim 40$ to ~ 50 mas, at which point it appears to cross the the southern limb, and then to cross back to the northern limb at $r > 65$ mas. Much of the jet also appears to be "limb-brightened" in that one or both edges are brighter than the jet center. For example, at 40 mas from the core the jet is brightest along both edges, and weaker along its centerline. And again at 65 mas it is similarly bright at both edges. In other places, for example at 50 and 75 mas from the core, it appears bright only along the northern limb. The brightness centroid of the jet also appears to oscillate from side to side. Many of these characteristics can be interpreted in terms of helical, filamentary structures within the jet or wrapped around its surface (Königl and Choudhuri 1985; Hardee 1987; Owen, Hardee, and Cornwell 1989; hereinafter OHC89).

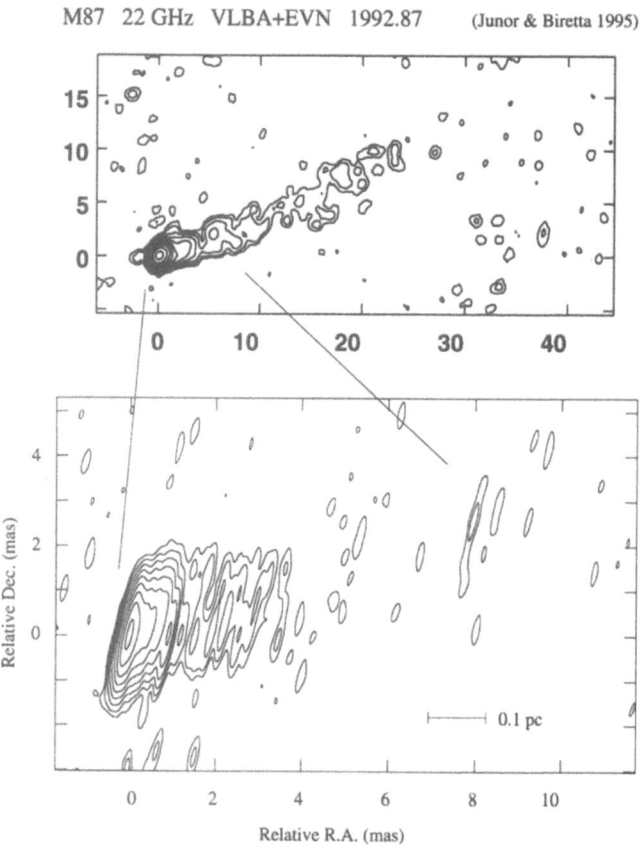

Fig. 3. VLBA+EVN image of M 87 sub-pc-scale jet at 22 GHz from epoch 1992.87. North is up. Contours are at −2, −1, 1, 2, 4, 8, 16, 32, 64, 128, 256, and 512 mJy/beam. (a.) Image made from tapered (u, v) data; resolution 1.1 × 0.9 mas FWHM at PA= −14°. (b.) Full resolution image with 1.3 × 0.14 mas FWHM at PA= −13° beam. North is up. (Junor and Biretta 1995)

Global VLBI observations at 22 GHz show that these characteristics continue down into sub-parsec scales (Spencer and Junor 1986; Junor and Biretta

1995; 100 GHz observations by Bääth *et al.* 1992 have higher resolution but the dynamic range is rather low). Figure 3.a shows a tapered image with ~ 1 mas resolution. Filamentary structure, limb-brightening, and side-to-side oscillation are again evident. The full resolution image (Fig. 3.b) also shows evidence of oscillation and linear features, though the dynamic range is somewhat poorer than the tapered image. Evidence of filamentary features, limb-brightening, and side-to-side oscillation are also seen in the kpc-scale jet (OHC89); these characteristics are present on scales from 0.1 to 1000 pc.

The jet appears to be well-collimated on scales from < 1 pc to several kpc, and there is a systematic trend for the jet's opening angle to decrease with core distance. Models fit to the (u,v) data of Fig. 3.b within 0.1 pc of the core give a full-width-quarter-maximum opening angle of $\phi \sim 23° \pm 8°$, while Fig. 2.a indicates a value near 13° at 1.5 pc. Farther out, we find $\phi \sim 9°$ at tens of pc (Fig. 2), and $\phi \sim 6°$ at 1 kpc. It is also interesting to consider the location of the jet's origin, as defined by the cone's apex on the smallest scales. If we extrapolate the brightness peaks along the jet's edges in Fig. 3.a back toward the core, we find that the cone's apex lies within about 1 mas of the bright, unresolved core. Hence, in this simple picture of a cone with constant (or monotonically contracting) opening angle, the jet's initial collimation must take place on a scale smaller than about 0.1 pc.

The jet's kinematics appear to be complex on all scales. As we will discuss in the next section, both stationary and superluminal features have been seen in the kpc-scale jet (Biretta, Zhou, and Owen 1995; hereinafter BZO95). The fastest features move outward at ~ 2c and are located in knot D about 200 pc from the core, while other features in this knot, and also in knots B and C (1200 to 1500 pc from core), appear stationary (Table 1).

Table 1. Apparent velocities of jet features relative to the core.

Feature	Core Dist. (pc)	Apparent Speed (v/c)	Reference
S2	0.013	0.01 ± 0.01	Junor & Biretta 1995
S3	0.04	0.03 ± 0.03	Junor & Biretta 1995
N2	1.6	-0.03 ± 0.02	Biretta & Junor 1995
N1	5	0.02 ± 0.04	Biretta & Junor 1995
D	200	-0.2 ± 0.1 to 2.5 ± 0.3	BZO95
F	700	0.9 ± 0.2	BZO95
A	1000	0.41 ± 0.02 to 0.61 ± 0.02	BZO95
B	1200	-0.2 ± 0.2 to 1.2 ± 0.1	BZO95
C	1400	0.11 ± 0.04	BZO95

Similar complexity is seen on the pc-scale. Figure 4 shows a sequence of 18 cm images of the pc-scale jet. The two most prominent features, N1 and N2, both appear to be roughly stationary. Figure 5 shows position measurements for N2. While the early data give evidence for outward motion at $\sim 0.3c$ (Reid *et al.* 1989), this motion does not continue at later epochs. Apparently the position of N2 wanders by a few mas on timescales of a few years, but is fixed on longer timescales. A linear fit to the data gives a speed $v/c = -0.03 \pm 0.02$, though it is a rather poor fit. Similarly, N1 has a very slow speed of $v/c = 0.02 \pm 0.04$.

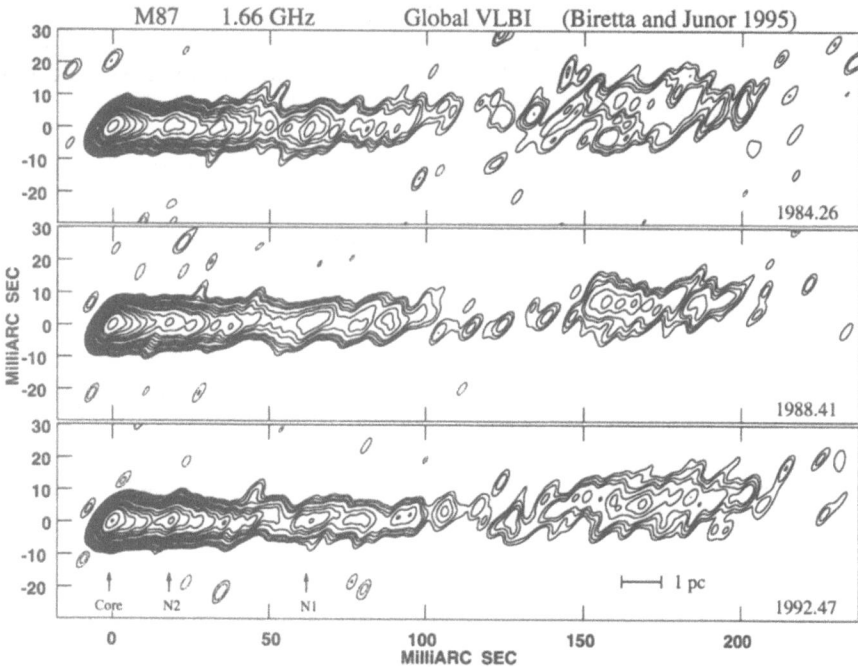

Fig. 4. Global VLBI images at 1.66 GHz for epochs 1984.26, 1988.41, and 1992.47. Prominent features N1 and N2 are labeled. All images have same 7.1×2.7 mas FWHM at PA= $-13°$ Gaussian restoring beam. Contours are at -2, -1.4, 1.4, 2, 3, 4, 7, 10, 14, 20, 30, 40, 70, 100, 140, 200, 300, 400, 700, and 1000 mJy/beam. The images are rotated to put the jet on the x-axis (Biretta and Junor 1995).

Careful examination of Fig. 4 also reveals many weaker features or local maxima in the brightness distribution. These features are very difficult to identify and track between epochs. They either change speed, flash on and off, or move large distances during the four-year sampling period. For example, any features moving at 2.5c, the largest speed seen in the kpc-scale jet, would travel the entire length of the visible pc-scale jet in only eight years. It is

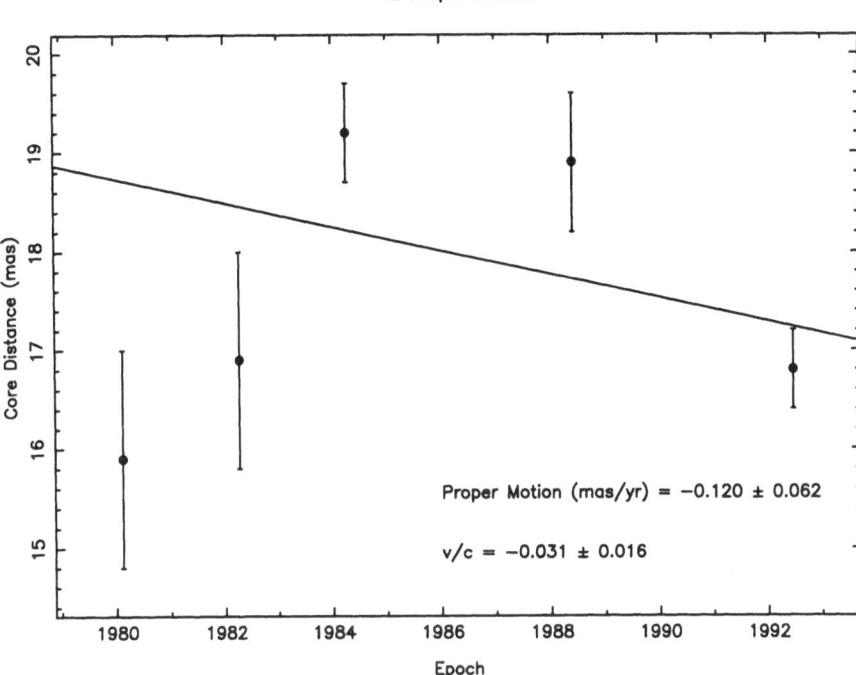

Fig. 5. Core distance vs. epoch for feature N2 in the pc-scale jet. Data are taken from 1.66 GHz VLBI images which are convolved to 8 mas resolution. The fitted line corresponds to motion at $(-0.031 \pm 0.016)c$ (Biretta and Junor 1995).

natural to ask whether these weaker features merely reflect errors in the images. This seems unlikely; the contrast between these local maxima and the surrounding emission in many cases is 30 to 100 times the expected thermal noise $(0.3\ mJy)$. As with the pc-scale jet, the most prominent features in the sub-pc-scale jet also appear to be nearly stationary. Models fit to 22 GHz data at epochs 1992.45 and 1992.87 give speeds of $0.03c$ or less for features S2 and S3, which are both within 0.04 pc of the core (Table 1).

Even though the prominent features are stationary in position, large flux variations are seen. Figure 6 shows 18 cm brightness profiles along the jet at 8 mas resolution. Even though N2 is stationary, it appears to brighten by 60% between 1988 and 1992. In fact, the entire inner 30 mas of the jet brightens by amounts ranging from 20 to 60%. If this brightening is interpreted as the result of an outburst in the core, it implies an apparent propagation speed of at least $2c$ for the jet material. The available 22 GHz data also indicate 30% variations in the core and jet flux over 5 months (1992.45 to 1992.87).

As previously mentioned, the path of the jet shows evidence for side-to-side oscillation. It is interesting to ask whether the path of these oscillations

M87   Intensity Profiles Along Jet   (1.66 GHz, 8 mas beam)

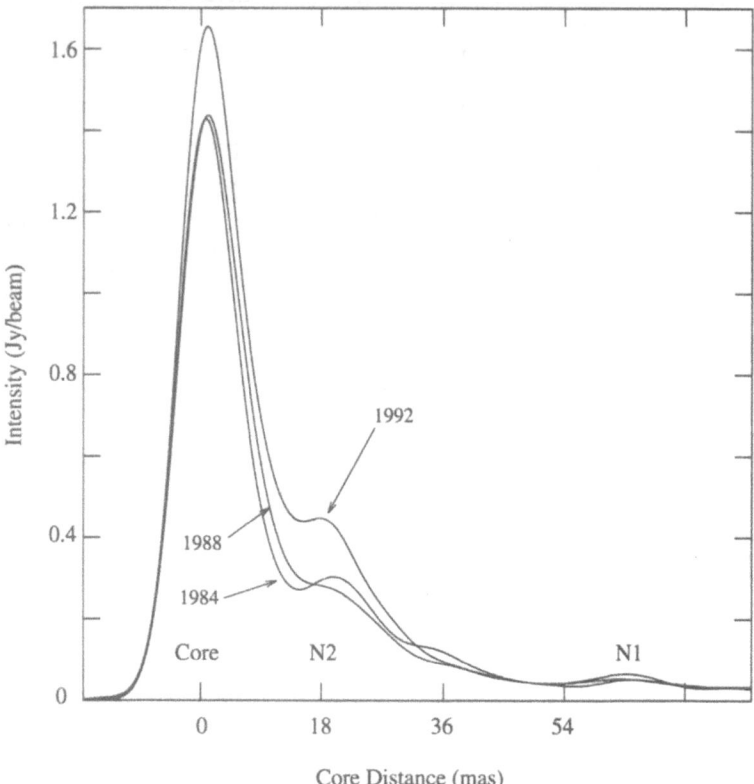

Fig. 6. Intensity as function of distance along jet at 1.66 GHz for three epochs. Core distance vs. epoch for feature N2 in the pc-scale jet. Data are taken from 1.66 GHz VLBI images which are convolved to 8 mas resolution. The fitted line corresponds to motion at $(-0.031 \pm 0.016)c$ (Biretta and Junor 1995).

is fixed or moves outward. The former result would suggest instability or external forces as the cause of the oscillations, while the latter result would suggest precession-like effects. Examination of Fig. 4 shows a strong tendency for the oscillation pattern to remain fixed. For example, between $r = 25$ and 40 mas from the core, the jet consistently runs south of the mean centerline. Similarly, it always runs north of the centerline between $r = 40$ to 55 mas.

In the remainder of this section we discuss some implications of these observation of the pc-scale jet. Because of its proximity, M 87 presents a unique opportunity to image an AGN at small linear scales, and thus to study the jet during its initial formation. From the 22 GHz observations, it is apparent that most of the jet collimation occurs on scales < 0.1 pc, which corresponds to about 300 Schwarzschild radii for a putative 2×10^9 solar mass black hole. It thus appears likely that the initial collimation is provided by electromagnetic processes which are tied to the black hole and accretion disk (Sakurai 1987; Li, Chiueh, Begelman 1992). The interstellar medium might provide some collimation on larger scales (*e.g.* Wiita and Siah 1986), but is probably not the dominant agent.

One of the most interesting parameters of the pc-scale jet is the bulk flow velocity. Models that attempt to unify the FR-I sources with BL-Lacs predict relativistic flows in the cores of FR-Is (Browne 1983; Urry, Padovani, Stickel 1991). While there is no single piece of compelling evidence supporting relativistic flow in the core of M 87, there is considerable circumstantial evidence for this. Superluminal motion (v up to $2.5c$) is seen 200 pc from the core, which strongly suggests relativistic flow on those scales, and it is easy to imagine that flow originating in the nucleus. The most prominent features in the pc-scale jet appear to be stationary, but these may merely represent standing shocks or stationary obstructions in a rapid flow. The kpc-scale jet displays superluminal features adjacent to stationary ones, so one cannot dismiss the possibility of relativistic flow based on the stationary features. The pc-scale jet also contains numerous features that cannot be tracked between available epochs, and rapid motion ($v_{apparent} > 2c$) is a viable explanation of their behavior. We have also seen that the inner 30 mas (2.4 pc) of the jet brightens in < 4 years, which can be taken as evidence of flow at $> 2c$. Finally, the usual limit on the jet/counter-jet brightness ratio (> 150, Reid *et al.* 1989) implies flow with Lorentz $\gamma > 2$.

An alternate picture is one in which the pc-scale jet is slow, but then accelerates between the pc- and kpc-scales. Such an acceleration might occur as the internal energy of a relativistic plasma is converted to bulk energy at a nozzle (Daly and Marscher 1988). However, the morphology of the jet and its opening angle are remarkably similar on these two scales. There is no obvious evidence for nozzle effects between these scales, or candidates for the site of such a nozzle. And other properties, such as the flux evolution and jet / counter-jet ratio, would still remain unexplained.

If one accepts that the M 87 pc-scale jet contains a relativistic flow, it is interesting to ask why it is different from the BL Lacs and quasars, where the most prominent features are superluminal. A possible answer may lie in M 87's orientation and limb-brightened appearance. Constraints from motions in the kpc-scale jet, and from the morphology of knot A, together suggest the jet is oriented about $\theta \sim 43°$ from the line of sight (Biretta, Zhou, and Owen 1995). Deprojection of the circumnuclear gas disk also gives this orientation, if one presumes the disk is normal to the jet (Ford *et al.* 1994). For such an angle, regions with bulk flow $\gamma \sim 3$ have no Doppler boosting, and for larger γ the emission is actually beamed away from the observer. We also note that the limb-brightened appearance suggests much of the emission arises from the outer surface of the jet, where the external medium might cause instabilities and standing shocks, along with lower flow speeds. Hence, M 87 and its superluminal cousins may be intrinsically similar, but appear different due to orientation effects: in M 87 ($\theta \sim 43°$) the visible emission is dominated by slower material ($\gamma \lesssim 3$) near the jet's surface where the speed and orientation give little Doppler beaming, whereas in BL Lacs and quasars ($\theta < 10°$) the visible emission is dominated by highly boosted material in a faster ($\gamma \gtrsim 5$), more coherent flow near the jet's center.

Work is underway to observe M 87 at closely spaced epochs ($\lesssim 1$ month intervals) with the VLBA. This should greatly clarify the nature of the rapidly evolving features in the jet, and show whether they in fact indicate relativistic flow in the nucleus. Future mm and space VLBI of M 87 will further improve resolution, and present a unique opportunity to image the "engine" within a few Schwarzschild radii of the putative black hole.

3   The Inner Jet: Between Parsec Scales and $\sim 1\,\mathrm{kpc}$

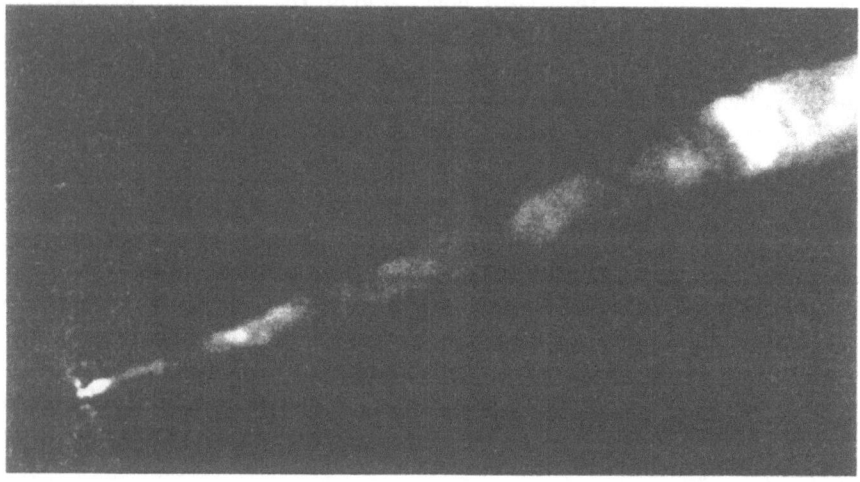

Fig. 7. Greyscale image of inner jet from nucleus (left) to knot A (right) observed at 15 GHz with 0.''1 resolution. The bright point source at the nucleus has been subtracted in this display. North is up, and the images is $\sim 13.''6$ or $\sim 1.06\,\mathrm{kpc}$ wide. From OHC89.

Next we consider the region from roughly 0.2 arcseconds out to knot A. We will refer to this region as the "inner jet." As we will see, its morphology and kinematics appear distinct from those seen farther from the nucleus. The most prominent features in this region are four quasi-periodic knots usually called D, E, F, and I. These knots were first identified in radio by Owen, Hardee, and Bignell (1980), and later confirmed with better resolution by Biretta, Owen, and Hardee (1983).

The first image to clearly show the internal structure of the knots, and to detect the weak inter-knot emission, was the 15 GHz VLA image at 0″.1 resolution by Owen, Hardee, and Cornwell (1989 OHC89; Fig. 7). This image still remains one of the best radio images of the jet. It clearly illustrates one of the key features of the inner jet — that it appears as a cone with remarkably straight sides and an opening angle of ~6° (FWQM). To a large extent, this cone is defined by a faint diffuse emission which appears to fill the jet. Limb-brightening is another key feature of the inner jet. There are many places where the structure is brighter at either or both limbs than in the jet center (*i.e.* limb-brightened). For example, in knot E, between knots E and F, and again between F and I, the edges of the jet are brighter than the jet center. There are also regions where one limb is particularly bright, such as in knots D and E. Filamentary or extended linear features are also evident in this region. For example, it is possible to follow the brightness ridgeline through knot D along a line which starts at the northern limb on the eastern end of the knot, crosses to the southern limb near the middle of the knot, and then returns to the northern limb as one continues toward the western end of the knot. Pieces of similar features are also apparent in knots E and F.

The magnetic field structure of the emission regions may be inferred from polarization images. The polarization structure of the 20″ jet has been studied in detail (OHC89; Fig. 8; also see Biretta, *et al.* herein). The magnetic field runs roughly parallel to the jet axis throughout the inner jet region. In general, wherever filamentary, linear, or elongated features are apparent, the field always runs along the length of the feature. Examples of such regions are seen in knots D and E, and in the brightest region of knot I. Fractional polarizations are typically 15 to 30 percent throughout the jet, though some regions, particularly near the edges of the jet, show polarization as high as 50 to 70 percent. These latter values approach the maximum possible for synchrotron emission, and indicate highly ordered magnetic fields. The polarization of the inter-knot regions can be seen more clearly in the recent 8 GHz image by Zhou, Owen, and Biretta (1999; Fig. 9). Again the fields are directed predominantly along the jet, and the highest fractional polarizations occur near the jet limb in knots E, F, and I.

Motions and variability have been detected in the kpc-scale jet by comparing high dynamic range 15 GHz VLA images observed at many epochs between 1982 and 1994 (BZO95; Zhou 1998). Small changes in the positions of features were measured using a two-dimensional cross-correlation technique (Biretta, Owen, and Cornwell 1989). Because of the high sensitivity of this technique, and the high dynamic range of the images, it is possible to measure position changes to an accuracy better than 1 percent of the 0.15 arcsecond resolution.

The highest speeds and most dramatic changes are seen for this inner jet region, and in particular over the first 300 pc of the jet, encompassing

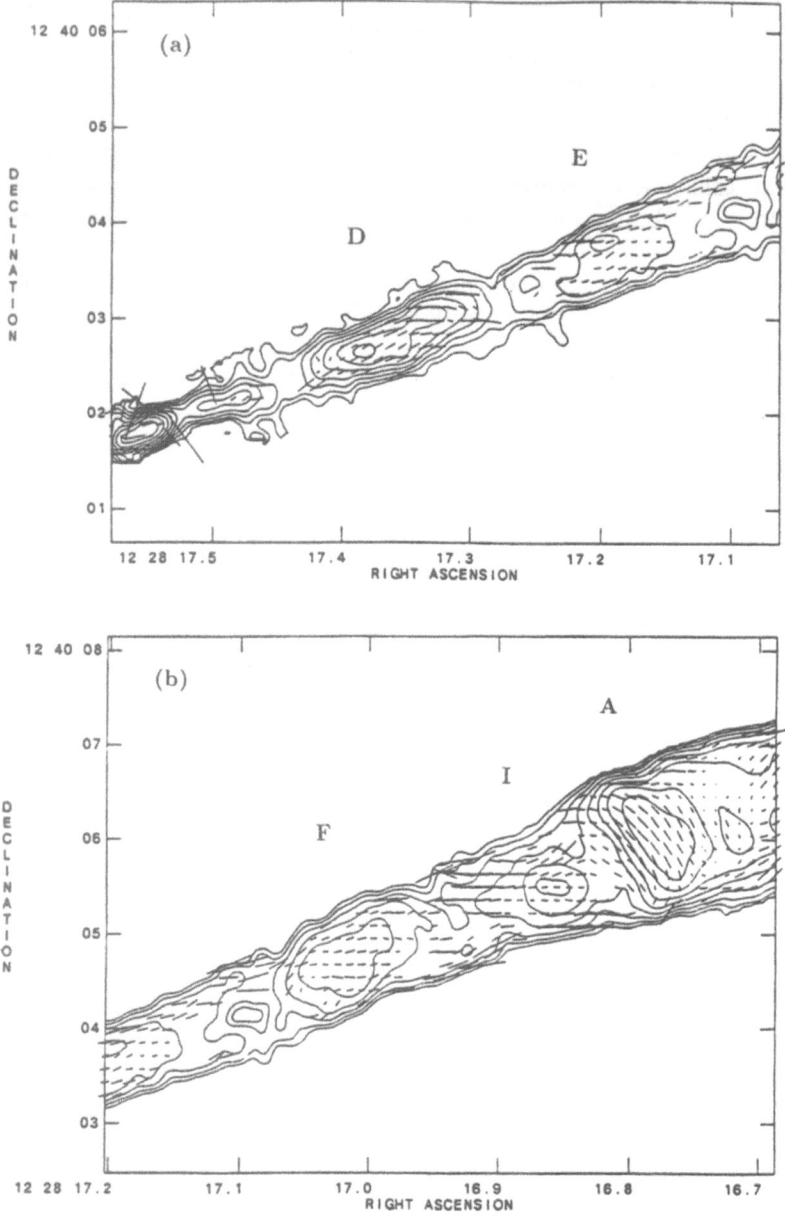

Fig. 8. Contour plots of jet with vectors showing the local magnetic field direction at frequency 15 GHz. The magnetic field lines are predominantly along the jet until knot A. A vector length of $1''$ corresponds to 216% polarization. Contours are spaced at factors of two in intensity beginning at 1.58 mJy/beam, and the resolution is $\sim 0''\!.15$ (a) Nuclear jet and knots D and E. (b) Knots E, F, and I. From OHC89.

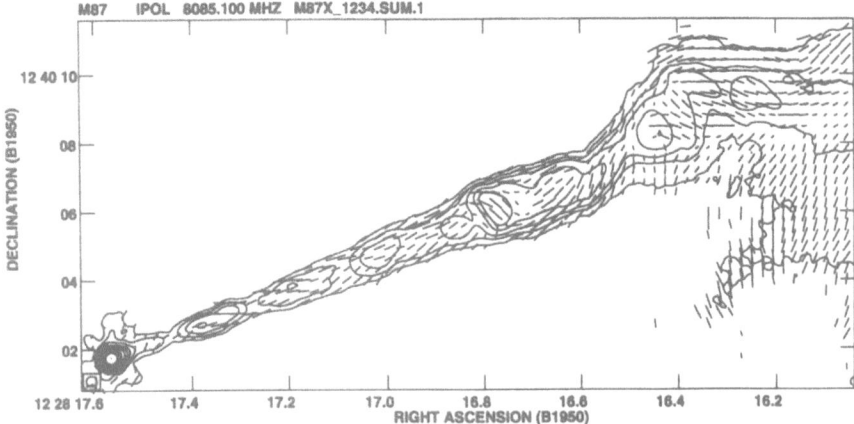

Fig. 9. Contour plots of jet with vectors showing the magnetic field direction from high-dynamic range frequency 8 GHz observation. The magnetic field lines are parallel to the jet direction except in knots A and C. A vector length of $1''$ corresponds to 62% polarization. From Zhou, Biretta, and Owen 1999.

the region from the nucleus to knot D. Figure 10 shows the entire jet with vectors superposed indicating the speed and direction of motion for some of the brighter knots. Typical speeds are in the range $0.5c$ to c. For the inner jet, the motions are predominantly parallel to the jet axis. It is also possible to measure speeds for smaller features within the knots, and this is illustrated in Fig. 11 for knot D.

The three features within knot D present an interesting situation. The eastern-most feature (DE) appears to be nearly stationary (speed $-0.2c \pm 0.1c$), while the middle and western features (DM and DW) display superluminal motion at $(2.5 \pm 0.3)c$ and $(2.1 \pm 0.2)c$, respectively. One interpretation is that the eastern feature represents some stationary obstruction in the flow, while the faster speeds may be more characteristic of the fluid flow. The fast speeds imply motion with the Lorentz factor $\gamma \gtrsim 3$ in the standard beaming model.

More recently Zhou (1998) has noticed that much higher speeds are obtained for DW if one looks at the position of its peak brightness, rather than cross-correlating the entire feature. Figure 12 shows contour maps of the first few arcseconds of the jet at five epochs spanning 9 years. Rapid motion of the flux peak in DW is quite evident, and Fig. 13 shows formal fits to its position as a function of time. The velocity component along the jet axis is $v_x = (5.1 \pm 0.8)c$, and implies $\gamma \gtrsim 5$; this is similar to nearby features in the optical jet (see Biretta *et al.* herein). DW also has a significant component of motion across the jet and toward the south with $v_y = (-0.9 \pm 0.2)c$. Apparently it is moving at an angle of about $10° \pm 2°$ to the jet axis, and is following

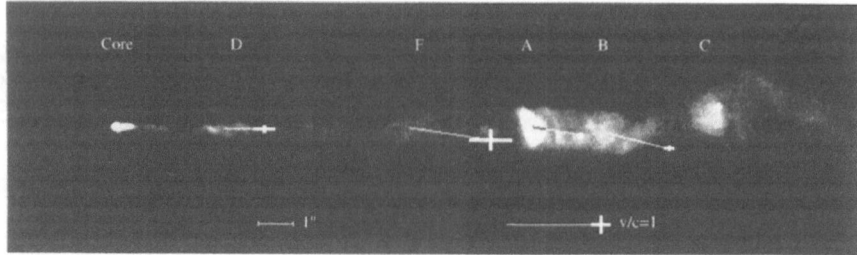

Fig. 10. Observed velocity vectors for entire knots D, F, A, B, and C superposed on an image of the jet. Motion is toward the heavy cross at the end of each vector, which also indicates the 1σ uncertainty on each component of the motion. The image is rotated to put the jet along the x-axis. From BZO95.

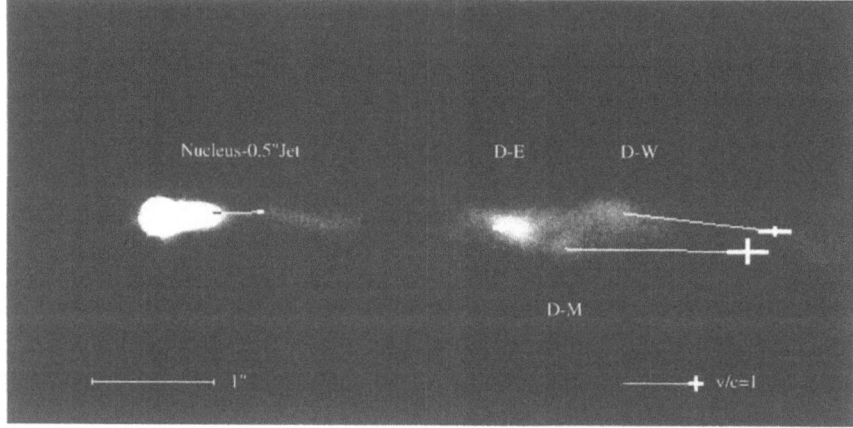

Fig. 11. Observed velocity vectors for features within knot D. (See Fig. 10 for details.) From BZO95.

the local ridgeline of the jet. This is suggestive of motion along a helical path within the jet, though other interpretations are certainly possible.

Figure 12 also shows dramatic evolution in the region labeled "H" about 1 arcsecond from the nucleus. This corresponds roughly to the optical region HST-1 (Biretta *et al.* herein). At the early epochs there are only two features, but by 1991 a third appears (labeled "H-E") which then brightens considerably by 1993. Rapid variability of extended features can be used to argue for relativistic Doppler factors. A simple extrapolation of the jet opening angle gives a size for H-E of 6 pc; either the emission region is much smaller than this size, or the Doppler factor exceeds ~ 5. Efforts so far to directly measure the sizes of these features have been inconclusive (Zhou 1998).

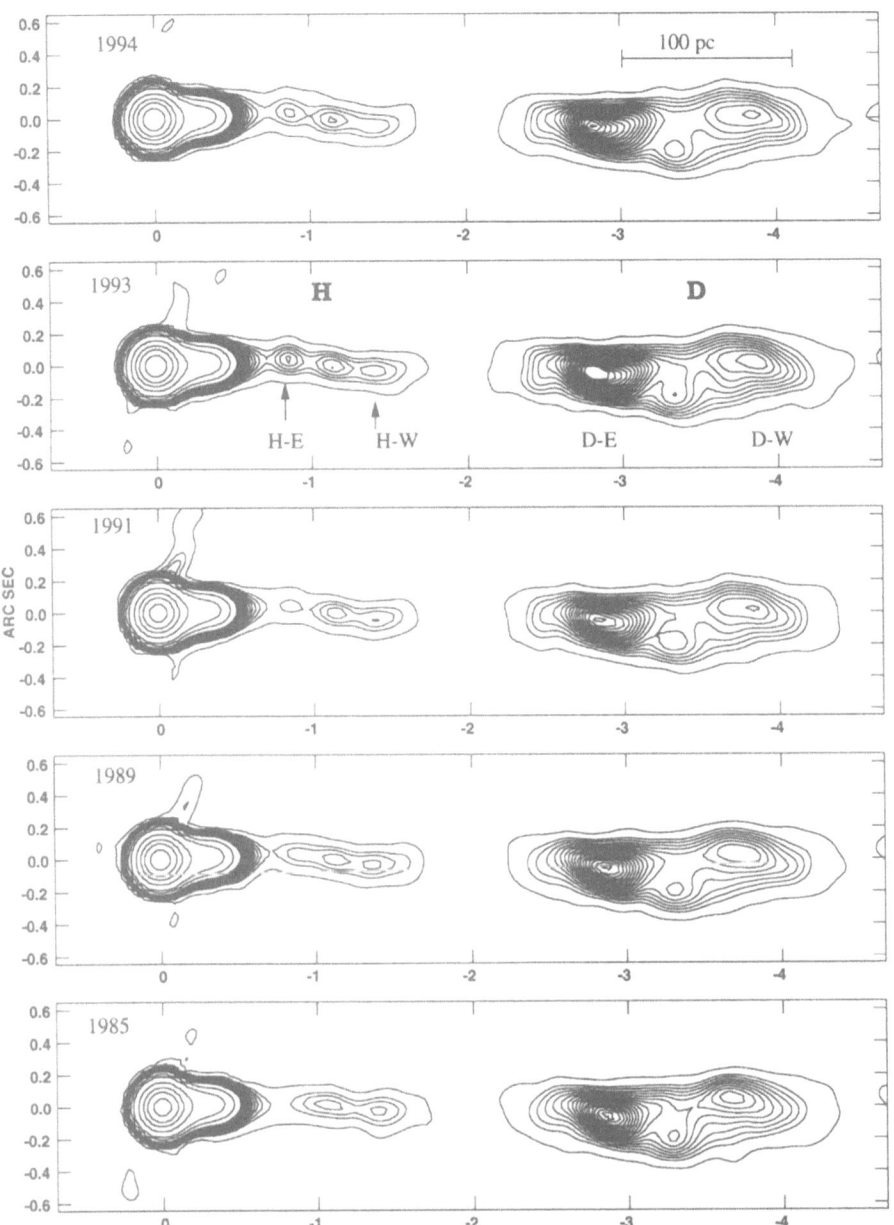

Fig. 12. Contour plots of the region from the nucleus to knot D for epochs 1985, 1989, 1991, 1993, and 1994. Note the formation of feature H-E, as well as rapid motion of the flux peak of DW. The images have been rotated to put the jet along the x-axis. From Zhou 1998.

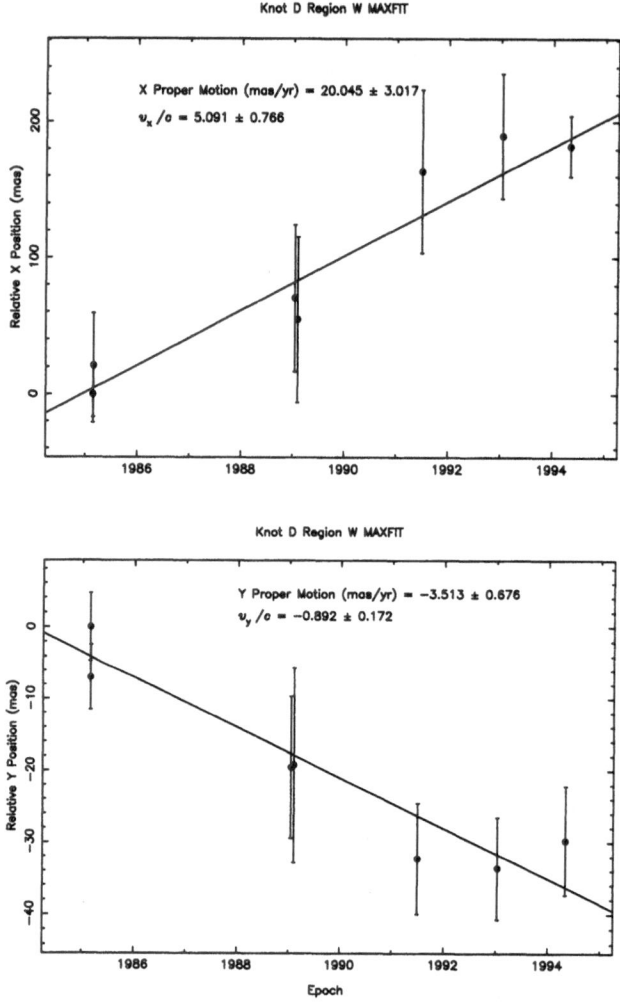

Fig. 13. Position vs. time for the brightness peak of feature DW. Top panel shows component of motion along the jet axis indicating a speed of $(5.1 \pm 0.8)c$. Bottom panel shows position change in direction normal to the jet axis, and indicates a speed of $(-0.89 \pm 0.17)c$. The Feb. 1985 position is defined as zero. From Zhou 1998.

4   Transition Region: 1 to 1.5 kpc from the Nucleus

We now consider the region of the jet containing knots A, B, and C. At both radio and optical wavebands the morphology of the jet changes dramatically at knot A (Fig. 14). While the inner jet is quite straight, in this "transition" region the entire jet appears to bend and oscillate from side to side. At knot A it is deflected towards the south; in knot B it is deflected back towards the north.

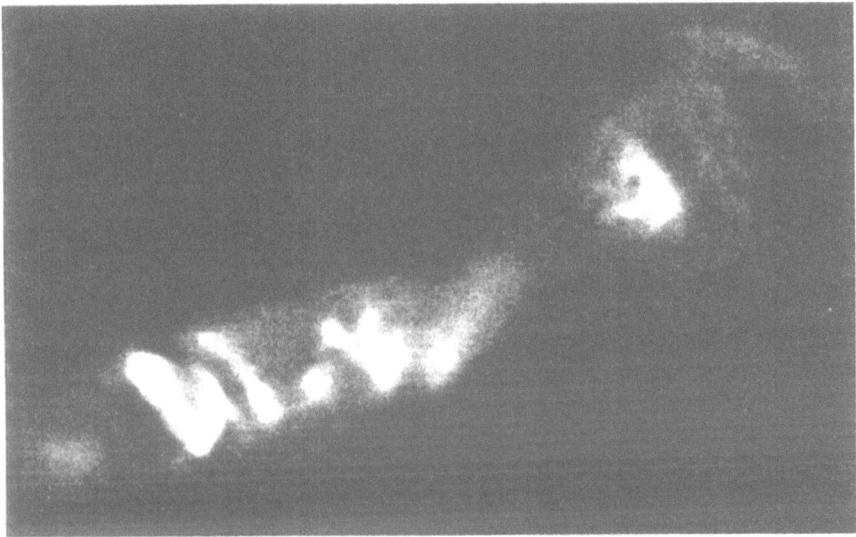

Fig. 14. Greyscale image for knots A, B, and C observed at 15 GHz with 0.''1 resolution. North is up. From OHC89.

The expansion angle of the jet also changes dramatically at knot A. Between the nucleus and knot A the (radio) jet appears as a uniformly expanding cone with an opening angle of $\sim 6°$. In this transition region, however, the jet appears to become nearly cylindrical, perhaps even with some decrease in the diameter between knots A and B. Between knots B and C, the jet seems to narrow considerably and then expand again.

The brightest regions of both knots A and C both are linear features running nearly perpendicular to the jet axis. These are suggestive of shocks in the flow (Biretta, Owen, and Hardee 1983), but have also been modeled as filaments wrapped around the jet surface (OHC89). The magnetic field in both these regions runs along the transverse feature (normal to the jet axis; Fig. 15). This is consistent with the shock interpretation, since compression would tend to orient a random magnetic field so that the resulting field would be parallel with the shock front. Careful examination also shows several other bright linear features oriented at large angles to the jet axis in knots A and B; and again the magnetic field lines appear to follow these features. These morphologies are different from the inner region of the jet, where the knots are either amorphous, or contain features elongated at relatively small angles to the jet axis.

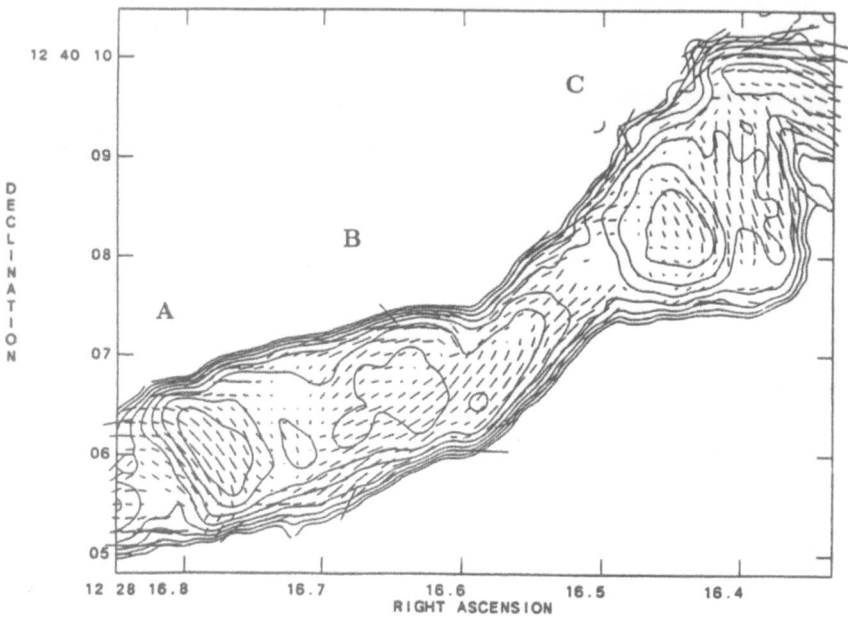

Fig. 15. Contour plots of jet with vectors showing the local magnetic field direction at frequency 15 GHz. The magnetic field lines are predominantly along the jet, except in knots A and C. A vector length of 1″ corresponds to 216% polarization. Contours are spaced at factors of two in intensity beginning at 1.58 mJy/beam, and the resolution is ~0″.15. From OHC89.

Another interesting morphological feature of this region is a narrow dark lane which runs roughly along the centerline of the jet. It is most apparent in the region between knots A and B, and there is perhaps some evidence of it continuing into knot C. This dark lane led OHC89 to hypothesize a low-emissivity region at the jet's center. Such a region might allow low-loss propagation of energetic particles from the nucleus to the outer knots, and hence solve the particle lifetime problem posed by the optical emission (*cf.* Biretta 1993).

We see a complex velocity field in this region, but with a tendency towards lower speeds than the inner jet, and larger misalignments between the jet axis and velocity vector. Most features have speeds in the range 0.4c to

1.2c, though one feature is essentially stationary (region B-SE with speed $(-0.2 \pm 0.2)c$; Figs. 16 and 17). Knot A has the most accurate speed measurement, yielding a value $v_x = (0.509 \pm 0.015)c$ (Fig. 18). Knot C, the outermost prominent knot, appears to be nearly stationary with speed $(0.11 \pm 0.04)c$. While the inner jet shows several features with speeds in the range 2c to 5c, the maximum observed speed in the transition region is only 1.2c. Many features in this region show significant motion transverse to the jet axis, with up to $\sim 21° \pm \sim 4°$ between the jet axis and velocity direction. This is the same region in which the jet begins to bend and oscillate, indicating a connection between morphology and kinematics. This is in contrast to the inner jet where motions are directed straight away from the nucleus.

We note that Zhou (1998) has looked for accelerations in the jet motions, and finds evidence for modest acceleration in only knot B and no other regions. It is not clear whether the acceleration in knot B represents a true acceleration or merely some small structural change inside the knot.

The matter content of jets has been one of the key questions surrounding these objects, and early efforts were made to detect internal Faraday rotation and depolarization. Proton - electron jets might show Faraday effects if the fields and particle densities were high enough, while an electron - positron jet would show no effect. Owen, Hardee, and Bignell (1980) reported early evidence for Faraday rotation $RM \sim 200$ rad m^{-2} in knot A from which they estimated an electron density 0.1 cm^{-3}. However, more recent observations show such values are also characteristic of the western radio lobe against which the jet is projected (Owen, Eilek, & Keel 1990; Zhou 1998). Apparently the observed Faraday rotation is due to a foreground screen which covers both the jet and the lobe, and any rotation in the jet istself must be very small.

5   Outer Jet: 1.5 to 30 kpc from the Nucleus

Beyond the transition region the jet continues, but now with strong side-to-side oscillations and bending (Fig. 19). Limb- and edge-brightening effects remain conspicuous on these larger scales. About 30″ (2 kpc) from the nucleus, there appear to be two bright filaments of emission along what might be the edges of the outer jet. Just beyond these scales the jet begins to bend strongly toward the south and merges into the diffuse western lobe.

While no "counter-jet" is visible within 20″ of the nucleus, there is clearly a diffuse two-sided "lobe" structure on these scales. On the western side of the source the jet appears to merge smoothly into a diffuse, filamented, lobe structure. In the eastern lobe (or counter-jet side) there are numerous arcs of emission which run perpendicular to the source axis; these are suggestive of expanding, edge-brightened shells. The filaments in the lobes are similar to those seen in the jet on mas and arcsecond scales, and have been studied in detail by Hines, Owen, and Eilek 1989. We note that the lobes, while low in surface brightness compared to the jet, completely dominate the source flux

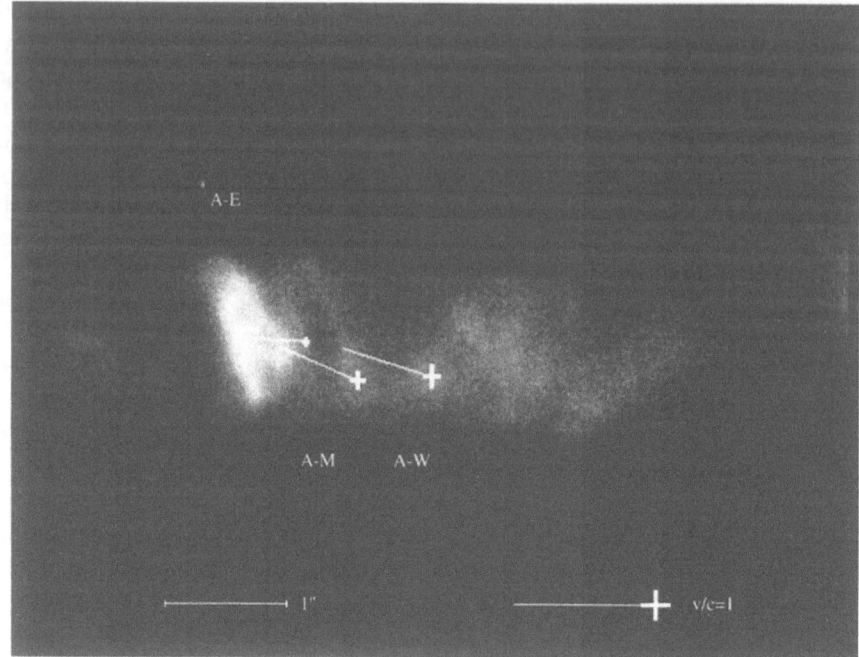

Fig. 16. Observed velocity vectors for features within knot A. Motion is toward the heavy cross at the end of each vector, which also indicates the 1σ uncertainty on each component of the motion. The image is rotated to put the jet along the x-axis. From BZO95.

at all radio frequencies. There also appears to be an exclusion between the radio lobes and the optical line emitting filaments; the optical line emission is mostly along and just beyond the northern and eastern boundaries of the radio lobe region (Ford and Butcher 1979; Jarvis 1990; Sparks, Ford, and Kinney 1993).

At even larger scales, two-sided jet-like structure is seen extending 10's of kpc from these lobes. The appearance on these larger scales, however, is more that of buoyant plumes, than that of a well-collimated "jet." Low-resolution VLA $\lambda = 20$cm D-array images show the western lobe (jet side) is connected to a plume which continues southward 30 kpc (Fig. 20; papers herein; also Böhringer et al. 1995). The eastern lobe (counter-jet or "unjetted" side) has a similar plume which extends eastward 20 kpc.

6   Summary and Discussion

We have divided the jet into three distinct regions based mostly on morphological considerations. The inner jet, including the pc-scale and first kpc, is

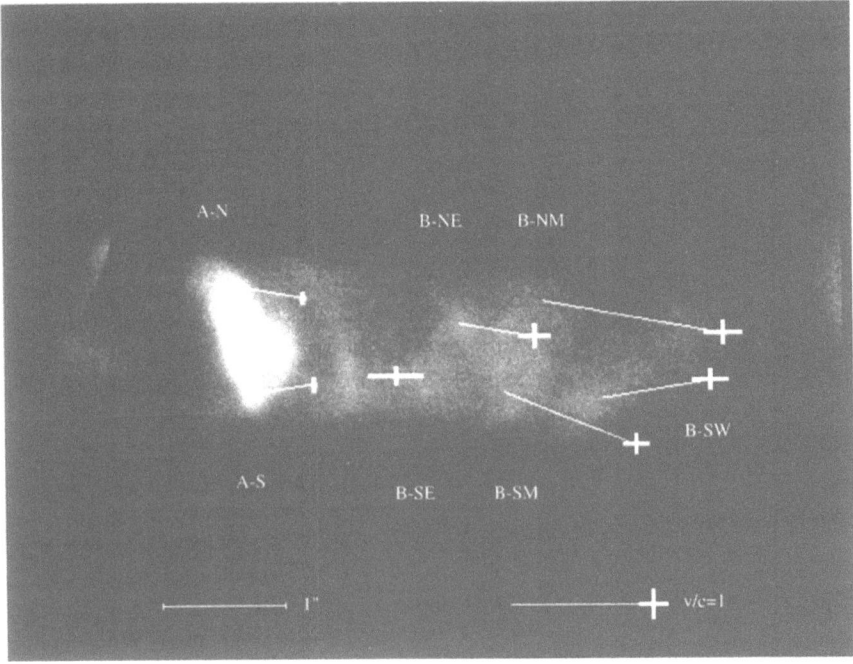

Fig. 17. Observed velocity vectors for features within knots A and B. See Fig. 16 for details. From BZO95.

consistent with a relatively straight, well-collimated cone. The opening angle of the cone appears to decrease somewhat from ∼ 23° on pc-scales to about ∼ 6° on the kpc-scale, but otherwise has straight sides. There is evidence for side-to-side oscillations of the jet ridge-line, but these may well be variations internal to a straight-sided cone. Radio emission in this region is often limb-brightened, while the optical emission is concentrated along the jet axis (*cf.* Biretta *et al.* herein). The magnetic field is primarily along the jet, though it sometimes follows linear features which make some small angle to the jet axis. A wide range of proper motions are seen; some features are stationary ($v < 0.1c$), while the fastest motions range between $2.5c$ and $5c$. The velocity vectors tend to run parallel to the jet axis, though in at least one region the motion follows a linear feature which makes a small angle to the jet axis (similar to the B field).

In the transition region, which includes knots A, B, and C, the jet suddenly ceases to expand in a simple cone and becomes a filled cylinder whose diameter fluctuates. Both the radio and optical emission extend across the entire jet. In this region the jet as a whole begins to bend and oscillate from side-to-side. Bright transverse, shock-like features are seen in both knots A and C. In each of these the magnetic field runs along the "shock front" (per-

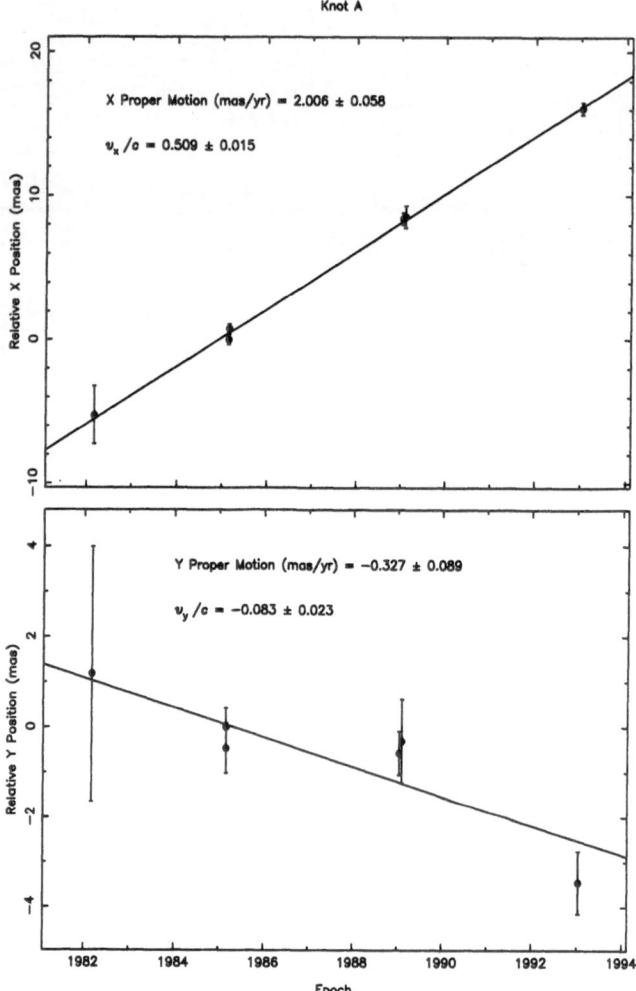

Fig. 18. Position vs. time for knot A. Top panel shows component of position change along the jet axis, with zero defined as the Feb. 1985 position; bottom panel shows position change in direction normal to the jet axis. From BZO95.

pendicular to the jet axis), as expected for compression in a shock. The fastest motions seen so far in this region are only 1.2c, and transverse motions are fairly common.

In the outer jet, beyond knot C, large bends and oscillations are seen, and the jet is often ill-defined. The jet merges into the western lobe, which at larger distances extends outward as a bent and distorted "plume" which can be followed for ~30 kpc towards the south. While there is no evidence of a well-collimated jet on the east side of the nucleus, a diffuse eastern lobe is

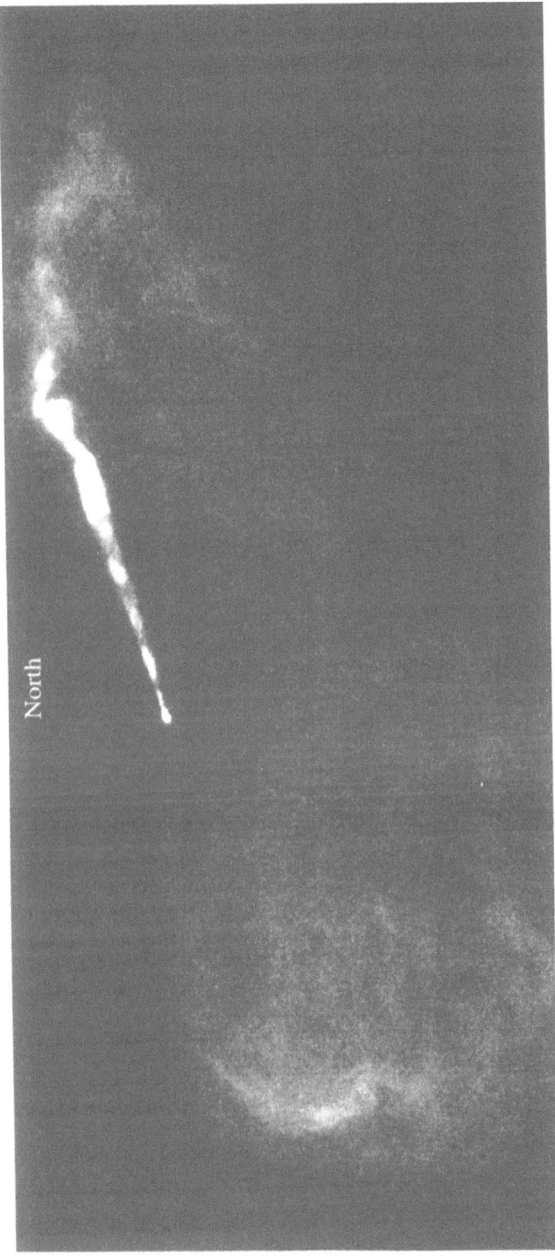

Fig. 19. High dynamic range image of M 87 jet and lobes made with the VLA at 15 GHz. The nucleus is near the center; total extent of the visible source is 75″ or about 6 kpc. The resolution is 0″.15, and intensity scale is logarithmic with white corresponding to ∼20 mJy/beam. Approximately 40, 6, and 2 hrs. of observation were used in the VLA A, B, and C arrays, respectively. North is indicated. From Biretta 1993.

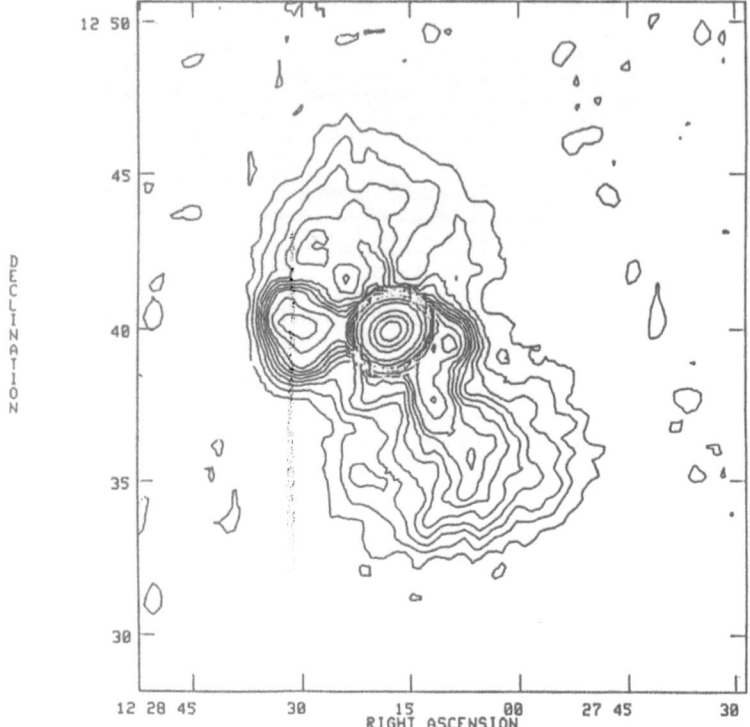

Fig. 20. Extended radio structure in M 87 observed with the VLA D array at 1.5 GHz. The entire lobe region of Fig. 19 is contained in the bright, nearly unresolved region near the image center. Jet-like structures (or "plumes") are seen extending eastward and south-westward from the lobe region. The resolution is ∼100 arcseconds, and the brightest region has an intensity of 90 Jy/beam. North is up; the image is 94 kpc wide. From Biretta 1993; original image from F. Owen.

present, and it similarly has a "plume" which can be traced at least 20 kpc to the east.

For many years one of the key issues regarding extragalactic jets was the flow velocity on kpc-scales (*e.g.* Bridle and Perley 1984). M 87 is the first kpc-scale jet with proper motions measurements, and from these results it appears fairly clear that the flow speed is relativistic, at least through the inner jet and transition region. The very fastest speed seen in the radio data, $5c$ in region DW (Zhou 1998), implies flow with a Lorentz factor $\gamma > 5$. This is quite consistent with predictions of unified models, which propose FR-I radio sources (such as M 87) are simply misdirected BL Lac objects (Browne 1983; Urry, Padovani, and Stickel 1991). Recently Heinz and Begelman (1997) have shown that relativistic speeds could also ameliorate the particle lifetime problem posed by the optical emission on kpc-scales.

It is less clear whether relativistic speeds persist into the outer jet region (beyond 1.5 kpc). The presence of two-sided structure beyond 2 kpc suggests the outer jet speed is at least not very relativistic. Moreover, knot A appears to have a significant impact on the jet properties, suggesting a large change in Mach number (although Bicknell and Begelman 1996 have modeled it as a relatively weak shock). We note that similar overall morphologies are seen other many other FR-I sources, where a relatively faint, one-sided jet ("one-sided jet base") is seen on scales to a few kpc, beyond which two-sided structure dominates (Eilek, *et al.* 1984; Bridle 1986). It seems likely all these sources are relativistic over the first few kpc.

The complexity seen in the kinematics (stationary features adjacent to fast ones) is likely due to differences between the bulk motion and "pattern" speeds of the visible features (*e.g.* Lind and Blandford 1985). Many of the jet features probably represent standing waves or shocks associated with obstructions or changes in the external atmosphere. One might also argue that the fast speeds could represent patterns due to "scissor" effects (crossing features whose intersection moves rapidly while the actual matter moves slowly), but the complete lack of inward motion towards the nucleus argues strongly against such models.

It is interesting to consider why the pc-scale radio features should be stationary in M 87 *vs.* superluminal radio cores, if both contain relativistic motion. Part of the answer may lie in the angle of the jet axis to the line of sight. From the motion of knot A and fast features in knot D, and from relativistic aberration constraints on the appearance of knot A, one can show the jet must be oriented $\sim 40°$ to the line of sight (Biretta, Zhou, and Owen 1995; though faster speeds seen now by Zhou 1998 and Biretta, *et al.* herein may argue for angles $\sim 20°$). This is much larger than that derived for superluminals, and causes a Doppler boost $\sim$unity (or less for large γ) for fast jet material. Hence, radio emission from fast moving material may be strongly diluted by slower material in M 87, and so the dominant features appear stationary. We note that low proper motions are also seen in the cores of other FR-I sources (Giovannini *et al.*, 1998).

A key problem regarding extragalactic jets is their composition — are they composed of electron-positron or electron-proton plasmas? As mentioned before, efforts to detect internal Faraday rotation have been unsuccessful, though arguing at least for low densities of protons. Recently Reynolds *et al.* 1996 have used the radio spectrum and kinematic constraints to argue for an electron-positron plasma, though better VLBI data are needed on the nucleus before an electron-proton jet can be eliminated.

Another key problem remains that of the jet production process deep within the galaxy nucleus. The jet is well-collimated on scales < 0.1 pc, as we saw from the VLB observations. This implies that the region forming the jet must be very small, and must collimate by ~ 300 Schwarzschild radii (assuming 10^9 solar mass black hole – Ford *et al.* 1994; Harms *et al.* 1994). It seems likely that electromagnetic processes must dominate the initial collimation. Mechanisms such as external pressure gradients cannot provide much collimation on these scales.

In the future, we expect that mm VLBI observations and space-based VLBI will allow us to follow the jet inwards closer to the engine, and thereby set stronger constraints on the collimation process. Rapid-monitor VLBA observations might clarify the pc-scale kinematics, and show whether fast (though weak) features can be tracked. Continued monitoring with the VLA (and HST) should provide a detailed velocity field for much of the jet. Eventually we hope to compare observed velocity fields with numerical models and add a fourth dimension to our modeling and understanding of jets.

References

Baade, W. and Minkowski, R. 1954, ApJ, 119, 215

Bääth, *et al.* 1992, A&A, 257, 31

Bicknell, G. V., & Begelman, M. C. 1996, ApJ, 467, 597

Biretta, J. A., Owen, F. N., and Hardee, P. E. 1983, ApJ, 274, L27

Biretta, J. A., Owen, F. N., and Cornwell, T. J. 1989, ApJ, 342, 128

Biretta, J. A., 1994, in Astrophysical Jets, eds. D. Burgarella, M. Livio, C. P. O'Dea, (Cambridge Univ. Press, Cambridge, UK), pp. 263.

Biretta, J.A., and Junor, W., 1995, Proc. Natl. Acad. Sci., 92, 11364

Biretta, J.A., Zhou, F., Owen, F.N., 1995, ApJ, 447, 582 (BZO95)

Böhringer, H., Nulsen, P. E. J., Braun, R., and Fabian, A. C., 1995, MNRAS, 274, L67

Bolton, J. G., Stanley, G. J., and Slee, O. B. 1949, Nature, 164, 101

Bridle, A. H. 1984, AJ, 89, 979

Bridle, A. H. 1986, Can. J. Phys., 64, 353

Bridle, A. H. and Perley, R. A. 1984, ARA&A, 22, 319

Browne, I. W. A. 1983, MNRAS, 204, 23p

Cohen, M. H., Moffet, A. T., Shaffer, D., Clark, B. G., Kellermann, K. I., Jauncey, D. L., and Gulkis, S. 1969, ApJ, 158, L83

Daly, R. A., and Marscher, A. P. 1988, ApJ, 334, 539

Eilek, J. A., Burns, J. O., O'Dea, C. P., and Owen, F. N. 1984, ApJ, 278, 37

Fanaroff, B. L. and Riley, J. M. 1974, MNRAS, 167, 31p

Ford, H. C., *et al.* 1994, ApJ, 435, L27

Ford, H. C. and Butcher, H. 1979, ApJS, 41, 147

Fraix-Burnet, D., *et al.* 1990, AJ, 101, 88

Giovannini, G., Cotton, W. D., Feretti, L., Lara, L., and Venturi, T. 1998, ApJ, in press.

Hardee, P. E., 1987, ApJ, 318, 78

Harms, R. J., Ford, H. C., Tsvetanov, Z. I., Hartig, G. F., Dressel, L. L., Kriss, G. A., Bohlin, R. C., Davidsen, A. F., Margon, B., Kochhar, A. K. 1994, ApJ, 435, L35

Heinz, S. & Begelman, M. C. 1997, ApJ, 490, 653

Hines, D. C., Owen, F. N., and Eilek, J. A. 1989, ApJ, 347, 713

Jarvis, B. J. 1990, A&A, 240, L8

Junor, W., and Biretta, J. A. 1995, AJ, 109, 500

Keel, W. C. 1988, ApJ, 329, 532

Kellermann, K. I., Clark, B. G., Cohen, M. H., Shaffer, D. B., Broderick, J. J., and Jauncey, D. L., 1973, ApJ, 179, L141

Königl, A. and Choudhuri, A. R. 1985, ApJ, 289, 173

Li, Z.-Y., Chiueh, T., and Begelman, M. C. 1992, ApJ, 394, 459

Lind, K. R. and Blandford, R. D. 1985, ApJ, 295, 358

Liu, F. K. and Xie, G. Z. 1992, A&ASS, 95, 249

Owen, F. N., Eilek, J. A., and Keel, W. C. 1990, ApJ, 362, 449

Owen, F. N., Hardee, P. E., and Bignell, R. C. 1980, ApJ, 239, L11

Owen, F. N., Hardee, P. E., and Cornwell, T. J. 1989, ApJ, 340, 698 (OHC89)

Reid, M. J., Biretta, J. A., Junor, W., Spencer, R., Muxlow, T. 1989, ApJ, 336, 125

Reid, M. J., Schmitt, J. H. M. M., Owen, F. N., Booth, R. S., Wilkinson, P. N., Shaffer, D. B., Johnston, K. J., and Hardee, P. E. 1982, ApJ,263, 615

Reynolds, C. S., Fabian, A. C., Celotti, A., & Rees, M. J. 1996, MNRAS, 283, 873

Sakurai, T. 1987, Publ. Astron. Soc. Japan, 39, 821

Schreier, E. J., Gorenstein, P., and Feigelson, E. D. 1982, ApJ, 261, 42

Sparks, W. B., Ford, H. C., and Kinney, A. L. 1993, ApJ, 413, 531

Spencer, R. E. and Junor, W. 1986, Nature, 321, 753

Tonry, J. L. 1991, ApJ, 373, L1

Turland, B. D. 1975, MNRAS, 170, 281

Urry, C. M., Padovani, P., and Stickel, M. 1991, ApJ, 382, 501

Wiita, P. J. and Siah, M. J. 1986, ApJ, 300, 605

Zhou, F., 1998, Ph.D. thesis

Zhou, F., Owen, F., and Biretta, J. A., 1999, (in preparation)

VLA 7mm Images of the M 87 Jet

Frazer Owen[1] and John Biretta[2]

[1] National Radio Astronomy Observatory
Socorro, NM 87801 USA
[2] Space Telescope Science Institute
Baltimore, MD21218 USA

Abstract. We report observations of M 87 with the VLA at 7 mm in the A-array. The image with 35 mas resolution shows the inner jet structure out to 400 mas from the core and as well has the unresolved leading edge of knot A.

1 Observations

In order the increase the resolution and frequency coverage over which the M 87 jet has been observed we used the VLA in the A-array at 43 GHz (7 mm) to image the source at 35 mas resolution. The observations used the 13 antennas equipped with 7 mm receivers in 1996. Every 30 minutes the core of M 87 was used as a pointing calibrator keep the antennas centered on the source. The data were self-calibrated and MEM images were constructed using the programs in the AIPS package.

2 Results

In Fig. 1 we show the core and the 400 mas jet. This image shows the same sort of jet structure we have previously seen on smaller and larger scales. Along with the "knots", the image shows the displacements of the centroid of the emission away from the jet axis which probably indicates the same sort of helical/edge brightened structure seen elsewhere in the jet.

In Fig. 2, we show the knot ABC complex. The signal-to-noise is low, so one can jet make out the overall structures seen at other wavelengths. However, the sharp feature at the leading edge of knot A does stand out. The width of this feature must be less than the resolution of 35 mas and thus smaller than 3 pc. If this structure extends all the way through the jet and is narrow because it is seen in projection, this implies an alignment with the line-of-site of better than 2 degrees and a planar structure with a thickness to length ratio of better than 30/1. Alternatively, perhaps this is a surface feature.

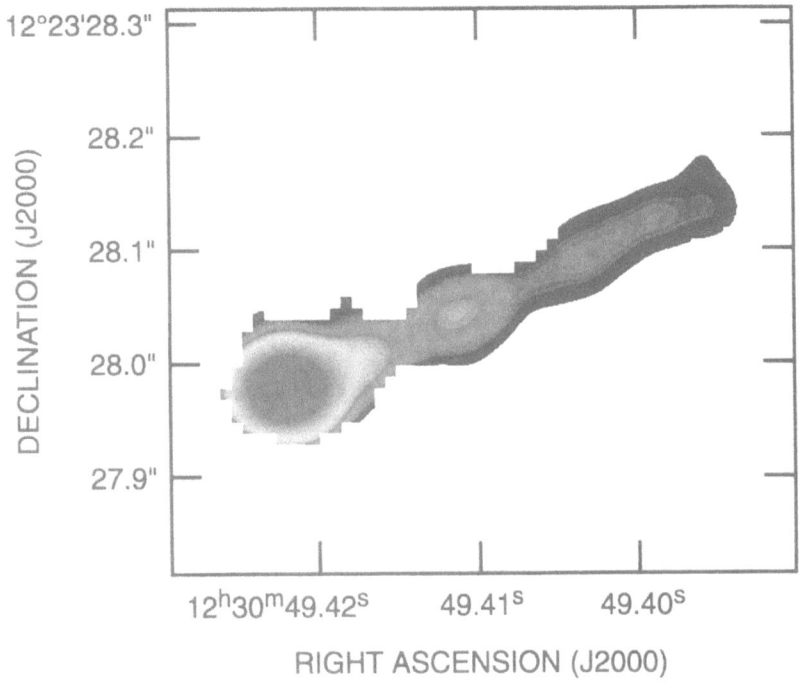

Fig. 1. 7 mm image of the M 87 core

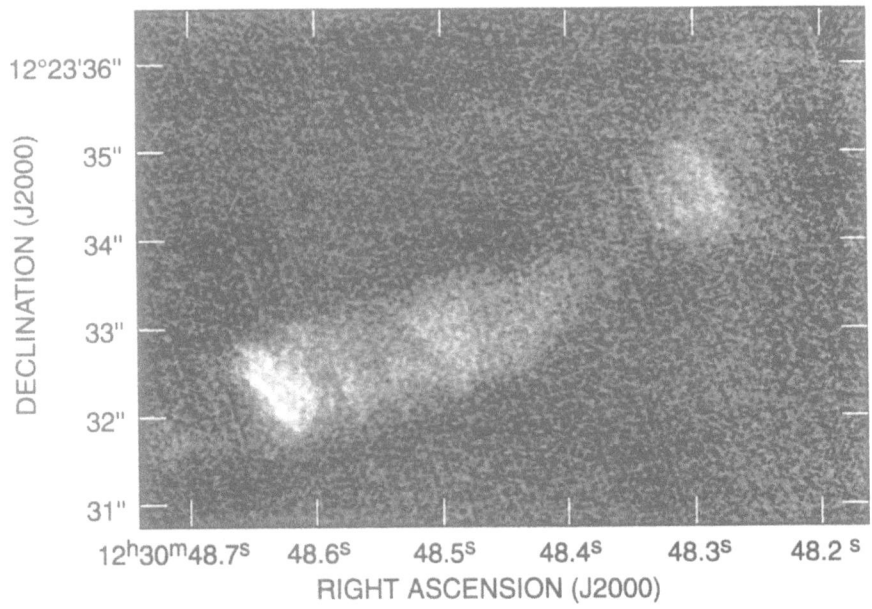

Fig. 2. 7 mm image of the M 87 knot ABC complex

High-Frequency Observations and Spectrum of the Jet in M 87

Klaus Meisenheimer

Max-Planck-Institut für Astronomie
Königstuhl 17
D69117 Heidelberg
Germany

Abstract. This review collects the current fund of ground-based observations at $\nu > 10^{14}$ Hz which allows to determine the overall synchrotron spectrum of the jet in M 87 very accurately. Beyond knot D (at $3''$ distance from the core), the spectrum is characterized by a straight power-law $S_\nu \sim \nu^{-0.66}$ which cuts off steeply at some frequency ν_c The optically observed spectral variations along the jet are caused by changes in the cutoff frequency between $\nu_c = 4 \times 10^{15}$ Hz and $\nu_c = 7 \times 10^{14}$ Hz, while the power-law index between 10^{10} and 10^{14} Hz stays amazingly constant at $\alpha_{PL} = -0.66 \pm 0.02$. We demonstrate that an ubiqitous particle distribution function $N(E) \sim E^{-2.32}$ which cuts off steeply at a maximum energy $E_c = 10^6 m_e c^2$, together with local variations of the magnetic field strength, accounts well for both the observed variation of ν_c and the apparent brightness distribution along the jet (*i.e.* the "knots"). The global particle spectrum might even be maintained into the inner lobes — albeit the eastern, "un-jetted" side seems to contain a steeper particle spectrum $N(E) \sim E^{-2.65}$. Thus, permanent particle re-acceleration seems required all along the jet and in the inner lobes.

Recent high-resolution observations with the HST (spatial scale $\lesssim 0''.1$) seem to support this general behaviour of the kpc jet beyond knot D. However in the inner 300 pc from the core, rapidly evolving optical features with deviating synchrotron spectra are found. A careful analysis of their time-dependence (on time-scales comparable to the synchrotron loss-time $\tau_{syn} \simeq 10$ years) might eventually help to understand the nature of the particle acceleration process. The X-ray emission from the jet is most likely of synchrotron origin. It seems to arise from compact substructures where locally the maximum particle energy is boosted above its steady state value to $E_c > 5 \times 10^6 m_e c^2$.

1 Introduction

The jet in M 87 stands out from all extragalactic radio jets in three respects: Its extra-ordinary high surface brightness in both the radio and the near-infrared to optical windows which are easily accessible form the ground. Its relative proximity to our Galaxy which allows observations with rather high spatial resolution ($1''$ corresponds to approximately 100 pc), and last not least its location in the northern hemisphere which makes it observable from Europe and Northern America where until very recently the largest and most advanced telescopes where located. From these prepositions it is hardly surprising that observations of the jet of M 87 played a key role in shaping our concepts on the phenomenology and physics of extragalactic radio sources in general. In this it is only paralleled by Cygnus A (see Carilli & Harris 1996), the other northern radio source of extraordinary surface brightness. The radio lobes and hot spots in Cygnus A, however, are undetected in the optical window.

Thus in the "early optical era", before the advent of the radio interferometer telescopes, the optical visibility of the jet in M 87 allowed important insights into the dominant emission mechanism in radio jets — synchrotron radiation — long before the ubiquity and diversity of the radio jet phenomenon was even realized. We will highlight some of the important conclusions from this era in Sect. 2.

Even when during the eighties and nineties it could be proven for several more extended radio sources that their synchrotron spectra which dominate the radio frequencies do extend to the high frequency regime of optical observations, none of these sources came close to M 87 in either surface brightness or accessible spatial resolution. Thus the overall synchrotron spectrum of the jet in M 87 and its variation from place to place is still the best known by a large margin[1]. The ground-based results are reviewed in Sect. 3 and a first interpretation is given in Sect. 4, in order to set the stage for the observations with higher resolution (VLA and HST at $0\rlap{.}''1$ resolution). Deviating from the oral contributions in Ringberg, we include here the high frequency observations and synchrotron spectra of the inner lobes as Sect. 5. The question how the HST observations may change the global picture are addressed in Sect. 6. Section 7 discusses the X-ray observations and their interpretation. The concluding Sect. 8 summaries the actual status and identifies the next step to be taken, in order to solve the mystery of particle acceleration in this key object for extragalactic radio sources.

2   Spectrum and emission mechanism
— the historical perspective

To most astronomers of the nineties the jet in M 87 is one of the hundreds of extragalactic radio jets which have been detected by modern interferometric radio telescopes (see *e.g.* the early review by Bridle and Perley 1984). However, its singular position in shaping our picture of extragalactic jets can only be appreciated if one realizes that this jet was first identified on *optical photographs*, dating back almost half a century before the advent of imaging radio astronomy: In 1918, Curtis reported the existence of "a curious ray" in the inner region of M 87. This finding beard little consequences until the

---

[1] The jet of the quasar 3C 273 is the only other jet readily observable from radio to X-rays and is described in detail *e.g.* by Röser *et al.* 1997

fifties when Shklovsky (1953) proposed that the unusual coincidence of radio and optical radiation from the Crab nebula could be understood in terms of a radiation process which only recently had been detected in laboratory experiments accelerating electrons trapped in a magnetic field to energies $E_{tot} \equiv \gamma m_e c^2$, well above the rest mass energy $m_e c^2 = 511 \, \text{keV}$. The emerging *"synchrotron radiation"* is characterized by a smooth and broad frequency distribution and high linear polarization up to 75%. Indeed, the polarization of the Crab's light could be demonstrated within a few years (Oort & Walraven, 1956). At that time radio polarimetry was still a task nobody could dream of.

At about the same time it was detected that M 87 is also a source of intense radio emission. During a conference in Manchester (1955) it was suggested that the peculiar "ray" of optical emission might be related to the radio source and the synchrotron radiation process could be responsible for both. Within two years Burbidge (1956) laid out the basic implications of synchrotron radiation in astrophysical sources and W. Baade (1957) demonstrated by optical (photographic) polarimetry that the "ray" in M 87 shows indeed linear polarization of $\gtrsim 20\%$. Thus the exceptional optical brightness of the M 87 jet allowed to establish the emission mechanism of extragalactic radio jets, well before they were identified as a widespread class of astronomical objects.

The *optical* observations of the M 87 jet dominated the investigations until the mid-seventies (the advent of radio-interferometers with full imaging capabilities). Therefore, we would like to refer to this period as "early optical era". In 1959, Hiltner obtained the first quantitative measurement of the optical polarization using an aperture photometer. Subsequent photographs of the jet clearly revealed its "knotty" fine-structure.

Although the short loss-time of the ultra-relativistic electrons which radiate synchrotron light was already noted by Burbidge, the full extend of the dilemma which arises from the fact that M 87 emits optical synchrotron radiation at $\geq 1 \, \text{kpc}$ from the core was rigorously pointed out not before Felten (1968) who demonstrated that particle acceleration processes in the core can hardly be responsible for the optical synchrotron emission of the jet ("Felten's dilemma") and stated that there are only two viable solutions: (i) a relativistic plasma flow, the time dilation within which could weaken the synchrotron loss-time argument, or (ii) a local process of particle acceleration to ultra-relativistic energies. However, at that time no promising theory of particle acceleration in jets had been developed.

The dusk of the "early optical era" loomed when Turland (1975) published the first radio map (taken with the Cambridge 1 mile telescope) which indicated that the radio-optical correspondence is extremely tight in the jet of M 87. He also proposed the first overall spectrum of the synchrotron emission from the jet, which is in amazing agreement with current results (see Fig. 1, below). At about the same time, the first radio maps of Cygnus A with $\lesssim 2''$ resolution (Hargrave & Ryle 1974) had revealed "hot spots" of extreme surface brightness which again seemed to demand an *in situ* acceleration process and triggered the general concept of "beams" or jets as drivers of extragalactic radio sources (Blandford & Rees 1974, Scheuer 1974). It was also the hot spots of Cygnus A for which Bell (1978) developed the theory of "diffusive shock acceleration" ("first order" Fermi acceleration) which seems to be able to accelerate electrons to the required energies.

The first radio maps of M 87 with a resolution comparable to seeing-limited optical images (Laing 1980, Owen *et al.* 1980) confirmed the 1:1 correspondence between optical and radio morphology in the jet and showed — for the first time — that the radio emission is polarized similar to the optical radiation. This provided the unambiguous proof that the emission is dominated by optically thin synchrotron radiation in the entire frequency range between 1 GHz and 10^{15} Hz. Although it is appropriate to refer to the eighties and early nineties as "radio era" of the investigations of the jet (see below), there were three crucial observations during this period which have been obtain at frequencies $\nu > 10^{14}$ Hz:

(1) The near-infrared photometry of the brightest parts of the jet (knots A+B+C) by Stocke *et al.* (1981) indicated that the overall synchrotron spectrum is characterized by a radio-to-NIR power-law which steepens or cuts off at around 3×10^{14} Hz. The authors proposed the (still valid) concept of a power-law electron energy spectrum which abruptly cuts off at some maximum energy E_{max} at which the synchrotron losses balance the acceleration gain.

(2) In 1982, Nieto and Lelièvre obtained an optical image with about $0\rlap{.}''7$ resolution which revealed that the knots break up in substructures. Their image played a important role in driving the hydro-dynamic simulations of extragalactic radio jets which developed around that time (Norman *et al.* 1982, Falle & Wilson 1985).

(3) Schlötelburg *et al.* (1988) published a very accurate map of the optical polarization of the jet, the implication of which became obvious only later when the radio maps of Owen *et al.* (1989) showed an almost identical radio polarization (*cf.* comparison in Meisenheimer *et al.* 1989b).

Nevertheless, the observational studies of the jet were dominated by the VLA, the resolution of which ($\lesssim 0\rlap{.}''15$) was almost 10 times better than that obtained by typical optical images. The new radio maps (Hines *et al.* 1989, Owen *et al.* 1989 revealed a wealth of filamentary fine-structure in the jet and the lobes. The most important result of these observations regarding the topic of this review was the detection of edge-brightening in the inner kpc jet ($3''$ to $5''$ from the core) which led to the concept of a jet consisting of an empty "spine" and a "mantle" of enhanced synchrotron emissivity. Based on their radio maps, Owen *et al.* proposed to bypass "Felten's dilemma" by invoking a channel of low field strength along which electrons could stream from the core without losses. The high surface brightness of synchrotron radiation would have to be explained by some kind of diffusion of the ultra-relativistic electrons ($E \gtrsim 10^5 m_e c^2$) into the mantle which contains the strong magnetic field inferred from equipartition arguments.

VLBI maps of the inner 20 pc of the jet (Reid *et al.* 1989) made it clear that the collimation to opening angles of a few degrees persist to $\lesssim 1$ pc from the core and that the filamentary radio structure of the jet can be traced inwards to at least 10 pc from the core. In addition, the multi-epoch VLBI observations led to the first detection of proper motions in the jet. A highlight to this "radio era" has certainly been set by the work of Biretta, Zhou and Owen (1995) who used a set of VLA maps spanning an observing period of more than 10 years to demonstrate that also the kpc jet of M 87 exhibits detectable proper motions which show an astonishing variety in apparent speeds (< 0.1 to $> 2c$) and orientation with repect to the jet axis. This was a first hint that on small scales ($\lesssim 10$ pc) the steady-state approach to the jet spectrum might be inappropriate.

In the meantime, the first HST images of the jet became available (Boksenberg *et al.* 1992). Although their significance for the jet spectrum was rather limited – due to the complicated point-spread-function of the HST before refurbishment – it became evident immediately that a fully operational HST could provide a completely new insight into details of the jet spectrum. Thus another "modern optical era" was about to dawn. The first results of this era are presented in the current review and the article by Biretta *et al.* (on page 210).

3   The radio-to-optical spectrum of the jet

The starting point of the present article is set by Biretta & Meisenheimer (1993) who reviewed the radio and optical properties of the jet in M 87 as they were known in 1992. Their conclusions concerning the optical properties and the overall spectrum can be summarized as follows:

- At a resolution of about 1"the radio and optical *morphology* and *polarization* of the jet are almost identical.
- The *optical spectral index* α_{opt} varies very smoothly along the jet.
- The *overall spectrum* between 10^9 and 10^{15} Hz seems to be characterized by a straight power-law $S_\nu \sim \nu^\alpha$ with $\alpha \simeq -0.65$ which "cuts off" at frequencies $\nu_c \gtrsim 10^{14}$ Hz. The cutoff spectrum follows the theoretically expected synchrotron spectrum of a power-law electron energy distribution with a rather pronounced cutoff at some maximum energy $E_c = \gamma_c m_e c^2$.
- There is a close relationship between the local *cutoff frequency* ν_c (as derived from α_{opt} and a universal spectral shape) and the local surface brightness B_ν which can be explained by an almost constant maximum energy γ_c together with variations of the local magnetic field strength.
- The observed X-ray flux (Biretta *et al.* 1991) does not fit onto the extrapolation of the radio-to-optical spectrum. Thus the origin of the X-rays remained unclear.

In order to determine the the high frequency spectrum more accurately, Neumann, Meisenheimer and Röser observed M 87 in two near-infrared (NIR) bands (K' at $\lambda = 2.1\,\mu m$ and H at $\lambda = 1.65\,\mu m$) with the near-infrared camera MAGIC at the Calar Alto 3.5 m telescope. The fairly large field provided by the NICMOS 256 × 256 array allows an accurate determination of the background due to the starlight of M 87 by exactly the same isophot fitting procedure which has been used for analysing the optical CCD data (see Meisenheimer *et al.* 1996, further details of the NIR observations are given in Neumann (1994) and a forthcoming paper by Neumann *et al.* 1999).

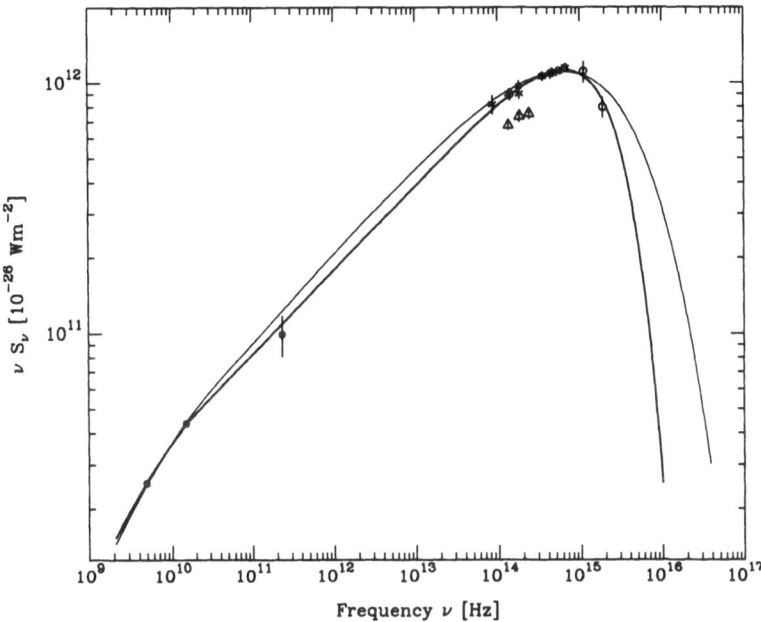

Fig. 1. Overall synchrotron spectrum of the brightest part of the jet in M 87 (knots A+B+C). Radio flux values are from Hines *et al.* (1989) and Owen *et al.* (1989). The mm flux is from Meisenheimer *et al.* (1996). The near infrared measurements by Stocke *et al.* (1981), Killeen *et al.* (1984) and Neumann *et al.* (1999) are shown as *, $\triangle$ and •, respectively. In the optical, various flux measurements by Perez-Fournon *et al.* (1988), Keel (1988) and Meisenheimer *et al.* (1996) have been included. The UV points refer to the IUE measurments by Perola & Tarenghi (1980). Continuous lines refer to two different model spectra (*cf.* text).

Adding the new NIR measurements into the overall spectrum (Fig. 1) confirms that the integrated spectrum of the brightest knots ABC can perfectly be described by the synchrotron spectrum from a powerlaw electron energy distribution $N(\gamma) \sim \gamma^{-2.32}$ which cuts off either exponentially or even steeper at some maximum energy γ_c. Although formally the "perfect" cutoff (*i.e.*

$N(\gamma) = N_0\gamma^{-2.32}$ at $\gamma \leq \gamma_c$; $N(\gamma) \equiv 0$ at $\gamma > \gamma_c$) seems to provide a much better fit to the spectrum than the exponential cutoff it should be noted that this relies entirely on the UV points measured with IUE (Perola & Tarenghi 1980) which have large uncertainties. First results of the HST observations indicate a somewhat smoother cutoff. The new observations confirm the early NIR photometry by Stocke *et al.* (1981) and rule out the "broken powerlaw" spectrum proposed by Biretta *et al.* (1991) on the basis of incorrect measurements between $2.2\,\mu m$ and $850\,nm$.

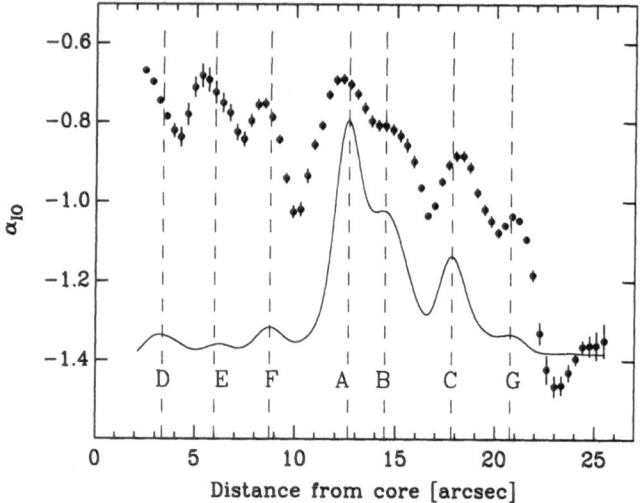

Fig. 2. NIR-to-optical spectral index α_{IO} as derived from a linear fit through the flux measurements at $\lambda = 2.1$, 1.65, 0.85, 0.65, and $0.45\mu m$ (see Neumann *et al.* 1999). For reference the brightness profile along the jet is shown. Capital letters label prominent knots.

The spatial analysis of the jet spectrum can now be based on 9 images with $\lesssim 1''$ resolution:

- the 2 cm VLA map (kindly provided by Frazer Owen),
- three images at $\lambda = 2.1$ and $\lambda = 1.65\,\mu m$ (Neumann *et al.* 1999).
- five optical images at $\lambda = 0.85$, 0.65, and $0.45\,\mu m$ (Meisenheimer *et al.* 1996).

All images have been convolved to the same "effective" beam and aligned to $\leq 0\!''\!05$ (see Meisenheimer *et al.* 1996 for details). The effective beam of $1\!''\!3$ (FWHM) is set by the image with the worst seeing ($1\!''\!1$ FWHM) and the pixelation of the optical data ($0\!''\!35$). Photometry is done on a grid with $0\!''\!35$ spacing, the central points of which follow the ridge line of the optical jet.

Although the curvature of the spectra (*cf.* Fig. 1) is evident when comparing the spectral index between 2.1 μm and 0.85 μm with that between 0.85 and 0.45 μm, we show in Fig. 2 the *average* NIR-to-optical spectral index α_{IO} as derived from a straight powerlaw fit through all measurements between 2.1 and 0.45 μm. It essentially corresponds to the spectral index at $\langle \lambda \rangle \simeq 0.90$ μm. It is obvious from Fig. 2 that the occurance of a flat spectral index α_{IO} is well correlated with local peaks of the jets surface brightness (*i.e.* the "knots").

With the flux measurements in B, R, I, H and K′ the high frequency part of the spectrum is sufficiently well covered to allow fits of synchrotron spectra at every grid point along the jet. The fitted model spectra are assumed to arise from electron energy distributions with a "perfect cutoff" at some maximum energy γ_c. Accordingly there are three free parameters:

- the powerlaw slope $\alpha_{PL} = (1 - q)/2$, where q is the slope of the electron distribution: $N(\gamma) \sim \gamma^{-q}$,
- the cutoff frequency ν_c, and
- a normalization at some frequency, *e.g.* $B_{15\,GHz}$ = brightness at $\nu = $ 15 GHz.

Surface brightness measurements on all 9 images mentioned above are included, that is the degree of freedom is $n_{free} = 6$. Satisfactory fits (with $\chi^2/n_{free} < 1.5$) could be obtained at every grid point.[2]

The parameters of the fitted synchrotron spectra (Fig. 3) show that the powerlaw index α_{PL} stays amazingly constant around $\alpha_{PL} \simeq -0.66$. Consequently, all observed variations of the high frequency spectral index α_{IO} are solely caused by variations of the cutoff frequency ν_c from place to place. As seen in the upper panel of Fig. 3 the cutoff frequency varies between $\nu_c \geq 8 \times 10^{15}$ Hz (upstream of knot D) and $\nu_c = 5 \times 10^{14}$ Hz (downstream of the outermost knot G). However in the region between knot D and knot G (projected distance 1.7 kpc), for which ν_c is well determined, it does not change by more than a factor of 5. This is extremely astonishing since for a typical magnetic field strength of about 30 nT and a jet inclination w.r.t. the line of sight of 45° one would expect that synchrotron losses change ν_c by factors ≥ 4 on a scale of $L_{syn} < c\tau_{syn}/\sqrt{2} = 42$ pc (0″.4).

4   Interpretation of the jet spectrum

The dilemma posed by the low apparent synchrotron losses becomes even more evident if one converts the observed run of ν_c into the corresponding cutoff energy $\gamma_c = E_c/m_e c^2$ of the underlying particle distribution according to

---

[2] Spectral fits where obtained with a "perfect" cutoff in the electron distribution. Assuming a smoother cutoff — as *e.g.* predicted by shock acceleration — does not increase χ^2 significantly, but leads to higher values of ν_c.

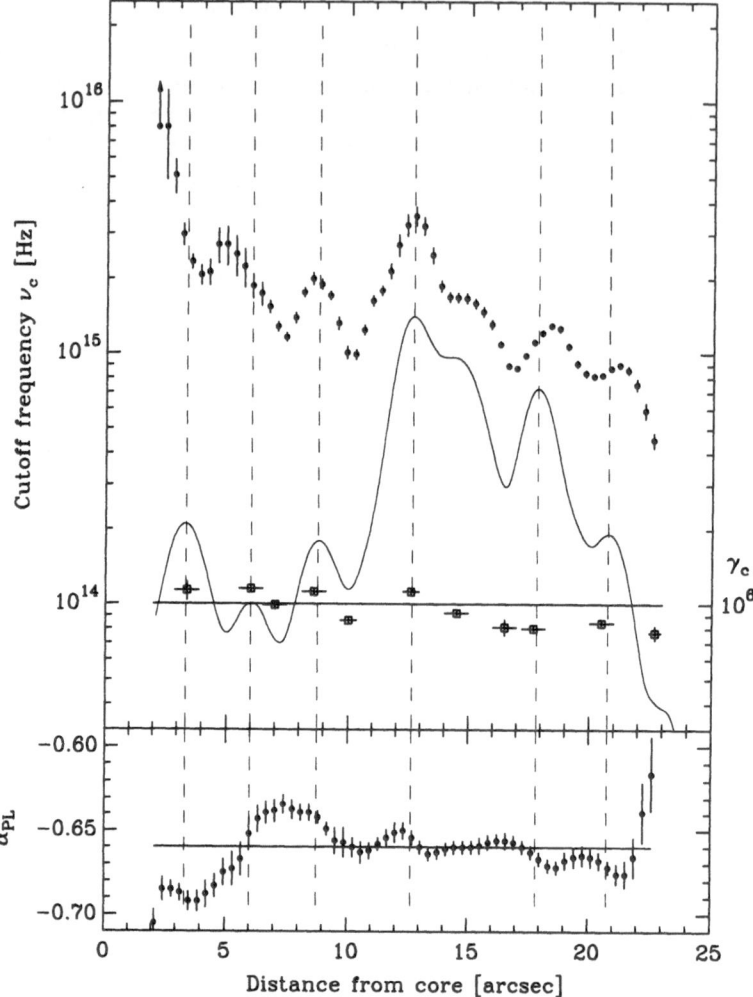

Fig. 3. Parameters of the synchrotron model fits for every grid point along the ridge of the jet. *Top panel:* Cutoff frequency ν_c. For reference a *logarithmic* representation of the brightness profile at $\lambda = 2\,\mathrm{cm}$ is given. The maximum energy γ_c inferred from ν_c and a minimum energy estimate of the field is shown be □ (refer to labeling on the right). *Bottom panel:* Best-fit power-law index α_{PL} in the range 10^{10} to $10^{14}\,\mathrm{Hz}$.

$$\nu_c = 42\,Hz\,|\mathbf{B}|\,\sin\theta\,\gamma_c^2 \qquad (1)$$

Since a direct measurement of $B_\perp = |\mathbf{B}|\sin\theta$, where θ is the angle between **B** and the line-of-sight, is not possible, one has to rely on the standard "minimum energy" estimate[3] for deriving the field strength (*cf.* discussion below). Such estimates of the field strength have been derived by Meisenheimer *et al.*

---

[3] Often referred to as "equipartition magnetic field" in the literature.

(1996), based on the morphology and surface brightness of the jet on 2 cm VLA maps with $0''.1$ resolution.

Using their values for $|\mathbf{B}|$ ($10 \ldots 70\,\mathrm{nT}$) and a jet orientation of $45°$w.r.t. the line-of-sight yields the run of γ_c along the jet which is displayed in the central panel of Fig. 3. Obviously the maximum energy γ_c of the relativistic electrons (positrons?) does not change by more than 30% along the entire length of the jet between knots D and G. This is in dramatic conflict with the expectation from synchrotron theory which would predict that γ_c decreases by a factor of 2 within only 40 pc. As already pointed out by Felten (1968, see above) there are, in principle, two ways to explain the absence of detectable synchrotron losses in the particle spectrum:

(A) The synchrotron losses are balanced all along the jet by a re-acceleration process which is able to maintain $\gamma_c \simeq 10^6$.

(B) The relative change $\Delta\gamma/\gamma_c$ is severely overestimated by the assumption of an equipartition magnetic field, because either

    (1) the majority of electrons is transported in a "loss free channel" with low magnetic field (*cf.* Owen *et al.* 1989) from which the synchrotron-bright regions of high magnetic field are supplied with electrons at γ_c, or

    (2) the "minimum energy" magnetic field over-estimates the true field strength by at least an order of magnitude.

Since hypothesis (B-1) requires a two-fluid jet which would be extremely unstable against relaxation towards a force free plasma configuration, only (B-2) seems viable. This alternative has been explored by Heinz & Begelman (p. 229, these proceedings) who show that a magnetic field of $2.8\,\mathrm{nT}$ (*i.e.* $10\times$ lower than the minimum energy field derived by Meisenheimer *et al.* 1996), combined with mildly relativistic jet speeds (Lorentz factors 2-3) could indeed yield the unexpected low loss rate $\Delta\gamma/\gamma_c$ observed. However, three elements of circumstantial evidence seem to argue against this bypass around "Feltens dilemma" (see above):

(1) In hotspots of radio galaxies, Meisenheimer *et al.* (1989, 1997) found evidence that the minimum energy estimate is correct within a factor 2 or so.

(2) Lowering the field strength by a factor of 10 below its minimum energy value implies that the total energy stored in relativistic particles is increased 30-fold. This would make the energy flow in relativistic particles (and fields) comparable if not larger than the total kinetic jet power derived for radio galaxies of comparable radio luminosity.

(3) The apparent morphology of the M87 jet seems to be dominated by magnetic filaments which show up both in high synchrotron brightness and well ordered magnetic fields (*cf.* the 2 cm VLA maps presented by Owen *et al.* 1989). It is hard to imagine how an energetically negligible field could imprint such dominant signatures onto the morphology.

Because of these obvious shortcomings of alternative (B), we would like to explore the implications of alternative (A) in more detail:

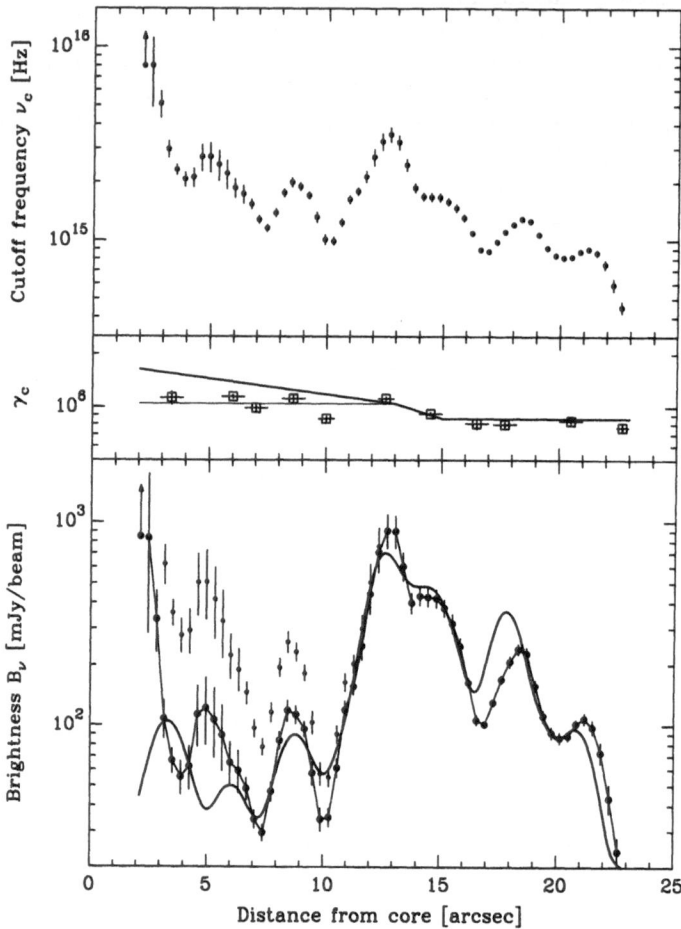

Fig. 4. Model of the brightness distribution B_ν along the jet. *Top panel:* measured values of ν_c. *Middle panel:* γ_c as inferred from the minimum energy fields ($\square$) and the two functions which have been used for the run of γ_c. *Bottom panel:* The observed brightness B_ν^{obs} at $\nu = 15$ GHz together with the predicted $B_\nu^{\mathrm{mod}} \sim |\mathbf{B}|^{2.2}$ for the step function $\gamma_c = 1.05 \times 10^6 \to 8.6 \times 10^5$ (small symbols) and for the fine-tuned function for γ_c ($\bullet$) which is shown as thicker line in the middle panel.

Ignoring the nature of the (unknown) re-acceleration process for the moment, one could ask the question whether an almost constant maximum energy plus local variations of the magnetic field strength would be sufficient to explain *both* the run of ν_c and that of the local surface brightness B_ν

along the jet. To this end, one should compare the run of $B_\perp$ (as derived from the observed ν_c, equation (1) and $\gamma_c \simeq 10^6$) with the observed surface brightness B_ν^{obs} along the jet. Standard synchrotron theory would predict $B_\nu \sim |B|^{2-\alpha\pm0.5}$, depending on field geometry and the possible causes for local field enhancements (*e.g.* adiabatic compression, MHD instabilities).

As it can be seen in Fig. 4, a brightness model $B_\nu^{mod} \sim |B|^{2.2}$ provides a reasonable fit to the observed profile B_ν^{obs} (at $\nu = 15\,\mathrm{GHz}$) beyond $\sim 10''$ from the core. The agreement between B_ν^{mod} and B_ν^{obs} can be extended almost to knot D, when we fine-tune the run of γ_c to

$$\gamma_c = 1.1 \times 10^6 e^{-(r-12.7'')/8''} \text{ at distances } 3'' < r < 15'' \text{ from the core, and}$$
$$\gamma_c \equiv 8.6 \times 10^5 \text{ at } r \geq 15'' \text{ (down-stream of knot B).}$$

This fine-tuning is well within the accuracy of the minimum energy estimate which critically depends on the simplified assumption, that all substructures are fully resolved at the 2 cm maps at $0\overset{''}{.}1$ resolution. From the good agreement between B_ν^{mod} and B_ν^{obs} (Fig. 4) and the constancy of α_{PL} along the jet we draw two conclusions:

– The particle energy distribution is essentially unchanged along the entire length of the jet between knots D and G:
  $N(\gamma) \sim \gamma^{-2.32}$ for $3 \times 10^3 \lesssim \gamma < \gamma_c$
  $N(\gamma) \to 0$ for $\gamma > \gamma_c$, with $\gamma_c \simeq constant$ (see above).
– Local variations of the magnetic field strength govern *both* the apparent run of the cutoff frequency and the brightness distribution along the jet.

This consequence of rigorously following alternative (A) does not only imply that the concept of a "permanent" re-acceleration process along the main body of the kpc jet in M87 is fully consistent with the observations but furthermore puts extremely strict boundary conditions onto the nature of any such process:

(i) The putative re-acceleration operating in the jet of M87 maintains the electron spectrum to extremely high accuracy in both the knots and the inter-knot regions (which differ by at least a factor of 5 in magnetic field strength).

(ii) Re-acceleration seems to occur "on the spot" (*i.e.* on scales $\ll 100\,\mathrm{pc}$).

It is hard to imagine how stochastic acceleration processes in various shock fronts along the jet could meet this criteria — instead a self-regulating process operating on scales < 10 pc seems to be required by the observations.

Admittingly, the nature of the required process is completely unknown at present. But it seems that MHD processes ("reconnection" and acceleration in current sheets, see Lesch & Birk 1998) are able to meet at least criterion (ii). Whether or not they are able to meet the greatest challenge — the constancy of α_{PL} — has to be investigated by a detailed analysis how an initial powerlaw spectrum is modulated by the conversion of magnetic field into particle energy. Nevertheless, for the first time since the fifties we seem to have reached the situation again, when the accuracy of the observations challenges theory beyond its present imagination.

5   High frequency observations and synchrotron spectra of the inner lobes

Shortly after the last Ringberg workshop on extragalactic radio jets, two groups reported the detection of a patch of optical continuum emission in the eastern — "un-jetted" — lobe of M 87 (Stiavelli *et al.* 1992, Sparks *et al.* 1992). This patch coincides perfectly with the brightest ridge of radio emission in the inner eastern lobe (Fig. 5), the bow-like shape of which runs parallel to the outer radio contours of the lobe (Hines *et al.* 1989). According to the nomenclature invented by Hines *et al.* we will refer to this feature as "patch θ". The positional coincidence and the fact that its optical spectral index $\alpha_{opt} \simeq -1.3$ seems somewhat steeper than the radio spectrum of θ ($\alpha_{rad} = -0.84$) already indicated that the optical patch might represent the high frequency part of the synchrotron spectrum of θ. Early near-infrared observations by Neumann *et al.* (1995) and our new, deeper H, K' observations which have been obtained to measure the shape of the jet spectrum accurately (see above) combined with the VLA map of Hines *et al.* result in a well determined radio-optical spectrum of patch θ which is perfectly fitted by a synchrotron model spectrum with $\alpha_{PL} = -0.82$ cutting off at the maximum frequency $\nu_c = 3\,10^{14}$ Hz (see Fig. 6). In disagreement with Stiavelli *et al.* who interpreted patch θ as leading hot spot of the "counter-jet" (made invisible by relativistic feebling), Neumann *et al.* proposed to regard θ as the brightest part of the synchrotron emission of the eastern lobe which just happens to exceed the optical detection limit ("tip of the iceberg"). Our deeper NIR images impressively confirm this interpretation since they allow us to trace the patch to the level of the next lower radio contour (see Fig. 5).

Before discussing the implications of optical synchrotron emission from the eastern, "un-jetted" lobe any further we should direct our attention to the inner western lobe beyond the optical jet: Both our deep K' and H images show very clearly that there is indeed some detectable NIR light emerging from the lobe (Fig. 5). Unfortunately, our current B,R,I data do not reach deep enough to detect this lobe emission and establish its spectral shape beyond any doubt. However as demonstrated in Fig. 6 one might well shift the synchrotron spectrum of the jet (knots A+B+C) by a factor of 23 in frequency (*i.e.* to $\nu_c = 1\,10^{14}$ Hz) in order to account perfectly for the measured radio and NIR flux values of the western lobe.[4] In order to estimate the maximum energy γ_c of the electrons (positrons) radiating in the western lobe, we again have to rely on the minimum energy argument from which one gets $|B| \simeq 10\,$nT. This yields $\gamma_c \simeq 5 \times 10^5$ (western lobe) which is amazingly close to $\gamma_c = 8 \times 10^5$ inferred for the outer jet down-stream of knot B (*cf.* Fig. 3). Obviously the particle re-acceleration to $\gamma_c \approx 10^6$ does not cease at the end of the collimated jet but is maintained well into the inner lobes of M 87.

---

[4] On a very low surface brightness level the NIR emission might even be traced around the sharp bend at the outer edge of the western lobe.

Fig. 5. Comparison betwen a deep K' image of M 87 (greyscale, starlight subtracted) and the radio map at 5 GHz (isocontours, Hines *et al.* 1989). Note the brightest patch θ in the eastern lobe and the low brightness emission coinciding with the radio maximum of the western lobe.

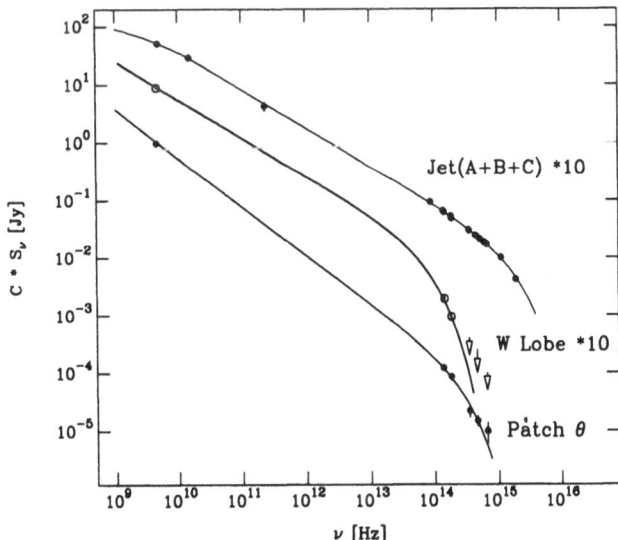

Fig. 6. The radio-optical spectra of the western lobe (o) and patch θ in the eastern lobe (•) in comparison with the spectrum of the knots A+B+C. Note that the spectra of the western lobe and knots A+B+C are scaled by 10 in order to avoid confusion.

Similar arguments for patch θ in the eastern lobe yield a field strength of $13\,\text{nT}$ and $\gamma_c(\theta) = 7.5 \times 10^5$. Thus also the "un-jetted", eastern lobe has to be supplied by enough (kinetic) energy to keep partical acceleration to $\gamma_{max} \simeq 10^6$ going. One possibility for providing this would be an intrinsically symmetric "counter-jet" which is made invisible by its unfavorable Doppler boosting away from the line-of-sight. On the other hand, the powerlaw index $\alpha_{PL} = -0.82$ seems to be characteristic for the *entire* eastern lobe (*cf.* Salter *et al.* 1989). This and the large opening angle of patch θ (when viewed from the core) make it more likely that it corresponds to the patch of NIR emission in the western lobe and the two sides of M 87 are intrinsically different. In this case one would have to regard the high maximum energy γ_c in the eastern lobe as further indication of the ubiguity of effective partricle acceleration in the magnetized plasma ejected from the core of M 87, independant of whether or not a highly collimated, high surface brightness jet is formed. This view would certainly challenge the current paradigma of line-of-sight orientation being the main cause for observed asymmetries in extragalactic radio sources.

6   News from HST — some preliminary results

Deviating from our oral contribution at Ringberg which presented the published pre-COSTAR results of the HST observations, we will concentrate here only on HST results concerning the high frequency spectrum of the jet and particle acceleration. A complete review of the current HST results is given by Biretta *et al.* (on page 210). Specifically, we will focus on the following questions:

(1) Is the close correlation between surface brightness and optical spectral index maintained on scales $\lesssim 0\rlap{.}''1$ (10 pc, *i.e.* $< 1/10$ of the ground-based studies)?

(2) Are there localized and short-lived phenomena present in the HST observations (which for the first time probe scales of the order of $c\tau_{syn}$ for $\gamma \simeq 10^6$) which could directly point to sites of effective particle acceleration?

Although the final analysis of the optical spectral index (including the full spectral coverage $340\,\text{nm} < \lambda < 814\,\text{nm}$, matched to exactly the same resolution and aligned to $\lesssim 20\,\text{mas}$) has still to be carried out, it is obious from the preliminary results (*e.g.* Fig. 5 in Biretta *et al.*), that the general behaviour of the spectral index does indeed conform to the expectations from the analysis presented in Sects. 4 and 5: The fainter regions of the jet (especially the inter-knot regions) and some regions close to the jet boundary have very steep optical spectra $\alpha_{opt} < -1.2$. This certainly enhances the impression (noted by various authors, *e.g.* Sparks *et al.* 1996) that the inner optical jet (out to knot F) is more central brightened than the radio jet for which pronounced edge brightning has been reported at some places (Owen *et al.* 1989). At the present state of analysis it seems not entirely clear, whether the above proposed model for the "natural" correspondence between ν_c and B_ν is able to account fully for the apparent differences in radio and optical jet morphology on $0\rlap{.}''1$ scale. However, the pronounced morphological difference which is apparent *e.g.* between knot E and F (see Fig. 7 in Biretta *et al.*) seem to make a careful analysis of the pointwise jet spectrum very rewarding (*i.e.* using a grid of $0\rlap{.}''1$ spacing), since it could reveal those subtle changes in either α_{PL} or γ_c which will eventually be needed to pin down the locations of most effective particle acceleration and the flow pattern of the plasma into regions where synchrotron losses dominate the re-acceleration process. Nevertheless, it seems already clear from the smoothness of α_{opt} even on $0\rlap{.}''1$ resolution, that we are looking out for very *subtle* effects. Simple particle acceleration schemes which try to locate the re-acceleration in a dozen or so "knots" (strong shocks) along the jet are virtually ruled out by the HST data! Even if the deviation from a rather smooth "permanent" re-acceleration process is marginal and hard to trace down-stream of knot D (projected distance 300 pc from the core), the HST observations have now provided a wealth of data that this "quiet" jet behaviour does not persist all the way towards the core:

At the up-stream edge of knot D, the current HST data (observed in 1995*ff*) show a distinguished optical knot (D_1) with transverse field orientation which has no obvious radio counterpart on the 2 cm VLA maps observed in 1985. Further upstream (at $1\rlap{.}''5$ from the core) there exists now a new optical knot (called HST-1, see Biretta *et al.*) which is located in a region of low brightness on the radio maps from 1985. Furthermore, HST-1 rapidly ejects "bullets" of optical emission which travel down the jet with apparent speeds of up to $6c$ (see Fig. 10 in Biretta *et al.*). They are fainting on a time scale of about one year. Such behaviour can certainly not be regarded as smooth "permanent" re-acceleration process! The above mentioned finding that the $B_\nu \sim |\mathbf{B}|^{2.2}$ model completely breaks down for $r < 3''$ argues against D_1 and HST-1 being completely new features, but rather favours the interpretation that the jet physics in the innermost $300 \ldots 500$ pc from the core is indeed different from that in the kpc jet.

Sofar no convincing theory has been put forward to explain the extraordinary behaviour detected in the innermost $3''$ of the M 87 jet. We would like, therefore, to sketch a scenario which could perhaps account for both the fast variability in the inner jet and the constancy of the particle spectrum beyond $r = 5''$.

Our basic assumption is that an electro-magnetic acceleration process, driven by magnetic reconnection maintains the high energies of the particles along the jet of M 87 and perhaps into the lobes (Lesch & Birk 1998). In the outer regions $r > 5''$ where the jet has widened to $D_{jet} > 50$ pc (*i.e.* $> c\tau_{syn}$), the small relative size ($\ll 10$ pc) of the individual reconnection zones and integration along the line-of-sight make it extremely hard to resolve the fine-structure of the acceleration process. However, close to the core where the jet seems to be confined to a very small opening angle ($D_{jet} < 10\,pc$) we are able to witness the rapid variations which should be present in reconnection processes of an inhomogenous magnetic field structure: Bunches of ultra-relativistic particles ($\gamma \simeq 10^6$) are ejected along reconnecting field lines (the bullets ejected from HST-1 ?) since a component $\mathbf{E} \parallel \mathbf{B}$ is needed to generate effective acceleration. Alternatively, apparent superluminal motions could be explained by the progress of the reconnection zone which could appear to change its location faster than light — even for a sub-relativistic jet flow. The fading time scale of the "bullets" ejected by HST-1 matches the expectation for synchrotron losses of electrons with $\gamma \simeq 10^6$ in a field of several tens of Nanotesla, supporting the former interpretation. Further out, where the jet structure is more complicated (with strong transverse field components) the output of the reconnection zones has lost its unidirectionality and thus mainly results in a permanent re-boost of the particle energies.

We are rather confident that the wealth of high resolution data produced by the HST together with simultanous observations at the highest VLA frequencies (43 GHz) will eventually provide the data base which will allow us to decide between different acceleration processes and to study their properties in great detail.

7   How do the X-rays fit into the overall spectrum?

Recent X-ray data obtained with the ROSAT HRI (Neumann *et al.* 1997) confirmed the presence of X-ray emission from knot D and the vicinity of knot A which has already been reported by Schreier *et al.* (1982) on the basis of observations taken with the EINSTEIN HRI. The latest re-analysis of the EINSTEIN data has been published by Biretta *et al.* (1991) who claim also the detection of knot B and (perhaps) knot C. Due to problems in reconstructing the actual PSF of the ROSAT HRI observations (an uncorrectable jitter in the satellite pointing prohibits to recover the full resolution of $\lesssim 5''$ FWHM), the new observations add little to the EINSTEIN result on knot D. However, in contrast to the finding by Biretta *et al.* there is no indication for X-ray emission from knots B to C on the ROSAT maps presented by Neumann *et al.* (1997).

Moreover, the location of the — unresolved — ROSAT X-ray emission seems to be closer to the core than both the EINSTEIN peak and the radio-optical knot A (see Fig. 7). Also the ROSAT flux of knot A (epoch 1992) seems to fall short by 30% of the result of the EINSTEIN HRI (this fact was first noticed by Harris & Biretta, *priv.comm.* 1995). Due to the difference in detected quantum efficiency between the ROSAT and EINSTEIN HRI (the latter was able to detect harder photons), which could affect both the detected flux from the jet and the shape of the underlying background from the X-ray halo around M 87, it is not entirely clear whether the apparent

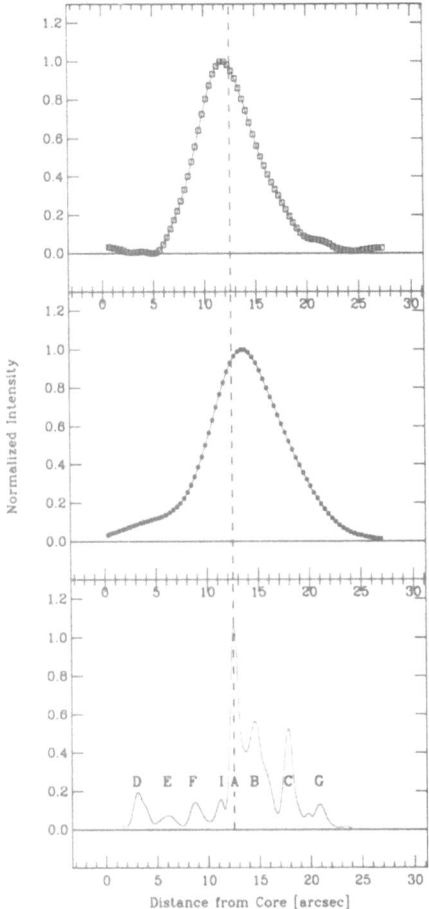

Fig. 7. Location of the X-ray emission near knot A as observed by the ROSAT HRI (from Neumann *et al.* 1997). *Top panel:*Tracing of the observed X-ray emission along the jet (background subtracted, resolution: 5.″5 FWHM). *Middle panel:* Tracing of the optical brightness profile convolved with the X-ray beam. *Bottom panel:* Tracing of the radio brightness at 0.″3 resolution.

difference indicates a true variability (rather likely, see discussion below) or can be attributed to differences in instrument performance and data analysis.

The nature of the X-ray emission. The local radiation field near or within knot A is too weak by at least an order of magnitude to make comptonization of low energy photons a promising process for generating the observed X-ray flux from the vicinity of knot A. Likewise thermal bremsstrahlung — albeit suggested by the apparent offset between the radio-optical synchrotron peak of knot A and the X-ray peak (see Fig. 7) — requires rather contrived assumptions about shock-enhanced high-density regions near knot A. Although it is hard to exclude this explanation for the X-ray emission on the basis of the current data, Neumann *et al.* argue that a synchrotron origin of the X-ray emission is most likely. In this scenario the offset between the location of the radio-optical knot A (at $0\overset{''}{.}3$ resolution, see Fig. 7) and the X-ray peak would have to be attributed to a combination of the uncertainty in the X-ray astrometry and the fact that only the upstream boundary of knot A (at $\sim 12\overset{''}{.}1$ from the core) emits synchrotron radiation up to X-ray frequencies.

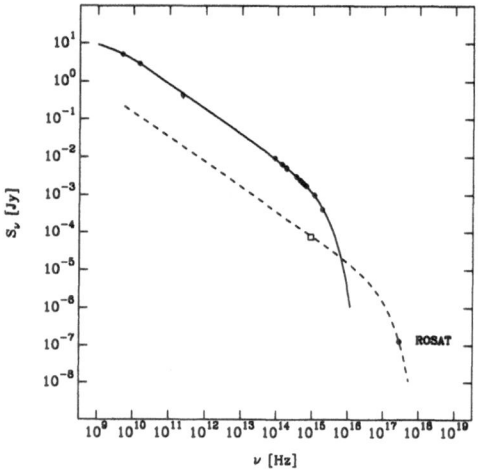

Fig. 8. Comparison between the radio-optical spectrum of the jet (knots A+B+C) with the X-ray flux mesured by the ROSAT HRI. Although the extrapolation of the spectrum of the entire region falls short of the X-ray flux by several orders of magnitude, a synchrotron spectrum with similar shape but higher cutoff frequency $\nu_c \simeq 1.5 \times 10^{17}$ Hz for the narrow feature at the leading edge of knot A ($\square$) may well account for the observed X-ray flux.

The X-ray emission interpreted as synchrotron radiation. The bluest and brightest part of the optical knot A is a narrow feature located at $r \simeq 12\overset{''}{.}2$ from the core (see Fig. 2 in Biretta *et al.*, these proceedings). This location is marginally consistent with the origin of the X-ray emission (Fig. 7). Since this substructure makes up for only 4% of the optical flux complex A+B+C, a possible connection between the radio-optical spectrum and the X-ray data would be the overall synchrotron spectrum displayed in Fig. 8. The required cutoff frequency is $\nu_c \simeq 1.5 \times 10^{17}$ Hz, well above the value inferred for knot A at $1\overset{''}{.}3$ resolution. Even if one allows for a local field enhancement by a factor of 1.4 above the average field strength inferred for knot A (*i.e.* $|\mathbf{B}| \simeq 100\,\mathrm{nT}$), a maximum energy $\gamma_c \gtrsim 5 \times 10^6$ would be required in order to explain the X-ray emission as high-frequency tail of the synchrotron spectrum. Although such a high γ_c would be in conflict with the above outlined model to explain cutoff frequency and brightness by local variations of the local magnetic field, it is not unreasonable to attribute an additional boost to extra-ordinary high energies to the strong shock front which apparently is located at the front edge of knot A (*cf.* the flip of the field orientation from parallel to perpendicular w.r.t. the jet axis). Obviously, the proposed synchrotron spectrum of the front feature of knot A is rather steep in the X-ray band between 0.1 and 2 keV. Thus, minor changes in either γ_c or the local field could change ν_c and consequently the flux in the X-ray band dramatically. Since the synchrotron lifetime for electrons with $\gamma = 5 \times 10^6$ in a field of 100 nT is only 3.5 years, rapid variability of the X-ray flux on a time-scale of years is not unexpected and could well account for the reported differences of the X-ray emission in 1979/80 (EINSTEIN observations) and 1992 (ROSAT observations).

On the other hand, the required shape of the synchrotron spectrum of the bright leading component is such that it could easily be verified by future observations and the careful analysis of already obtained HST images. Specifically the here proposed interpretation of the X-ray emission from knot A as high-frequency tail of the synchrotron spectrum predicts three properties:

(1) The *optical spectrum* of the front feature of knot A should be a straight powerlaw with $\alpha_{opt} = -0.66$ out to the highest observable frequencies.

(2) Most of the X-ray emission should emerge from the front feature of knot A at $r = 12\overset{''}{.}2$ from the core.

(3) The *X-ray spectrum* around 1 keV should be rather steep with $\alpha_x \lesssim -2$. While (1) should be confirmed on HST data, (2) and (3) will easily be tested by the AXAF observations soon to come. If (3) is falsified, we will have to think the origin of the X-ray emission over. In this case it will have to be interpreted either by thermal bremsstrahlung or synchrotron radiation with an even higher maximum frequency, emerging from substructures which have not been resolved yet on the HST images.

8 Summary

Very detailed studies of the overall synchrotron spectrum along the jet of M 87 from the ground have now reached the accuracy that the astonishing constancy of the underlying particle energy distribution $(N(E) \sim E^{-2.32}$, maximum energy $E_{max} = 10^6 m_e c^2)$ is establihed beyond any doubts. This purely observational fact is a great challenge to *all* current theories of particle acceleration in extended extragalactic radio sources. On top of this rather "quit" and smooth behaviour in the outer parts of the radio-optical jet, HST observations show rapid variations in the inner $3''$ of the jet. These observations may for the first time resolve the spatial and temporal scale of the acceleration process. Even if we are not able to measure the basic parameters of the plasma directly (density and composition, flow speed and magnetic field strength) the wealth and accuracy of the available data are such that theoretical interpretations are no longer hindered by the lack of data but seem to be faced with too accurate measurements which immediatly rule out the standard models. The observers have done their homework ! Now it is the turn of the theorists to promote our understanding of the jet in M 87 and extragalactic radio sources in general.

References

Baade, W. 1956, *Astrophys. J.* **123**, 550

Bell, A.R. 1978a, *Mon. Not. R. astr. Soc.* **182**, 147

Biretta, J.A., Stern, C.P. and Harris, D.E. 1991, *Astron. J.* **101**, 1632 (BSH91)

Biretta, J.A. & Meisenheimer, K. 1993: *The jet of M 87*. In: "Jets in extragalactic radio sources", proceedings of the second Ringberg workshop, eds. H.-J. Röser and K. Meisenheimer, Springer Verlag Heidelberg, Berlin, usw., p. 159

Biretta, J. A., Zhou, F. and Owen, F. N. 1995, *Astrophys. J.* **447**, 583

Blandford, R.D. and Rees, M.J. 1974, *Mon. Not. R. astr. Soc.* **169**, 395

Boksenberg, A. *et al.* 1992, *Astron. Astrophys.* **261**, 393

Bridle, A.H. and Perley, R.A. 1984, *Ann. Rev. Astron. Astrophys.* **22**, 319

Burbidge, G.R. 1956, *Astrophys. J.* **124**, 416

Carilli, C.L. and Harris, D.E. (eds.) 1996: "Cygnus A — Study of a Radio Galaxy", Cambridge University Press, Cambridge U.K.

Curtis, H.D. 1918, *Lick Obs. Publ.* **13**, 11

Falle, S.A.E.G. and Wilson, M.J. 1985, *Mon. Not. R. astr. Soc.* **216**, 79

Felten, J.E. 1968, *Astrophys. J.* **151**, 861

Hargrave, P.J. and Ryle, M. 1974, *Mon. Not. R. astr. Soc.* **166**, 305

Hiltner, W.A. 1959 *Astrophys. J.* **130**, 340

Hines, D.C., Owen, F.N. and Eilek, J.A. 1989, *Astrophys. J.* **347**, 713

Keel, W.C. 1988, *Astrophys. J.* **329**, 532

Killeen, N.E.B., Bicknell, G.V., Hyland, A.R. and Jones, T.J. 1984, *Astrophys. J.* **280**, 126

Laing, R.A. 1980a, *Mon. Not. R. astr. Soc.* **193**, 439

Laing, R.A. 1980b, *Mon. Not. R. astr. Soc.* **193**, 427

Lesch, H. and Birk, G.T. 1998, *Astrophys. J.* **499**, 167

Meisenheimer, K., Röser, H-J., Hiltner, P., Yates, M.G., Longair, M.S., Chini, R. and Perley, R.A. 1989a, *Astron. Astrophys.* **219**, 63

Meisenheimer, K., Röser, H.-J. and Schlötelburg, M. 1989b, in E. Meurs and R. Fosbury (eds.): *Extranuclear Activity in Galaxies*, ESO Proceedings Garching.

Meisenheimer, K., Schlötelburg, M. and Röser, H.-J. 1996, *Astron. Astrophys.* **307**, 61 206 237

Neumann, M. 1994: *PhD Thesis*, University of Heidelberg.

Neumann, M., Meisenheimer, K., Röser, H.-J. and Stickel, M. 1995: *Astron. Astrophys.* **296**, 662-664

Neumann, M., Meisenheimer, K., Röser, H.-J. and Fink, H.H. 1997: *Astron. Astrophys.* **318**, 383-389

Nieto, J.-L. and Lelièvre, G. 1982, *Astron. Astrophys.* **109**, 95

Norman, M.L., Smarr, L.L., Winkler,K.-H.A. and Smith, M.D. 1982, *Astron. Astrophys.* **113**, 285

Oort, J.H. and Walraven, T. 1956, *Bull. Astron. Inst. Netherlands* **12**, 285

Owen, F.N., Hardee, P.E. and Bignell 1980, *Astrophys. J. (Lett.)* **239**, L11

Owen, F.N., Hardee, P.E. and Cornwell, T.J. 1989, *Astrophys. J.* **340**, 698 (OHC89)

Perez-Fournon,I., Colina, L., Gonzalez-Serrano, J.I. and Biermann, P.L. 1988, *Astrophys. J. (Lett.)* **329**, L81

Perola G.C. and Tarenghi, M. 1980, *Astrophys. J.* **240**, 447

Reid, M.J., Biretta, J.A., Junor, W. Muxlow, T.W.B and Spencer, R.E. 1989, *Astrophys. J.* **336**, 112

Röser, H.-J., Meisenheimer, K., Neumann, M., Conway, R.G., Davis, R.J., Perley, R.A. 1997, *Reviews in Modern Astronomy* **10**, 253

Salter, C.J., Chini, R., Haslam, C.G.T., Junor, W., Kreysa, E., Mezger, P.G., Spencer, R.E., Wink, J.E. and Zylka, R. 1989, *Astron. Astrophys.* **220**, 42

Scheuer, P.A.G. 1974, *Mon. Not. R. astr. Soc.* **166**, 513

Schlötelburg, M., Meisenheimer, K. and Röser, H.-J. 1988, *Astron. Astrophys. (Lett.)* **202**, L23

Schreier,E.J., Gorenstein, P. and Feigelson, E.D. 1982, *Astrophys. J.* **261**, 42

Shklovsky, I.S. 1953, *Doklady Akad. Nauk. U.S.S.R.* **90**, 983

Sparks, W.B., Fraix-Burnet, D., Machetto, F. and Owen, F.N. 1992, *Nature* **355**, 804

Sparks, W.B., Biretta, J.A. and Machetto, F. 1996, *Astrophys. J.* **473**, 254

Stiavelli, M., Biretta, J., Møller, P. and Zeilinger, W.W. 1992, *Nature* **355**, 802

Stocke, J.T., Rieke, G.H. and Lebofsky, M.J. 1981, *Nature* **294**, 319

Turland, B.D. 1975, *Mon. Not. R. astr. Soc.* **170**, 281

HST Observations of the M 87 Jet

J. A. Biretta, E. Perlman, W. B. Sparks, and F. Macchetto

Space Telescope Science Institute
3700 San Martin Dr.
Baltimore, MD 21218, USA

Abstract. We present highlights and preliminary results of our HST observing programs on the M 87 jet. We confirm previously known similarities between the radio and optical jets, but also describe a wealth of differences and new structural details. We identify numerous compact optical features, as well as new structures in the shock-like knots A and C. We also find considerable evidence for structure in the jet as a function of distance from the jet axis — evidence is seen for radial structure in the radio / optical spectrum, in the optical / optical spectrum, and in the polarimetry. Our monitoring between 1994 and 1996 shows numerous features with superluminal speeds in the range $4c$ to $6c$, as well as the emergence of new superluminal components and rapid fading of old features.

1 Introduction

While many hundreds of extragalactic jets are known from their non-thermal radio emission, there are only a dozen or so with prominent optical emission. Of these few, the M 87 jet is by far the nearest, and is thus the "Rosetta Stone" for optical jet studies.

HST observations of the M 87 jet can provide critical insights to many key questions related to extragalactic jets (*cf.* reviews by Biretta 1993; Biretta and Meisenheimer 1993; Meisenheimer, Röser, and Schlötelburg 1996). One of the most important questions is that of energy transport — what exactly is being transported (electrons + positrons? protons?) — and how fast? What is the internal energy spectrum of the jet material, and how does it evolve? We see optical emission from short lived $\gamma \sim 10^6$ electrons at large distances from the nucleus — how exactly do they get there — is there a loss-less transport mechanism such as low B-fields or high-γs, or are they accelerated in-situ perhaps at shocks? What are the flow speeds, and how fast is the kinetic energy dissipated? Are relativistic effects important, and if so, what does this imply about objects at other orientations (*i.e.* unified models)? What is the internal structure of the jet, and what role do B-fields play in confining and collimating jets, and allowing them to propagate many kpc (Mpc in some cases) from the galaxy nucleus? Finally, study of well-resolved jets as in M 87 may give clues to the origin of non-thermal continuum emission from BL Lac objects and QSOs, and provide a physical (or at least spectral) connection to non-thermal X-ray and γ-ray emission in these objects.

HST is ideally suited for addressing many of these issues. HST can provide images with unsurpassed detail using the Faint Object Camera at UV wavelengths. Resolution of 30 milliarcseconds is easily obtained, and without dynamic range or spatial sampling issues usually associated with interferometer (u, v) coverage. This provides $\sim$ 50 resolution elements across the jet M 87, allowing imaging of the jet structure and shock-like knots with unprecedented detail.

Comparisons between the optical and radio images allows us to locate the highest energy particles, and can delineate the locations of particle acceleration sites at shocks. Measurement of the IR – optical – UV spectrum throughout the jet can show how the particle spectrum evolves, and mapping of the high-frequency spectral break can illustrate magnetic field variations and show what role they play in shaping the jet's brightness distribution.

Optical polarimetry can provide information on magnetic field direction and degree of order which is free of Faraday rotation effects. Moreover, systematic differences between the radio and optical flux distributions (the optical emission being concentrated on the jet axis) can be utilized to obtain 3-dimensional information about the field structure.

Finally, one of the most important goals is to study the jet's evolution and kinematics directly via regular monitoring. The resolution advantages of HST together with the greater compactness of features in the optical band make it ideal for detecting flux variations and measuring proper motions. And again, systematic differences from the radio may allow derivation of 3-dimensional information. Ultimately it may be possible to map the jet's velocity field and compare it against numerical simulations.

We have undertaken an extensive study of the M 87 jet with HST at nearIR, optical, and UV wavelengths, including polarimetry and annual monitoring. Here we present highlights and preliminary results of our study. We will assume a distance of 16 Mpc (Tonry 1991), for which $1''$ equals 78 pc.

2 Optical Jet Structure

High resolution images of the jet taken in the UV with the post-COSTAR Faint Object Camera reveal many details which were not apparent in previous images. Figure 1 shows the entire jet as seen by the FOC at $0''03$ resolution. Many structural details of the classical knots are readily apparent, as are many new compact structures. The first 200 pc of the jet — roughly the region between the nucleus and knot D — appears as a narrow chain of numerous bright, unresolved knots. The opening angle of the jet as defined by this chain of features is extremely narrow. The opening angle here is less than about $1°$, whereas a value near $6.5°$ is apparent for the rest of the jet beyond knot D. There are additional compact features between knots D and E, as well as bright, slightly resolved condensations (scale $\sim 0''1$) throughout knots D, E, F, and I of the "inner" jet.

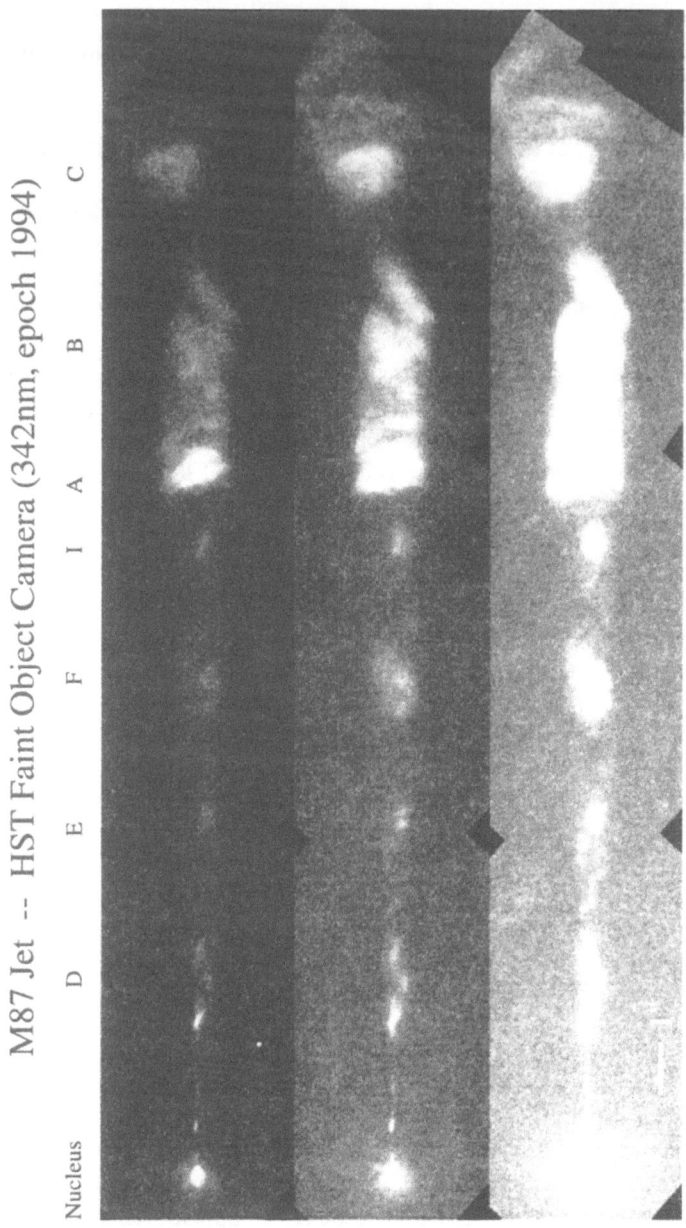

Fig. 1. HST Faint Object Camera image of the jet observed at wavelength 342 nm in 1994 with 0."03 resolution. Each image is a mosaic of three different FOC pointings. The three panels show the same image at different contrast settings. Knots are labeled with usual letter designations.

The FOC images also reveal many fine structures within the knots that are apparent for the first time. Knot A, in particular, shows a wealth of new detail. This is the brightest knot in the jet, and is thought to represent a shock in the flow (Biretta, Owen, and Hardee 1983; Biretta 1993). Evidence favoring this interpretation are the bright, thin line of emission normal to the jet axis which resembles a Mach disk; the change in the overall appearance of the jet at this point from a very faint jet interspersed with faint knots, to a bright, uniformly filled jet; the onset of side-to-side bulk oscillation of the jet downstream from knot A (suggestive of instability and reduced Mach number); and polarimetry which shows transverse B-fields indicating compression.

Previous radio images with $0\rlap{.}''1$ resolution showed a sharp unresolved "edge" at the eastern end of knot A. Even at ~ four times the best previous resolution, knot A still appears to contain significant unresolved emission (Figure 2). This region has previously been modeled as a disk of emission seen edge-on (Biretta, Owen, and Hardee 1983), but these latest images make it seem less probable that such a model can account for all the emission, due to the increasingly severe geometric coincidence required with the line-of-sight. It seems likely that some portion of the emission arises in a narrow linear (*i.e.* one-dimensional) feature, such as a narrow ring of emission (Biretta, Zhou, and Owen 1995), or a bright narrow filament wrapped on the jet surface, as suggested by Owen, Hardee, and Cornwell (1989; OHC89). Besides the bright, narrow feature at the leading edge of knot A, a second linear feature is apparent about $0\rlap{.}''5$ downstream (Figures 2 and 3).

The eastern edge of knot A also shows several faint structures which are symmetric about the jet axis, and which must be must be related to the impending "shock" in the bulk flow at knot A (Figure 3). There is a faint "bar" of emission centered on the jet axis and very nearly perpendicular to it. One may speculate that this is a shock in the high-speed flow at the jet center, and appears relatively faint due to beaming effects. There are also faint "fans" of emission that start at the jet edge adjacent to the bar, and then trail inwards to the jet axis as you go farther down the jet towards knot A. Finally there is a faint loop or "cap" of emission precisely on the jet axis before the brightest region of knot A. It seems almost certain that these features are all related to shocking of the bulk flow at knot A, and represent conical sheets seen in projection. These features have some similarity to those predicted in numerical models (*e.g.* Burns, Norman, and Clarke 1991).

Knot C, the other "shock-like" feature in the jet, also contains interesting new features. This knot is in many ways similar to knot A, both in terms of possessing a bright transverse linear feature, increased instability downstream, and transverse B-fields. The FOC image reveals linear features parallel to the jet axis, as well as a criss-cross pattern, both of which are symmetric about the jet axis.

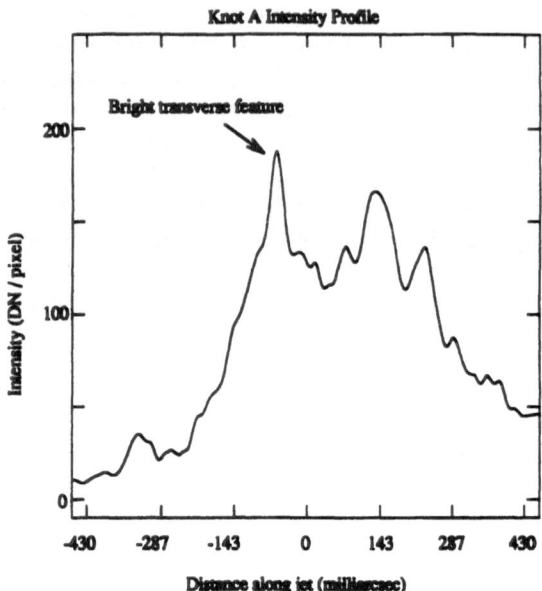

Fig. 2. Intensity profile along jet centerline through knot A. The bright, linear transverse feature is labeled. From HST / FOC $\lambda = 342\,$nm data. Resolution $0''\!.03$.

Figure 4 shows a comparison of the FOC images and VLA images taken at nearly the same epoch in 1994. At first glance the FOC and VLA images are similar. The brightest and largest of the knots (*i.e.* those labeled by de Vaucouleurs and Nieto 1979) are present in both bands and have similar internal structures (Biretta, Owen, and Hardee 1983; Biretta, Stern, and Harris 1991; Boksenberg *et al.* 1992). Closer examination begins to reveal systematic differences all along the jet. In the radio image, there is faint diffuse emission which appears to fill the jet from the nucleus to knot A, and tends to define the edges of a cone with an opening angle of about 6.5°. While traces of inter-knot emission are also present in the FOC image, it is much fainter relative to the knot emission. In addition, the optical emission in the knots is more condensed along the jet axis, as first noted by Sparks, Biretta, and Macchetto (1996) from pre-COSTAR FOC images. There are other detailed differences in knot structure, especially in and around knot E. The optical images show a bright pair of condensations in knot E (brightest two features in knot E), but these are only a relatively faint region in the radio image. Similarly there is a pair of unresolved features in the optical image between knots D and E, which again are not apparent in the radio image.

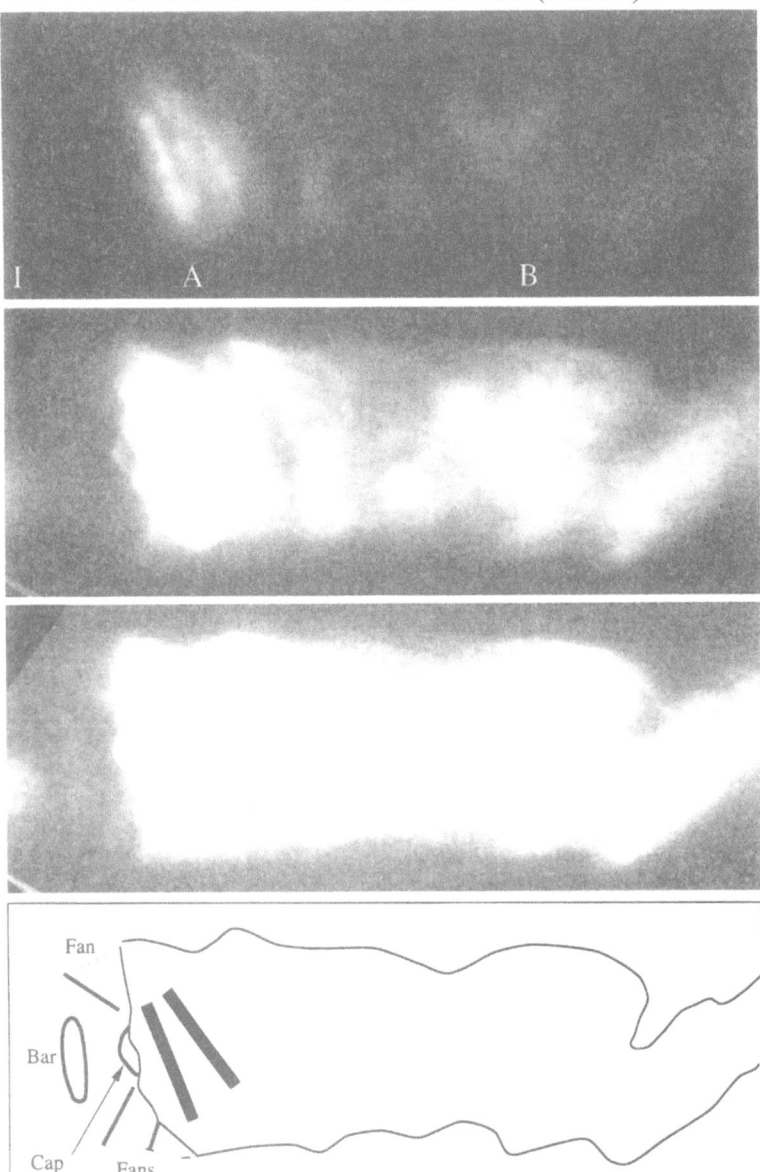

Fig. 3. HST Faint Object Camera image of knot A & B region. The three panels show the same image at different contrast settings. Newly identified features of knot A are labeled in the drawing at the bottom.

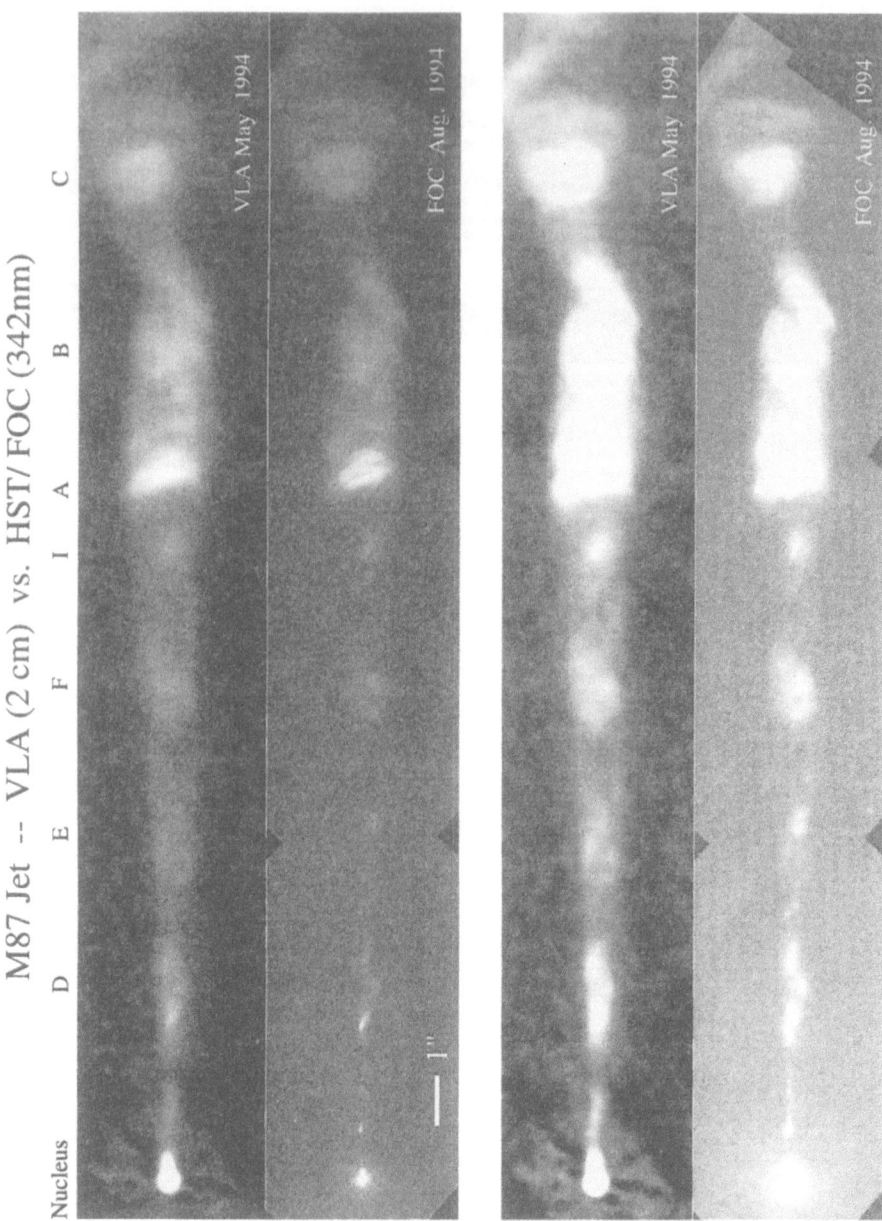

Fig. 4. Comparison of VLA and HST / FOC images of the jet from epoch 1994. The VLA image has 0.″15 resolution, while the HST/FOC image has 0.″03 resolution. The intensity scales are normalized so that the peak of knot A is equally bright in the VLA and HST displays. The top (left) panels show bright features, while the bottom (right) panels emphasize faint ones.

3   Optical Spectrum

It is also interesting to compare images within the optical band, as this delineates the locations of the highest energy synchrotron electrons. Figure 5 shows the spectral index derived between 555 nm and 814 nm WFPC2 images. Along the inner jet, the optical spectrum is flattest in the knots and near the jet axis with the values reaching -0.8 to -0.9 (defined in the sense that $S_\nu \propto \nu^\alpha$). Spectra tend to be several tenths steeper at the jet edges. This is clear evidence for radial structure within the jet, and suggests the presence of a channel containing high energy electrons on the jet axis. After the shock-like feature in knot A, this concentration on the jet axis is less apparent, and the jet becomes more uniformly filled with high-energy electrons. Between the knots of the inner jet (knots D to A), and between knots B and C, the spectra are appreciably steeper with values reaching -1.5 or beyond.

The flattest spectra appear associated with the brightest features — the eastern component of knot D and knot A. It is also interesting that while knot I seems undistinguished by either morphology or brightness, it also has a spectrum about as flat as adjacent knot A. This suggests there is already some enhancement of high-energy electrons before the bright feature in knot A, which is usually identified as the "shock." Work is underway to combine HST images at other optical and IR wavebands, so as to produce detailed spectra at all points in the jet. This will be useful in testing detailed models of the particle spectrum and B-field intensity within the jet.

The optical spectral index has been studied extensively by Meisenheimer, Röser, and Schlötelburg (1996) using ground-based images. One of their important results was that the optical spectrum was very similar throughout the jet, and this was used to argue in favor the magnetic field variations dominating the appearance of the jet. In particular, they note no significant variations perpendicular to the jet axis, and only relatively mild steepening between the knots. Our results are generally consistent with theirs, if we convolve our images to their 1.''3 resolution. However, the full resolution of HST reveals much greater spectral fluctuations as noted above. It is probably not surprising to see resolution effects, considering that the flat-spectrum regions are narrow, and that the knots are separated by only $\sim 2''$.

4   Optical Polarization

Polarimetry provides clues to the direction of the magnetic field within the jet, and hence is one of the most important diagnostics for deciphering its structure. The radio polarization has been well-studied by OHC89, and they find the field is primarily parallel to the jet axis, except in the bright transverse features in knots A and C, where the B-field runs along the transverse feature (*i.e.* normal to the jet axis). The parallel field in most of the jet can be understood as arising from shearing of a random field, while the perpendicular fields could arise from compression at shocks.

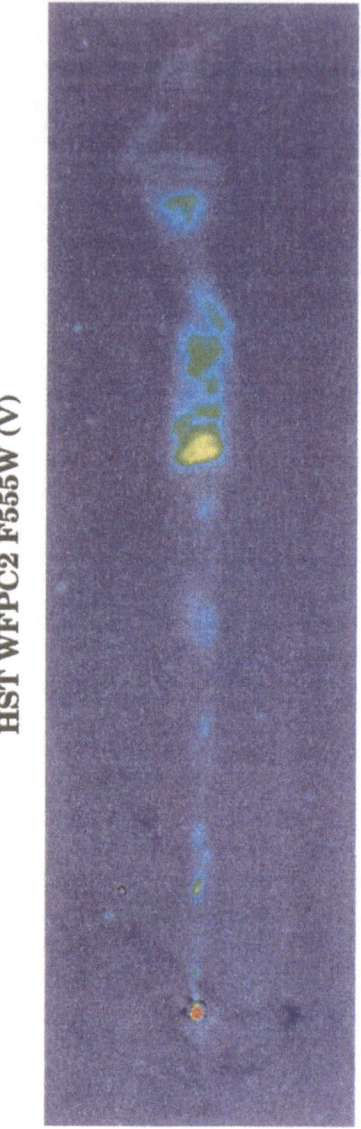

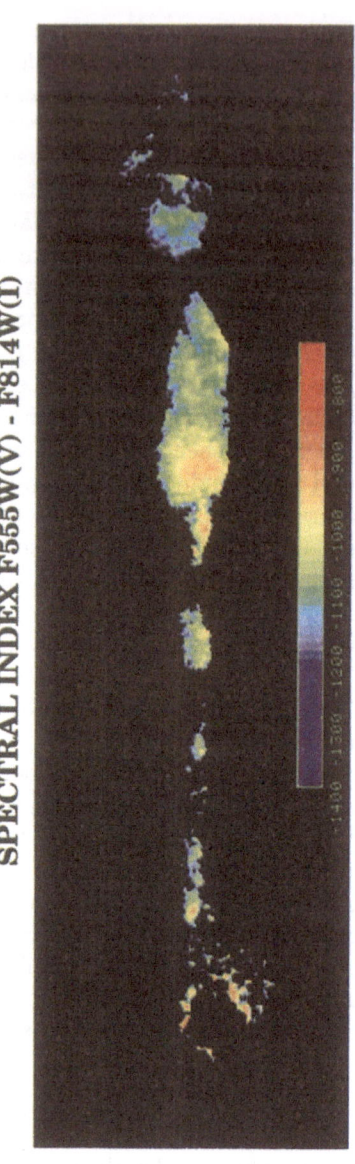

Fig. 5. Optical spectral index of the jet. Left panel shows HST / WFPC2 image of jet at $\lambda = 555$ nm; right panel shows spectral index α between $\lambda = 555$ nm and 814 nm. Inset shows color scale in units of milli-spectral index, defined in sense that $S_\nu \propto \nu^\alpha$. Dark intensity regions near nucleus are dust lanes; spectral mottling around nucleus is noise related to galaxy subtraction.

Optical polarimetry is especially interesting, because it can potentially tell us about the three-dimensional structure of the jet. As we have seen the optical emission is more concentrated on the jet axis, and hence there is an opportunity to obtain information about the magnetic structure at different depths within the jet.

Schlötelburg, Meisenheimer, and Röser (1988) have shown from ground-based data that the optical polarization is roughly consistent with that seen in the radio. Thomson *et al.* (1995) presented the first HST polarimetry using pre-COSTAR FOC data for knots A and B, and demonstrated an overall similarity with the radio results. Capetti *et al.* (1997) reported results of early WF/PC-1 polarimetry over the entire jet, and further confirmed the radio / optical similarities.

Here we present the first post-COSTAR HST polarimetry of the jet; these data are superior to previous data both in terms of sensitivity and resolution. These further confirm the overall similarity to the radio polarization, but also show interesting new features and new systematic differences from the radio.

Our observations were made using the corrective optics of WFPC2 at $\lambda = 555$ nm. The WFPC2 polarizer-quad filter was used, and the data were processed as described by Biretta and McMaster (1997). Figures 6 – 8 compare the radio and optical B-field images for adjoining regions of the jet covering the nucleus to knot B. The VLA radio image is at 15 GHz and is from epoch 1994.3, while the HST / WFPC2 optical image is at 555 nm from epoch 1995.4. Though the epochs are different, we expect only very small evolutionary changes throughout most of the jet.

In general terms our results agree with previous work. In both the radio and optical bands, the general tendency is for the field to run parallel to the jet axis, except for the bright, transverse, shock-like structures in knots A and C, where it runs perpendicular to the jet axis. Typical fractional polarizations are in the range 30% to 50%, and are comparable to those in the radio band. We do not find significant evidence for radio depolarization in the north-west region of knot A as mentioned by Thomson *et al.* (1995). Instead we find low polarization in both the radio and optical images.

Another interesting region is the "bar" located between knots I and A (Figure 3). This region appears highly polarized in the HST image (Figure 8) with the B-field running perpendicular to the jet axis, and is hence similar to the bright regions of knot A. This further supports a picture where the bar represents an initial shock or compression in the flow, which later develops forming the bright regions of knot A. The radio image is unpolarized in this region; this may be caused cancellation of the B-perpendicular emission in the bar by B-parallel emission closer to the jet surface (see discussion below of knot F and Figure 9). There may also be some blending between parallel emission in knot I and perpendicular emission in knot A.

Closer examination shows new areas with perpendicular B-fields. Both the eastern parts of optical feature HST-1 and knot D contain perpendicu-

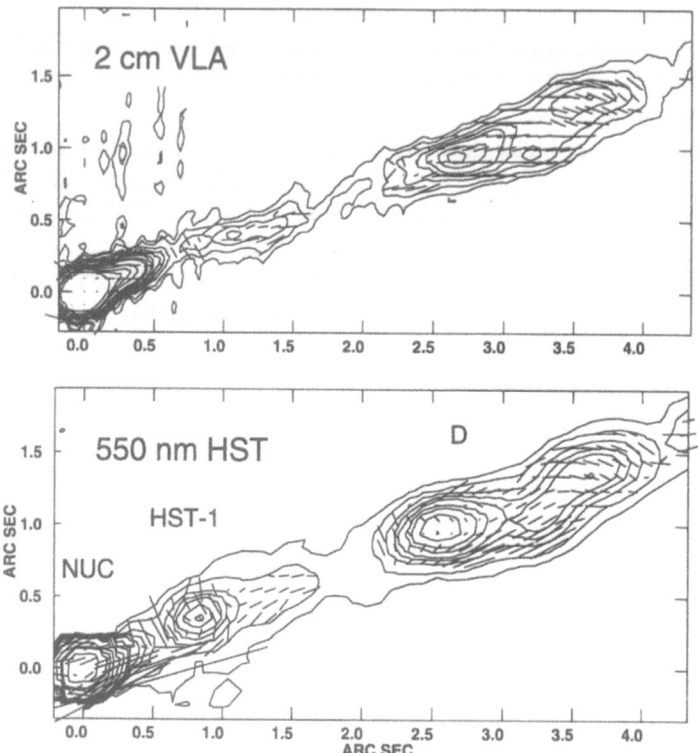

Fig. 6. Comparison of VLA and HST / WFPC2 polarization images of the first 4″ of the jet. The VLA data are from epoch 1994 while the HST data are from epoch 1995. Vectors indicate the B-field direction, and a vector length of 0.″33 corresponds to 100% polarization. The point-like nucleus has been subtracted from the VLA image.

lar B-fields, similar to knots A and C which are better resolved and show transverse shock morphology. Both HST-1 and D are similar to the knot A-B complex, in that they have bright emission with perpendicular B-fields on the upstream (eastern) side followed by fainter parallel B-field emission extending downstream (towards the west). The eastern region of HST-1 is also associated with the generation of new superluminal components (see next section).

Careful examination shows subtle radio / optical differences in knot D, and perhaps also in knots E and F. The optical image of D shows the field lines

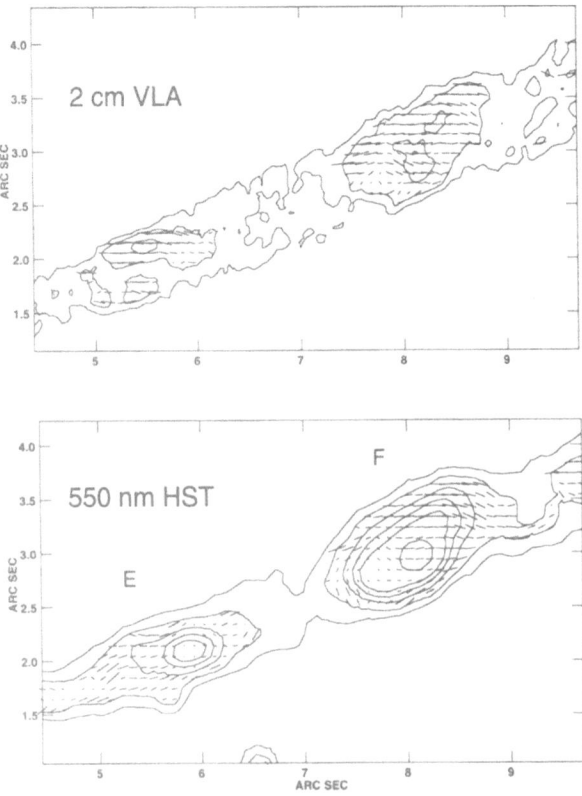

Fig. 7. Comparison of VLA and HST / WFPC2 polarization images of knots E and F. See Figure 6 for details.

turning from near-parallel to the jet axis at the knot center, to perpendicular at the upstream end of the knot. In contrast, the radio polarization remains parallel throughout this region. It is possible to interpret this in a simple picture where the B-field direction is a function of distance from the jet axis (Figure 9). As noted above, the optical emission tends to be concentrated on the jet axis, and these data show perpendicular B-fields on the upstream ends of the knot. On the other hand, the radio emission tends to be more concentrated near the jet surface, where shear against the external medium would cause parallel B-fields, thus accounting for the radio results. Such a picture explains the prevalence of parallel radio polarization, even where the optical results indicate perpendicular fields near the jst axis. This would

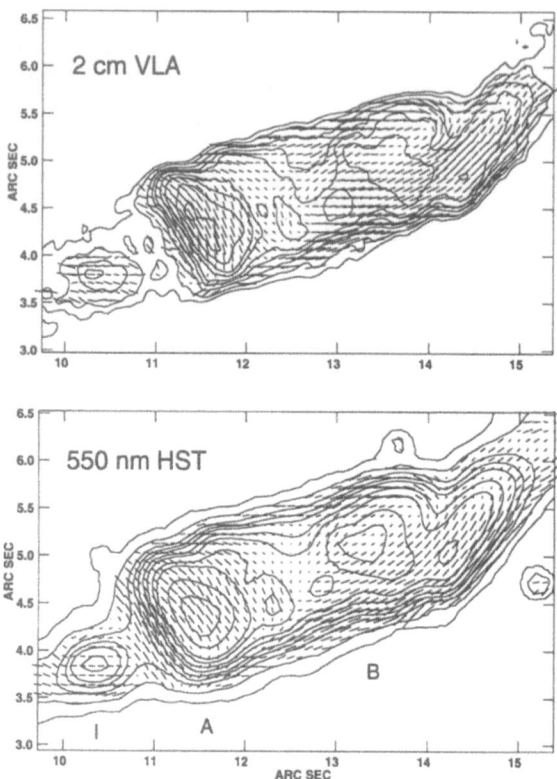

Fig. 8. Comparison of VLA and HST / WFPC2 polarization images of knots I, A, and B. See Figure 6 for details.

appear to give the first direct information about the internal B-field structure of the jet. We must caution, though, that some of the differences might also arise from structural evolution; the radio and optical epochs are not strictly identical, and knot D has shown changes on timescales as short as few years in HST and VLA monitoring (Zhou 1998). Better simultaneous observations are needed. The polarization results will be described more fully by Perlman *et al.* (1999).

Magnetic Field in M87 Jet and Knots

Jet Flow ⟶

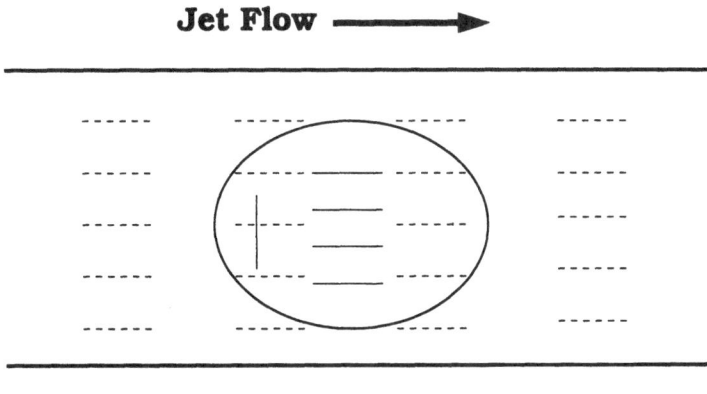

Key: ------ **Surface of jet, seen mostly in radio**

——— **Center of jet (knots), seen in optical**

Fig. 9. Possible model to explain differences between radio and optical polarization in knots D, E, and F.

5 Proper Motions and Evolution

One of the prime motivations for our program is to search for secular variations in the jet — including structural changes, flux variations, proper motions, and changes in polarization. This work can potentially give important clues to the velocity field within jets, about which very little is known from direct observation. Monitoring with the VLA at radio frequencies has detected motion in much of the jet with apparent speeds averaging near $0.5c$ (Biretta, Zhou, and Owen 1995; BZO95). The very fastest regions are in knot D and appear to move outward with speeds near $2.5c$. Due to the higher resolution of HST, as well as the greater compactness of the knots at optical frequencies, it seems likely that many more structural changes and motions might become apparent with HST. This has indeed been the case, and the results are extremely interesting.

One of the most optically active regions in the jet is the complex of compact knots about $1''$ from the nucleus which we have designated HST-1. Figure 10 shows a sequence of 342 nm FOC images from 1994 to 1996. The complex contains a bright eastern component which moves outward slowly ($0.4c$) and which appears to emit bright compact knots at speeds in the range $4c$ to $6c$ (typical uncertainty $\sim 0.5c$; Biretta, Sparks, and Macchetto 1999). The emitted components appear to fade with a half-life of about 1 year. While this image sequence is reminiscent of superluminal motion in compact radio sources, we emphasize that none of these components are the "core" in the usual sense; the nucleus is about 1″ (80 pc) away to the East. Apparently the eastern component must represent a slow-moving disturbance or shock in the flow, with the fast components being more indicative of the jet fluid speed. As seen in Figure 6, the slow eastern component has its B-field perpendicular to the jet, while the fast components are parallel.

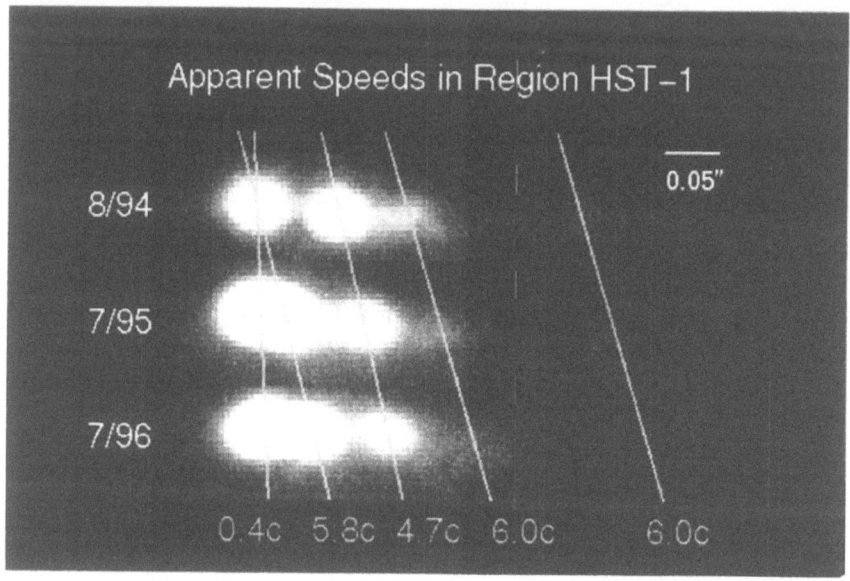

Fig. 10. Sequence of HST / FOC images of jet region HST-1 which is located about 1″ from the nucleus. The images are observed at $\lambda = 342$ nm at epochs Aug. 1994, July 1995, and July 1996. The slanting lines attempt to connect features between epochs, with apparent speeds given in units in c, the velocity of light.

Similar speeds are seen in knot D (Figure 11). The eastern-most region of knot D (labeled DE, about 200 pc from the nucleus) moves outward at 5.0c. Regions DM and DW are somewhat slower, with speeds of 4c and 2.6c, respectively. We also note rapid fading (half-life < 1 year) in features FOC3 and FOC4, which are located just east of knot D. Table 1 summarizes our preliminary speeds for several optical features in the jet. Measurement of speeds beyond knot D using the FOC is much more challenging, since images from multiple pointings must be mosaiced with extremely high accuracy; these analyses are still in progress.

The large speeds have a number of interesting implications within the context of the standard relativistic beaming model (Blandford and Königl 1979). The largest speeds give the strongest constraints, and here an observed

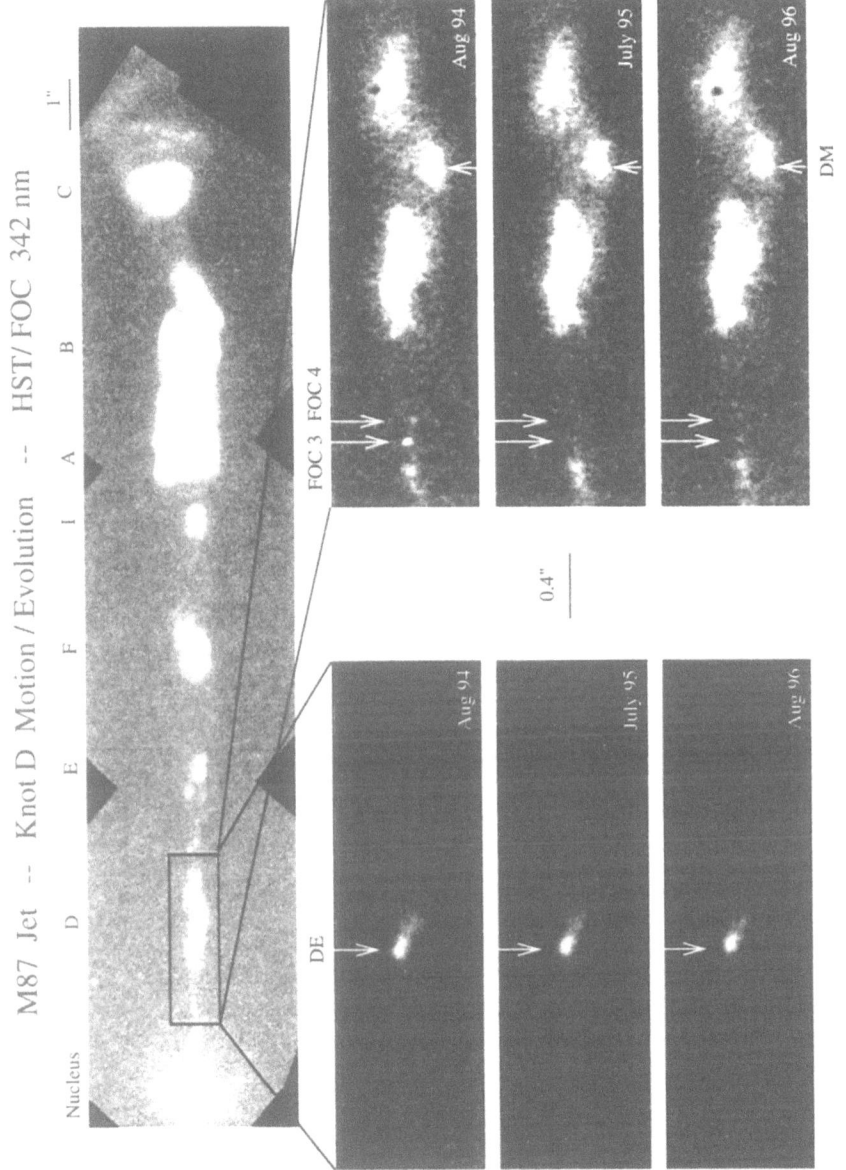

Fig. 11. Knot D region at 3 epochs. Top panel shows entire jet to provide orientation; lower panels show knot D at three epochs with two different contrast settings. Arrows are at fixed locations relative to galaxy nucleus, and can be used to judge motions. Features DE and DM (labeled) move outward at 4c to 5c, while FOC3 and FOC4 fade from view. Dark spots are instrumental Reseau marks on photocathode.

Table 1. Preliminary apparent speeds for optical jet features.

Region	Core Distance (pc)	Speed (v/c)
L	8	2.5
HST-1	80	0.4, 4.7, 5.8, 6, 6
FOC-1	120	2.1
D East	200	5.0
D Middle	240	4
D West	290	2.6

speed of $6c$ implies motion with a Lorentz factor $\gamma > 6$ and angle between the velocity vector and line of sight $\theta < 19°$. These two values represent extrema in the allowed parameter space where the other parameter (θ in the former case, and γ in the latter case) is most favorable. An example of a more general parameter set would be $\theta = 18°$ and $\gamma = 12$. These Lorentz factors are quite consistent with those predicted by unified models for FR-I sources such as M 87 (*e.g.* Urry, Padovani, and Stickel 1991).

The constraint on θ, however, is much smaller than previous estimates for the angle between the jet axis and line-of-sight for the M 87 jet. BZO95 estimated a line-of-sight angle of $42° \pm 5°$ based on the speeds and morphology of knot A, as well as the observed speeds in knot D. Ford *et al.* (1994) obtained similar estimates based on deprojection of the circumnuclear gas disk. In order to reconcile $\theta < 19°$ with these other estimates, and keep the velocity vector along the jet axis, the shock-like linear feature in knot A would have to be a highly oblique shock, and the axes of the gas disk and jet would have to be misaligned by approximately $20°$. Such a misalignment with the gas disk axis would not be highly unusual (Ford *et al.*, this volume). An alternative is that the velocity vector is misaligned with the jet axis by approximately $20°$. Such a situation could arise if the visible features moved on a twisted helical path, and there is some evidence for such structure in the morphology of knot D (see also OHC89; Conway and Wrobel 1995).

6   Summary

We have presented preliminary results of our HST study and monitoring of the M 87 jet. While there is overall similarity between the radio and optical jets, there are also clear systematic differences. The optical emission tends to be more concentrated on the jet axis than the radio emission, and the optical jet shows many more compact features. In addition, knots A and C show new structural details, many of which appear likely to be sheets of emission seen in projection. Further evidence for radial structure — that is, properties which depend on the distance from the jet axis — is seen in the optical spectral indices. The spectrum within the optical band is flattest on the jet axis in the knots, and steepens toward the jet edges and away from the knot centers. The magnetic field is generally aligned along the jet, except in the bright transverse features in knots A and C, and at the upstream ends of HST-1 and knot D, where the field runs perpendicular to the jet axis. There is evidence for radio / optical polarization differences in knot E and in the knot A "bar" region which may again be attributed to radial structure in the jet. The first $\sim 200\,\mathrm{pc}$ of the jet, and especially region HST-1, are found to be highly active. New superluminal features emerge, move with speeds up to $\sim 6c$, and fade from view with half-lives of 1 to 2 years.

We plan to continue our annual monitoring campaign, and add NICMOS data during 1998. The IR data should help constrain the spectral break-frequency, and show how it evolves as components brighten and fade in the jet. Work is underway on FOC 342 nm polarimetry images, which will double the resolution and show how the magnetic fields change as new components are born and fade. We are also looking for proper motions in the outer jet, though it is more technically challenging due to the larger distance from the position reference point (*i.e.* the nucleus). Eventually we hope to map the velocity field of the entire jet, and use this to study jet propagation and collimation, as well as kinematic properties of the luminous shocks within the jet.

References

Biretta, J.A, Owen, F.N., Hardee, P.E., 1983, ApJ, 274, L27

Biretta, J.A., Stern, C.P., Harris, D.E., 1991, AJ, 101, 1632

Biretta, J.A., Meisenheimer, K., 1993, *Jets in Extragalactic Radio Sources*, eds. H.-J. Röser, K. Meisenheimer, (Springer Heidelberg), 159

Biretta, J.A., 1993; *Astrophysical Jets*, eds. D. Burgarella, M. Livio, C. P. O'Dea, (Cambridge), 263

Biretta, J.A., Zhou, F., Owen, F.N., 1995, ApJ, 447, 582

Biretta, J.A., McMaster, M., 1997, WFPC2 ISR 97-11

Biretta, J.A., Sparks, W.B., and Macchetto, F., 1999, ApJ, in press

Blandford, R.D., Königl, A., 1979, ApJ, 232, 34

Boksenberg, A., *et al.*, 1992, A&A, 261, 393

Burns, J.O., Norman, M.L., Clarke, D.A. 1991, Science, 253, 522

Capetti, A., Macchetto, F., Sparks, W.B., Biretta, J.A., 1997, A&A, 317, 637

Conway, J.E., Wrobel, J.M., 1995, ApJ, 439, 98

de Vaucouleurs, G., Nieto, J.L., 1979, ApJ, 231, 364

Ford, H.C. *et al.*, 1994, ApJ, 435, L27

Meisenheimer, K., Röser, H.-J., Schlötelburg, M.,1996, A&A, 307, 61

Owen, F.N., Hardee, P.E., Cornwell, T.J., 1989, ApJ, 340, 698

Perlman, E., Biretta, J.A., Sparks, W.B., Macchetto, F., 1999, AJ, 117, 2185

Schlötelburg, M., Meisenheimer, K., Röser, H.-J., 1988, A&A, 202, L23

Sparks, W.B., Biretta, J.A., Macchetto, F., 1996, ApJ, 473, 254

Thomson, R.C., *et al.*, 1995, MNRAS, 275, 921
Tonry, J.L., 1991, ApJ, 373, L1
Urry, C.M., Padovani, P., Stickel, M., 1991, ApJ, 382, 501
Zhou, F., 1998, Ph.D. thesis

The Synchrotron Emission from the M 87 Jet

Sebastian Heinz[1,2], Mitchell C. Begelman[1,2]

[1] JILA
University of Colorado and National Institute of Standards and Technology
Campus Box 440
Boulder, CO 80309-0440
[2] Department of Astrophysical and Planetary Sciences
University of Colorado
Campus Box 391
Boulder, CO 80309
USA

Abstract. We show that the intensity changes and the spectral evolution along the M 87 jet are consistent with adiabatic changes to the particle momentum distribution function and the magnetic field. We use the deprojection of a 2 cm VLA intensity map by Sparks, Biretta, & Macchetto and spectral data by Neumann, Meisenheimer, & Röser to derive limits on the magnetic field and the total pressure. To explain the weakness of synchrotron cooling along the jet, the magnetic field strength must lie in the range 20 – 40 μG, departing from equipartition by a factor $\approx$ 1.5 – 5. Bulk Lorentz factors Γ_{jet} in the range 2 – 5 and inclination angles $\theta_{\mathrm{LOS}} \lesssim 25°$ lead to pressure estimates close to the ISM pressure. Extrapolation of our model back towards the core shows that particle acceleration to the observed spectral shape likely occurs at distances as far as 10 pc away from the core.

1 Introduction

The spectrum of the M 87 jet has been discussed at various other places in these proceedings already. Let us recapitulate the most striking features: the synchrotron emission of the jet follows a power-law spectrum of index $\alpha_R \approx 0.5$ in the radio, connecting to the optical data with a power law index $\alpha_{RO} \approx 0.65$, and steepening to α_O between 1.2 and 1.8 in the optical. This spectral shape seems to be conserved along the jet, both in the knots and the interknot regions. The radio-to-optical spectral index α_{RO} is very nearly uniform, while the optical spectral index α_O is anti-correlated with the intensity, *i.e.*, the optical spectrum is flatter in regions of higher intensity. The spectrum has been fitted by a power-law with a steep cutoff at high frequencies (Meisenheimer, Röser, & Schlötelburg 1996, Neumann, Meisenheimer, & Röser 1998, hereafter NMR). The cutoff frequency ν_c varies only weakly with distance from the core r, which is inconsistent with the strong decline in ν_c that would be expected if synchrotron cooling were present in a non-relativistic jet with equipartition B-field — after traveling a length of 2 kpc in a 300 μG (equipartition) field, the plasma would have cooled to exhibit a cutoff at $\nu_c \approx 10^{12}$ Hz, in contrast to the observed $\nu_c \approx 10^{15}$ Hz.

Several solutions to this dilemma have been proposed:

1. First-order Fermi re-acceleration in the knots. In this scenario, fine tuning is needed to produce the observed constant α_{RO}. Also, the lack of cooling between the knots is not explained by this model.
2. Particle transport in a loss-free channel in the interior. Radio brightness profiles across the jet suggest a limb-brightened emission, but a reinvestigation of the HST observations (Sparks, Biretta, & Macchetto 1996, hereafter SBM) shows that the optical emission is more concentrated to the inner regions of the jet.
3. On–the–spot reacceleration by a yet unknown process has the advantage of explaining all the observed features, but it invokes unknown physics to explain the apparent lack of cooling.

Here we propose a fourth way of explaining the observations by giving up the assumption of equipartition and taking the jet to be relativistic , with bulk Lorentz factors of $\Gamma_{\rm jet} \gtrsim 2-3$ and viewing angles $\theta_{\rm LOS} \lesssim 25°$. The small scale fluctuations of the cutoff frequency are produced by adiabatic compressions of the plasma in the knots, which are assumed to be weak shocks.

2 The Model

Our model rests on the hypothesis that Fermi acceleration is unnecessary to explain the fluctuations of the emissivity and the cutoff frequency along the jet. Given that small scale turbulence keeps the plasma isotropic in the fluid rest frame we can treat the compressions of the plasma in the knots (interpreted as weak, oblique shocks , Bicknell & Begelman 1996) as *adiabatic* . These adiabatic effects are readily combined with the effects of synchrotron cooling (Coleman & Bicknell 1988) to calculate the evolution of the particle momentum distribution along the jet, which can then be related to the observed spectrum and emissivity. For simplicity we assume a steady state injection of plasma by the central engine.

Since the physical basis of equipartition is weak, we consider the magnetic field strength B to be a free parameter. Depending on the orientation and the degree of disorder of the magnetic field, it will depend on the density to some power ζ: $B(r) \propto [\varrho(r)]^\zeta$, where ϱ is the proper particle density and r is the distance from the core. For a completely disordered magnetic field, $\zeta = \frac{2}{3}$, whereas for a homogeneous field the power depends on the orientation of the field with respect to the compression normal, with $\zeta = 1$ for an orthogonal orientation, $\zeta = 0$ for parallel orientation. We will take ζ to be of order $\frac{2}{3}$ throughout the rest of this paper, its actual value does not have a significant impact on the results of our calculations, however.

Because the shocks are believed to be weak and oblique , we take the fluid velocity to be constant, both in magnitude and direction (we will denote the bulk Lorentz factor of the fluid with $\Gamma_{\rm jet}$). For highly oblique shocks, the velocity component perpendicular to the shock front $v_\perp$ is small compared to the parallel component $v_\parallel$ and the velocity will not change significantly. Moderate changes in velocity would be easy to incorporate in principle; yet with our current ignorance of the velocity field and the shock parameters, such a level of detail is unwarranted. This is clearly one of the most serious simplifications of our model and will be removed in future studies. We further assume the knots to be moving with a constant velocity of $0.5\,c$.

With these assumptions in place we can use a volume emissivity map (calculated by deconvolving a 2 cm VLA radio image) provided by SBM to determine the magnetic field as a function of r up to a normalization constant B_0. Integrating the particle transport equation (*e.g.* Coleman & Bicknell 1988) we can then calculate the evolution of a particle of given injection momentum p_0 at some injection radius r_0 as a function of r, which allows us to predict the frequency evolution of spectral features, such as cutoffs or breaks , along the jet. Assuming a power-law particle momentum distribution with a spectral index of $\alpha = 0.5$ in the radio and a cutoff at optical frequencies (see NMR) we can compare our predictions of the cutoff frequency to the data by NMR to determine the two free parameters, B_0 and p_0 (for a more detailed description of the procedure we used see Heinz & Begelman 1997).

Figure 1 shows the observed cutoff frequency $\nu_{c,obs}$ (vertical bars) and the best fit curve (solid line) for a typical parameter set ($\Gamma_{jet} = 3, \theta_{LOS} = 25°$), which reproduces the scaling of the cutoff frequency reasonably well. The plot also shows the best fit cutoff frequency for an equipartition B-field , $\Gamma_{jet} = 1.1$ and $\theta_{LOS} = 25°$ (dashed line), which is clearly inconsistent with the data.

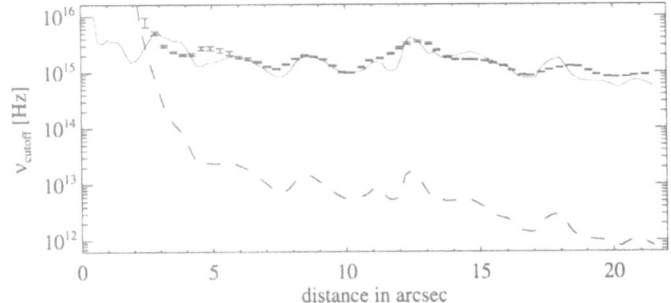

Fig. 1. Measured cutoff frequency along the jet from NMR (vertical lines with error bars). Solid line: best fit curve, calculated from the emissivity by SBM for $\Gamma_{jet} = 3$ and $\theta_{LOS} = 25°$; dashed line: best fit curve for equipartition B-fields for $\Gamma_{jet} = 1.1$ and $\theta_{LOS} = 90°$.

3   Results

Figure 2a (left) shows the best fit B-field averaged along the jet as a function of Γ_{jet} and θ_{LOS}. The values fall into the range $B \approx 20 - 40\,\mu G$ and are below equipartition by a factor of $\approx 1.5 - 5$. The impact a change in ζ or α_R has on the best fit parameters is much smaller than the formal uncertainty, which makes us confident that our assumption of constant ζ is not critical. A change in Γ_{jet} at knot A that satisfies the shock jump conditions produces a change in B of order 20%, also relatively modest.

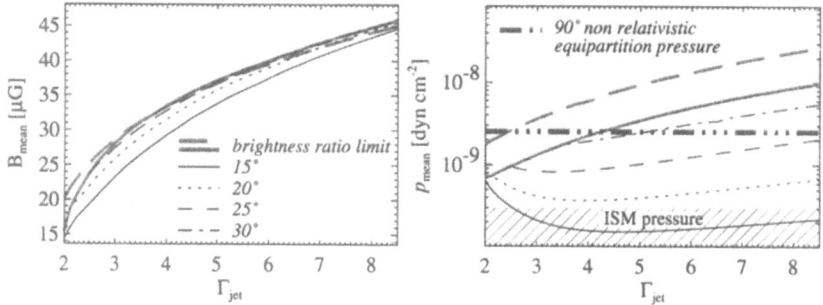

Fig. 2. (a) *Left:* B-field averaged along the jet between $0\rlap{.}''5$ and $22''$ as a function of Γ_{jet} for $\theta_{LOS} = 15°$ (solid line), $\theta_{LOS} = 20°$ (dotted line), $\theta_{LOS} = 25°$ (dashed line), and $\theta_{LOS} = 30°$ (dashed-dotted). The grey region indicates the jet-to-counterjet brightness ratio limit of 150 (dashed grey boundary to the left) – 380 (solid grey boundary to the right, Biretta 1993). (b) *Right:* Total pressure p_{mean} averaged along the jet as a function of Γ_{jet} for the same values of θ_{LOS}. For comparison, the dashed-triple-dotted line shows the equipartition pressure for a non-relativistic jet seen edge on (*i.e.* $\theta_{LOS} = 90°$). The hatched area shows the estimated ISM pressure (White & Sarazin 1988) in M 87. Labels according to (a) (*left*).

Figure 2b (right) shows the total (*i.e.*, particle + magnetic) pressure averaged along the jet corresponding to the best fit B-field from above. We assumed energy cutoffs at 10^7 and 10^{15} Hz and a spectral index of $\alpha = 0.5$ in between (to be consistent with SBM). The cutoff energies are essentially irrelevant in this case. The pressure lies between 2×10^{-10} and 10^{-8} dyn cm^{-2} for our parameter range and is out of equipartition by a factor of $0.8 - 3$. This has to be compared to the ISM pressure of $\approx 3 \times 10^{-10}$ dyn cm^{-2} (White & Sarazin 1988). Since the jet might well be embedded in an overpressured cocoon (Bicknell & Begelman 1996), it can easily be in pressure equilibrium with its surroundings.

Having an estimate of both $B(r_0)$ and p_0 we can try to determine where the actual particle acceleration to the observed power-law takes place. We assume that $B \propto r^{-\sigma}$. By taking $B \propto \varrho^{2/3}$ we can extrapolate backwards from r_0 to the distance r_{acc} at which the cutoff momentum reaches infinity, *i.e.* the minimum radius inside of which acceleration has to occur. In Fig. 3 we have plotted r_{acc} as a function of σ for $\theta_{LOS} = 25°$ and $\Gamma_{jet} = 3$. In the same figure we have plotted r_{acc} for the case in which B no longer scales like $\varrho^{2/3}$ — in this case we have taken $\varrho \propto r^{-2}$ (dashed line). The r_{acc}–curve is rather flat throughout most of the possible range for σ, which suggests that the most plausible value for r_{acc} falls between 1 and 10 pc. This is intriguingly far away from the central engine.

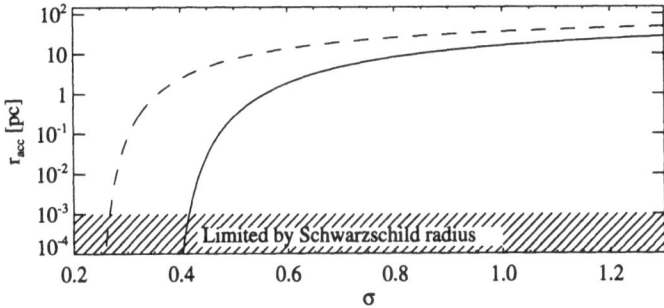

Fig. 3. Minimum acceleration radius r_{acc}, as a function of σ ($B \propto r^{-\sigma}$) for $\Gamma_{jet} = 3$ and $\theta_{LOS} = 25°$ with $B \propto \varrho^{2/3}$ (solid line). Dashed line: r_{acc} for the case of $\varrho \propto r^{-2}$. The hatched area indicates the limit set by 10 Schwarzschild radii for the $\approx 10^9$ M$_\odot$ central black hole.

4   Conclusion

We have shown that adiabatic compressions, coupled with a relativistic jet and sub-equipartition fields, can explain both the small-scale variations in brightness and cutoff frequency and the lack of a large-scale cooling trend, without requiring a pressure far in excess of the ISM pressure. The particle acceleration to relativistic energies likely operates out to distances $\approx 1-10$ pc from the core.

This work has been supported by NSF grants AST-95 29175 and AST-91 20599 and a Fulbright UP95/96/97 grant.

References

Bicknell, G.V., & Begelman, M.C. 1996, ApJ, 467, 597 (BB)
Biretta, J.A. 1993, in Astrophysical Jets. Space Telescope Science Institute Symposium Series Vol.6, D. Burgarella, M. Livio, & C. O'Dea (eds.) (Cambridge: Cambridge University Press), 263

Coleman, C.S., & Bicknell, G.V. 1988, MNRAS, 230, 497
Heinz, S., & Begelman, M.C. 1997, ApJ, 490, 653
Meisenheimer, K., Röser, H.-J. & Schlötelburg, M. 1996, A&A, 307, 61
Neumann, M., Meisenheimer, K., & Röser, H.-J. 1998, A&A, in preparation (NMR)
Sparks, W., Biretta, J.A., & Macchetto, F. 1996, ApJ, 473, 254 (SBM)
White, R.E., & Sarazin, C.L. 1988, ApJ, 335, 688

Theory of the M 87 Jet

Geoffrey V. Bicknell[1], Mitchell C. Begelman[2]

[1] Australian National University Astrophysical Theory Centre**.
[2] JILA
University of Colorado
Boulder, CO 80309
USA

Abstract. We summarize theoretical ideas on the dynamical properties of the jet in M 87 and its relationship to the inner lobe and adjacent emission line filaments. Shocks in one form or another and the Kelvin Helmholtz instability have long been held to be related to the emission from the M 87 jet and its shape respectively. We give a brief account of recent work which attempts to relate these two pieces of physics. We note that it is important to take relativistic effects on the appearance of shocks into account when attempting to determine the intrinsic obliquity of shocks. Our model naturally leads to consideration of the interaction of the jet-driven expanding inner lobes with surrounding dense gas. We have suggested that radiative shocks driven into clouds (no. density $\sim 1 - 10 \, \mathrm{cm}^{-3}$) are responsible for line emission from this gas. Outstanding problems are the confinement of the M 87 jet in the inner 100 pc and its initial approximately linear expansion.

1 Introduction

1.1 Crucial Points to Address in a Theory of the M 87 Jet

A comprehensive theory of the observational data on the M 87 jet, summarized in these proceedings by John Biretta, Frazer Owen and Klaus Meisenheimer, needs to address the following points:

- Jet sidedness — relation to the $V \sim 0.5c$ velocities inferred from proper motions
- The nature of the knots and their relationship to the twisting of the jet
- The apparent overpressure of the jet with respect to the ambient medium
- The relationship between the jet/inner lobe and the emission line filaments
- The initial constant expansion
- The synchrotron spectrum and the close similarity between radio (VLA) and optical (HST) images

Most of these points are addressed in this review and are mainly based on Bicknell & Begelman (1996). The synchrotron spectrum of the jet is addressed in Heinz & Begelman (1997) (see also Heinz & Begelman, this volume).

** The ANUATC is operated jointly by Mt Stromlo & Siding Spring Observatories and the School of Mathematical Sciences of the ANU.

1.2   Previous Models and Ideas

Many of the previous models of the M 87 jet have involved the idea that shocks are sites of particle acceleration and that these are responsible for the bright knots:

- Rees (1978) proposed that internal jet shocks result from variations in the flow velocity. This is still a feasible proposition for at least some of the knots since the M 87 jet is observed to vary significantly on many time scales (see Biretta, Tsvetanov, these proceedings)
- Blandford & König1 (1979) suggested that the shocks are bow shocks resulting from the interaction between the jet and interstellar clouds.
- Coleman and Bicknell (1985, 1988) extended the Blandford and Konig1 model to explain the large break in the synchrotron spectrum as due to cooling in an inhomogeneous post-shock region.

In addition to these shock models Biretta (1993) analyzed the M 87 jet data to estimate a bulk Lorentz factor, $\Gamma \gtrsim 3$ and an inclination to the line of sight $\theta \approx 40°$. Biretta also emphasized the importance of relativistic effects in interpreting the edge-on appearance of knot A. The papers of Hardee (1987a), Hardee (1987b) and Birkinshaw (1984), Birkinshaw (1991) have laid the groundwork for analyzing jet structure in terms of the relativistic Kelvin-Helmholtz instability. The paper by Hardee (1982) was the first to attempt to analyze existing M 87 jet images in terms in terms of the KH instability.

2   Theoretical Developments

2.1   Shock Proper Motion and Orientation

In 1994, we were motivated by the outstanding observational work of Biretta et al. (1995) who on the basis of 11 years radio observations of the M 87 jet, showed that the proper motions of brightest knots translate to "only" $0.5c$. Such a velocity is not fast enough to explain the brightness asymmetry between jet and assumed counterjet so that we either have to abandon this linchpin of relativistic jet physics or, keeping to the interpretation of the knots as shocks, address the implications of the difference of shock and underlying jet velocities. In our paper we adopted the latter view and proposed that the knots in the M 87 jet are oblique shocks induced by the helical body modes of the Kelvin-Helmholtz instability.

One of the more interesting effects associated with relativistically moving shocks is the apparent rotation of the shock front due to time retardation effects. This is depicted in Fig. 1. This rotation can be mathematically represented in the following way: Denoting the intrinsic (unit) normal by $\mathbf{n}$, the apparent (non-unit) normal by $\mathbf{m}$, the shock pattern speed divided by the speed of light by β_p, the component of this in the direction of the normal to the shock by β_n, the angle between the shock and observer directions by θ, and the unit vector in the direction of the observer by $\mathbf{k}_{obs}$, we have:

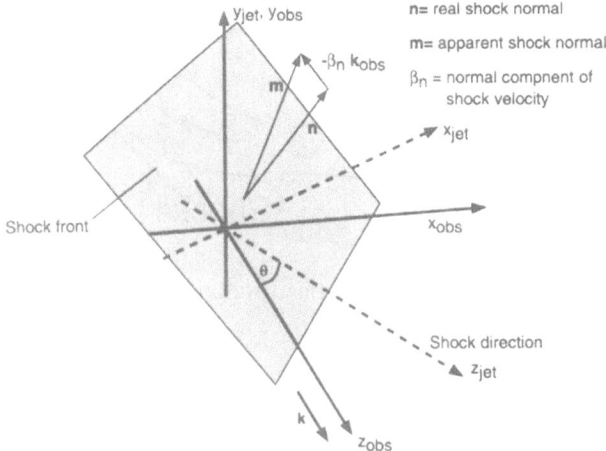

Fig. 1. The apparent rotation of the normal to a relativistic shock front resulting from time retardation

$$\mathbf{m} = \mathbf{n} - \beta_n \, \mathbf{k}_{\mathrm{obs}} \tag{1}$$

In a frame in which the z_{obs}-axis points towards the observer and the z_{jet}-axis points in the direction of motion of the shock, we have,

$$m_{z,obs} = -\sin\theta n_{x,\mathrm{jet}} + (\cos\theta - \beta_p)n_{z,\mathrm{jet}} \tag{2}$$

where the x_{jet}-axis is in the plane of z_{obs} and z_{jet} as shown in the figure. This particular equation shows why a relativistic shock tends to be viewed edge-on. When a relativistic jet is viewed at a small angle to the line of sight, $\sin\theta \to 0$ and $\cos\theta - \beta_p \to 0$ so that $m_z \to 0$, no matter what the intrinsic orientation of the normal.

Inverting the shock orientation relations, an apparent velocity of $\beta_a = 0.51$ for knot A and a jet inclination of $\theta = 30°$ implies an angle between the shock normal and the shock velocity of $\psi = 36°$; if $\theta = 40°$ then $\psi = 28°$. That is the apparently edge-on "shock" at knot A is oblique. Interestingly, the obliquity is not as marked as it would be if the shock were non-relativistic; for example, $(\beta_a, \theta) = (0, 30°) \Rightarrow \psi = 62°$. However, the shock *is* oblique and this has important ramifications for the dynamical properties of the shock which are discussed below.

2.2   Production of Shocks via the Kelvin-Helmholtz Instability

The helical Kelvin-Helmholtz instability twists a jet in different directions and we have suggested that the deflection of a supersonic flow results in shocks. Another feature which can assist the formation of shocks is the related distortion of the jet resulting from $m = 2$ modes (Massaglia *priv.comm.*).

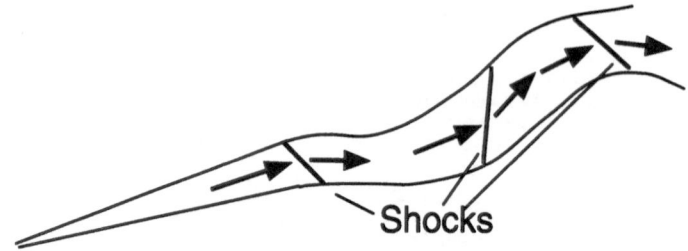

Fig. 2. Schematic of shock formation resulting from the KH instability

These grow at about the same rate as the $m = 1$ modes and indeed there is morphological evidence for a significant $m = 2$ component beyond knot A. To determine the growth rate of the $m = 1$ mode, we used the formulation of the dispersion equation due to Birkinshaw (1984) but with a relativistic equation of state of the form $w = \rho c^2 + 4P$ where w is the relativistic enthalpy, ρ is the mass density of cold (subrelativistic) matter and P is the pressure of particles, assumed to be dominated by the relativistic gas. The proportion of cold matter is parameterized by $\chi = \rho c^2 / 4P$. For example, in a jet with relativistic electron/positron pairs $\chi \approx 0$; if the electrons are relativistic and charge neutralized by cold protons $\chi \approx 130/\gamma_0$ where γ_0 is the lower cutoff in the electron distribution. The parameters defining the KH instability are the bulk Lorentz factor, Γ, the value of χ_{jet} and χ_{a}, the ambient value of χ.

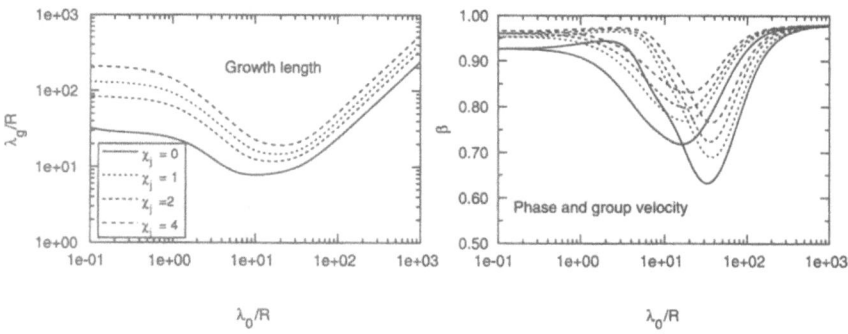

Fig. 3. Left panel: Growth wavelengths, λ_g, of the helical KH instability as a function of the wavelength, λ_0, for a $\Gamma = 5$ jet with $\chi_{\text{a}} = \rho_a c^2 / 4P = 10$. The different curves correspond to different values of χ_j. Right panel: Phase and group velocity for the same parameters.

Typical results for the dispersion relation for a $\Gamma = 5$ jet are shown in Fig. 3. Note that all curves show a pronounced minimum in the growth length, λ_g, corresponding to the maximally growing wave, at a specific value of the perturbation wavelength, λ_0. The minimum shifts to higher λ_g and λ_0 as either χ_{jet} or χ_a increases, although we only show the curves here specific to $\chi_a = 10$. A low χ_a environment makes a jet extremely unstable. (For an indication of this see below.) Since one expects the cocoon of waste products surrounding a jet to be of lower density than the jet itself this raises a problem. However, this can be resolved by the fact that mass injection by stars into the bubble can account for at least some of the required density. The rest can probably be accounted for by the incorporation of mass into the bubble from the surrounding cool filaments as they are swept up by the expanding inner lobe.

The jet and ambient densities also affect the velocity of propagation of the instability which determines the velocity of the resulting knot/shock wave. The right panel of Fig. 3 shows the phase and group velocities of the KH instability for $\Gamma = 5$ and $\chi_a = 10$, again with the curves corresponding to different values of χ_j. The well-defined minimum in the phase velocities corresponds to the maximum growth rate. Note that the group velocity and phase velocity have the same value at the minimum of the phase velocity. A shock that is produced by deflections in the jet resulting from the KH instability will move at the corresponding group velocity. The ambient value of χ needs to be quite high ($\sim 10 - 100$) in order to reproduce the velocities implied by the motions of the knots. More about this below.

The dual constraints of observed wavelength and proper motion of the knots are best examined via plots such as those in Fig. 4. The maximally growing wavelengths, λ_0/R and growth lengths λ_g/R are shown as functions of χ_a and λ_0/R is plotted against the corresponding group velocity. These plots are for a $\Gamma = 5$ jet. In the latter plot the range of these parameters implied by the observational data is also indicated by the oblique dashed curve. The interesting feature here is that the data imply a value of the parameter, $\chi_j \sim 1$, consistent with a jet dominated by electrons and positrons or an electron/proton jet with a fairly high low energy cutoff ($\gamma_0 \sim 100$). A similar result is obtained for all other Lorentz factors of interest.

3   Jet Dynamics

3.1   Insights from Rankine-Hugoniot Conditions

Given that the shocks in the jet are oblique, what extra insights does this give us into jet parameters? The plots (specific to knot A and also for $\Gamma = 5$) of various shock parameters in Fig. 5 tell an interesting story. (We solved the Rankine-Hugoniot shock conditions for cold matter plus relativistic plasma, utilizing the formulation due to Königl (1980).) First note (from the left hand panel) that for a moderate pressure jump, the shock inclination is of order $40° - 50°$. The lower end of this range is close to the shock inclination $\approx 36°$ inferred from the apparent edge-on appearance of knot A. (Note also that the obliquity is not well-determined.) The right hand panel of Fig. 5 shows

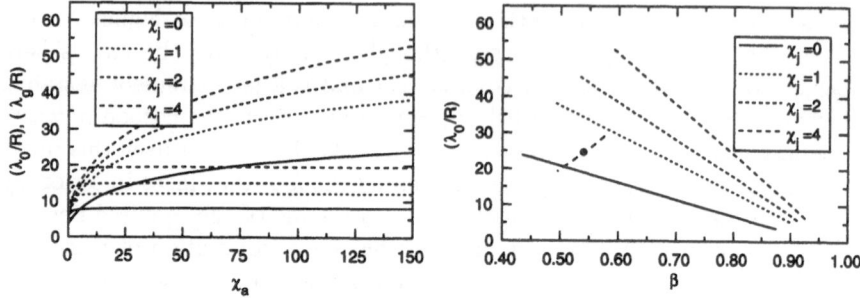

Fig. 4. Left panel: The wavelength and growth wavelength of the fastest growing mode as a function of the ambient value of $\chi = \rho c^2/4P$ for a $\Gamma = 5$ jet. Right panel: The relationship between the fastest growing wavelength and group velocity for the $\Gamma = 5$ Kelvin-Helmholtz instability. (The parameter χ_a varies along each curve.) The range of growth wavelength and shock pattern velocity implied by the data of Biretta, Zhou & Owen (1995) and $25° < \theta < 40°$ is indicated by the oblique dashed line. The filled circle represents $\theta = 30°$.

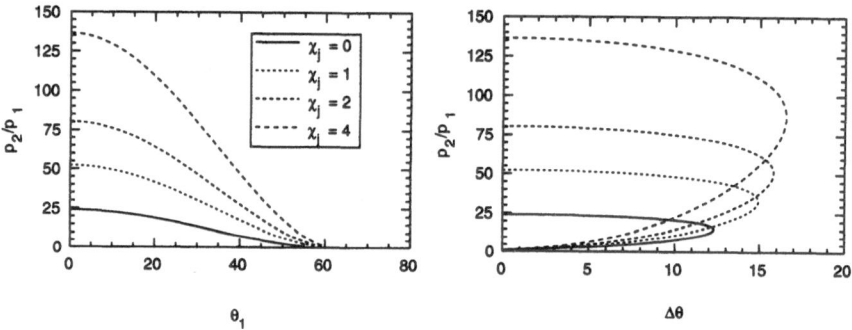

Fig. 5. Left panel: The shock pressure ratio (input parameters relevant to knot A) for a $\Gamma = 5$ jet as a function of the angle between the jet velocity and the shock normal (in degrees). Right panel: The shock pressure ratio as a function of the deflection at the shock (also in degrees).

the relationship between pressure ratio and deflection, showing that a modest pressure jump $P_2/P_1 \sim 10$ is possible for a small deflection $\sim 5° - 10°$. The observed deflection at knot A is approximately $6°$.

3.2   The Jet Energy Flux

For many reasons it is important to know the jet energy flux: This is the basic energy input to the surrounding inner lobe, which we modeled as an expanding bubble; the energy flux enters into calculations on the electron/positron or electron/proton nature of the jet (Reynolds et al. (1996)) and is also important as far as the large scale structure of M 87 is concerned, as has been discussed at this meeting. We therefore give a number of independent estimates of this parameter.

Direct calculation. The energy flux of a relativistic jet is

$$F_E = 4P_{\text{jet}}cA_{\text{jet}} \left[1 + \frac{\Gamma-1}{\Gamma}\chi_j\right] \Gamma^2\beta$$
$$\approx 1.4 \times 10^{43} \left[\frac{P_{\text{jet}}}{10^{-9}\text{dyn cm}^{-2}}\right] \left[1 + \frac{\Gamma-1}{\Gamma}\chi_j\right] \Gamma^2\beta \text{ ergs s}^{-1} \quad (3)$$

We have used fiducial parameters appropriate for knot A, where $p_{\text{min}} \approx 1.5 \times 10^{-9}$ dyn cm^{-2}. For the parameters that seem relevant for M 87, ($\Gamma \approx 3 - 5$; $\chi_j \approx 0 - 1$) this direct estimate of the energy flux easily implies a value of about 10^{44} ergs s^{-1}, even allowing for the fact that the jet velocity will be smaller than the pre-shock value in the immediate post-shock zone. The use of the minimum pressure here is not restrictive. The value used here is in fact the value that one obtains without taking relativistic effects into account. When these are incorporated in the calculation, the *minimum* pressure is less, but to be consistent with the pure hydrodynamical modeling we have employed, the actual value should be higher than this. The modeling of the spectral index variations undertaken by Heinz & Begelman (1997) (see also these proceedings) which assumes a sub-equipartition magnetic field, indicates an actual pressure possibly about a factor of 2 higher than the non-relativistic minimum pressure.

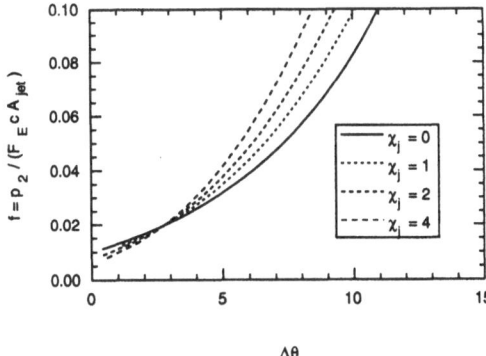

Fig. 6. The ratio of post-shock pressure to jet energy flux divided by cA_{jet} as a function of the jet deflection. Note that there is little dependence on the parameter χ. Also there is only a minor dependence on Γ.

Jet energy flux from shock dynamics. Assuming that knot A is a shock, the pressure times the area of the knot is an indirect measure of the jet momentum flux. Since, in a relativistic jet, the energy flux is approximately c times the momentum flux, then the knot pressure can be used to estimate the energy flux. Therefore, in figure 6 we show for a $\Gamma = 5$ jet, the relationship between the parameter $f = P_2/(F_E/cA_{jet})$ (calculated exactly) for a number of different values of χ_j and for $\Gamma = 5$. This relationship varies little with Γ in our favored range of Lorentz factors and, as can be seen from the diagram, little with χ_j. For a jet deflection of 6° the value of the parameter $f \approx 0.05$. Hence the jet energy flux from this method is:

$$F_E \approx f^{-1} \times P_2 \times cA_{jet}$$

$$\approx 0.05^{-1} \times \left[1.5 \times 10^{-9} \text{ dynes cm}^{-2}\right] \times c \times \left[\frac{\pi \times (140 \text{ pc})^2}{4}\right]$$

$$\approx 10^{44} \text{ ergs s}^{-1} \tag{4}$$

Lobe dynamics. The radius of the expanding inner lobe (see the following section) is consistent with a jet energy flux of approximately 10^{44} ergs s^{-1}. However, this is not as well determined as the previous two estimates since the lobe radius is relatively insensitive to the jet energy flux. On the other hand, if we consider the energy content of the lobe, a better constraint on the energy flux can be obtained. The predominantly relativistic energy content of an approximately spherical lobe of pressure P_{lobe} and volume V after a time t is given by:

$$3P_{lobe}V \approx \frac{5}{8}F_E t \Rightarrow F_E \approx \frac{8}{5} \times \frac{3P_{lobe}V}{t} = 4.9 \times 10^{43} \text{ ergs s}^{-1}, \tag{5}$$

for a radius of 1.2 kpc, $t = 10^6$ yr and $P_{lobe} = 1.5 \times 10^{-9}$ dyn cm^{-2}. This pressure is the largest minimum pressure of the filaments in the lobe. This calculation suggests an energy flux in excess of 5×10^{43} ergs s^{-1}.

4   The Dynamics of the Inner Lobe

The dynamical properties of the inner lobe fed by the M 87 jet are important for a number of reasons. The jet-driven expansion of this lobe provides a natural explanation for the juxtaposition of the emission line filaments and the inner radio lobe and also solves the long standing problem of why the jet is apparently overpressured with respect to the lobe.

4.1   The Basic Model

Modeling the lobe as a spherical bubble of radius R_{kpc} kpc, driven by a deposition of energy at a rate of $F_E = 10^{44}$ergs s^{-1} into a medium with a pressure of $10^{-10}p_{\text{ISM},-10}\,\text{dyn}\,\text{cm}^{-2}$ and a temperature of $10^7 T_7\,\text{K}$ for $10^6 t_6\,\text{yrs}$ we obtained for the radius, expansion velocity, V_{exp} and lobe pressure, P_{lobe},

$$R_{\text{kpc}} \approx 1.7 \left[\frac{F_{E,44}T_7}{P_{\text{ISM},-10}} \right]^{1/5} \tag{6}$$

$$V_{\text{exp}} \approx 600 \left[\frac{R_{\text{kpc}}}{t_6} \right] \text{km s}^{-1} \tag{7}$$

$$\frac{P_{\text{lobe}}}{P_{\text{ISM}}} \approx 3.4 T_7^{-1} \left[\frac{R_{\text{kpc}}}{t_6} \right]^2 \tag{8}$$

If we assume an age $t_6 \sim 1$ as derived by Turland (1975), and a jet energy flux $\sim$ a few $\times\, 10^{44}\,\text{ergs s}^{-1}$ then good agreement between the observed (1.2 kpc) and estimated radii is obtained for $p_{ISM,-10} \approx 5$ and $T_7 \approx 2$ as indicated by the M 87 cooling flow model of White & Sarazin (1988). The estimated age of 10^6 yrs is well worth revisiting in the light of higher resolution and more extensive radio data.

According to this bubble model the lobe is expanding at a velocity $\approx$ 750 km s^{-1} and its internal pressure $\sim 10^{-9}$ dyn cm^{-2}. Hence the internal bubble pressure is comparable to the pressures in the knots and is the natural confining pressure for them. There is no need to invoke a confining magnetic field on the kpc scale. However, as one extrapolates the pressure of the jet back to the core, using whatever model, this eventually exceeds the bubble pressure. The problem of confinement therefore becomes restricted to scales ~ 100 pc.

4.2   Excitation of Emission Lines

When an expanding bubble strikes dense clouds in the ISM, most likely originating from the cooling flow, radiative shocks are driven into them and shock-excited line emission results. As Ford & Butcher (1979) noted in their ground-breaking work on these filaments, the spectrum is typical of shock excited gas. The power, P_{exp}, supplied to the filaments (total area A_{fil}) by the expansion of the lobe, is given by

$$P_{\text{exp}} = P_{\text{ISM}} V_{\text{exp}} A_{\text{fil}} (1 + M_{\text{exp}}^2/3)$$
$$= 1.0 \times 10^{43} \left[\frac{P_{\text{ism}}}{5 \times 10^{-10}} \right] \left[\frac{A_{\text{fil}}}{8.4 \times 10^6 \text{ pc}^2} \right] \tag{9}$$

where M_{exp} the Mach number of the expansion and we have used the projected area of the filaments as a fiducial value for the total filamentary area A_{fil}. The total filamentary area could well be less than this since it is likely that the filamentary filling factor is less than unity. Nevertheless the estimated power comfortably exceeds the total emission line power $\sim 10^{42}\text{ergss}^{-1}$ estimated from the Hα luminosity $\approx 8.3 \times 10^{39}$ ergs s^{-1}.

What are the cloud densities implied by a shock excitation model? As the ISM is driven outwards by the expanding bubble, the stagnation pressure produced around each cloud $P_{st} \approx 3.5 \times 10^{-9} \mathrm{dyn\ cm^{-2}}$ and this drives a shock at velocity

$$V_{sh} \sim \left(\frac{3}{4}\right)^{1/2} \left[\frac{P_{st}}{\mu m_p n_{cl}}\right]^{1/2} \tag{10}$$

into a cloud of density n_{cl}. Hence for shock velocities $\sim 200\ \mathrm{km\ s^{-1}}$ (typical of the M 87 spectrum), the pre-shock density is given by

$$n_{cl} \approx 7 \left[\frac{P_{st}}{3.5 \times 10^{-9} \mathrm{dyn\ cm^{-2}}}\right] \left[\frac{V_{sh}}{200\ \mathrm{km\ s^{-1}}}\right]^{-2} \tag{11}$$

This is *not* the density in the recombination zone, where the density sensitive [SII] emission originates. As the gas cools it becomes denser until the increasing magnetic pressure prevents further collapse. If B_{cl} is the pre-shock magnetic field of a cloud, then the density, n_{rec}, in the recombination zone is determined by when the magnetic pressure becomes equal to the ram pressure of the shock in the shock frame, and is given by

$$n_{rec} \approx 190 \left[\frac{P_{st}}{3.5 \times 10^{-9} \mathrm{dyn\ cm^{-2}}}\right]^{3/2} \left[\frac{V_{sh}}{200\ \mathrm{km\ s^{-1}}}\right]^{-2} \left[\frac{B_{cl}}{10\ \mu G}\right]^{-1} \mathrm{cm^{-3}} \tag{12}$$

The magnetic field inferred from the substantial Faraday rotation measure across the inner lobes of M 87 (Owen et al. (1990)) $\approx 20 - 40\ \mu G$ so that the recombination density should be in the low density limit.

5   Discussion

We have outlined here the theory of the inner structure of M 87 which we proposed a couple of years ago. As we pointed out in the introduction, parts of this theory or similar ideas have been around for some time and what we have done is to attempt a synthesis of jet physics, lobe dynamics and emission line excitation in one model. As we have seen, all of these elements are potentially related, in part through the effects of relativity on the apparent obliquity of shocks. Possibly, some of the achievements of this model relate to our estimate of the parameter $\chi_j \sim 0 - 1$, our inference of an inner lobe which is denser than the jet, our solution for the confinement of the jet and the relationship of this to the emission line clouds enveloping part of the inner lobes of M 87. Much of this physics is relevant to many other radio-active AGN. Some of the observational basis for our model has been questioned at this meeting. In particular, is knot A really an edge-on shock or is it a filament? (See Owen, this volume.) The measurement of 6c proper motions (Biretta, this volume) also represents an interesting challenge. These could possibly be shocks traveling at a non-zero angle to the main jet flow. There are outstanding questions of jet dynamics: What is the reason for the approximately linear initial expansion of the jet and what confines it in the inner 100 parsec? What is the origin of the X-ray variability from the knots? (Harris, Biretta & Junor (1997); Harris, this volume.) There are also interesting links between this work and the work by Binney & Tabor (1995) (See also Binney, this volume.) Our estimate of a jet energy flux in excess of 10^{44} erg s^{-1} over a timescale of 10^6 yr suggests episodic activity, qualitatively consistent with their theory for episodic cooling flow and AGN activity. M 87, therefore, remains a fascinating laboratory for AGN physics and the theoretical issues are far from settled.

Chris Reynolds collaborated with us on § 3.2 of this paper. GVB would like to thank the organizers of this meeting and the Max Planck Society for financial support. MCB acknowledges support from NSF grant AST95-29175.

References

Bicknell, G. V. & Begelman, M. C. 1996, ApJ, 467, 597

Binney, J. & Tabor, G. 1995, MNRAS, 276, 663

Biretta, J. A. 1993, in Astrophysical Jets, ed. D. Burgarella, M. Livio, & C. P. O'Dea, Volume 6 of Space Telescope Science Institute Symposium Series (Cambridge: Cambridge University Press), 263

Biretta, J. A., Zhou, F., & Owen, F. N. 1995, ApJ, 447, 582

Birkinshaw, M. 1984, MNRAS, 208, 887

Birkinshaw, M. 1991, MNRAS, 252, 505 527

Blandford, R. D. & Königl, A. 1979, ApJ, 232, 34

Ford, H. C. & Butcher, H. 1979, ApJS, 41, 147

Hardee, P. E. 1982, ApJ, 261, 457

Hardee, P. E. 1987a, ApJ, 313, 607

Hardee, P. E. 1987b, ApJ, 318, 78

Harris, D. E., Biretta, J. A. & Junor, W., 1997, MNRAS, 284, L21

Heinz, S. & Begelman, M. C. 1997, ApJ, 490, 653

Königl, A. 1980, Phys. Fluids, 23, 1083

Owen, F. N., Eilek, J. A., & Keel, W. C. 1990, ApJ, 362, 449

Rees, M. J. 1978, September), MNRAS, 184, 61P

Reynolds, C. S., Fabian, A. C., Celotti, A., & Rees, M. J. 1996, MNRAS, 283, 873

Turland, B. D. 1975, MNRAS, 170, 281

White, R. E. & Sarazin, C. L. 1988, ApJ, 335, 688

Kelvin-Helmholtz Instabilities and Particle Acceleration in Jets

S. Massaglia, M. Micono, N. Zurlo, A. Ferrari

Dipartimento di Fisica Generale dell'Università
Via Pietro Giuria 1
I-10125 Torino
Italy

Abstract. We calculate the temporal evolution of distributions of relativistic electrons subject to synchrotron and adiabatic processes and Fermi-like acceleration in shocks. The shocks result from Kelvin-Helmholtz instabilities in the jet. Shock formation and particle acceleration are treated in a self-consistent way by means of a numerical hydrocode.

1 Introduction

Particle acceleration is a common requirement in many astrophysical contexts, such as the origin of the cosmic radiation, of the streams of relativistic particles from solar flares, and of the synchrotron emitting electrons in extragalactic radio sources and active galactic nuclei. These particles can be either directly detected, as for cosmic rays and solar streams, or their signature observed in form of non-thermal radiation. Among the physical mechanisms proposed for particle acceleration, Fermi-like processes are particularly attractive since they automatically lead to power-law particle distributions, as dictated by observations.

Astrophysical shocks are a favorite site of particle acceleration via Fermi-like mechanisms (see, *e.g.*, Bell 1978a,b; Drury 1983; Blandford and Eichler 1987; Achterberg 1990; Kirk, Melrose and Priest 1994).

Coming to extragalactic jets, HST observations in the optical band and X-ray observations from EINSTEIN and ROSAT satellites have shown that many jets emit non-thermal radiation extending from radio to X-rays, and that spectral and morphological features remain, in many instances, nearly constant along the jet from radio up to optical frequencies. This implies a constancy of the relativistic electron distribution function over at least five decades in energy, that needs to be interpreted (see discussions in Meisenheimer, Röser and Schlötelburg 1996 and Meisenheimer 1996 for M 87).

A possible clue could be the diffusive shock acceleration (DSA) by multiple shocks, as discussed in Melrose and Pope (1993) and Ferrari and Melrose (1997), under very general conditions.

In this paper we treat, in a consistent way and by means of a numerical hydrocode, the jet instabilities that yield shock formation, particle acceleration in these shocks, and the temporal evolution of the distribution function, subject to adiabatic effects and synchrotron losses.

The plan of the paper is the following: in the next section (Section 2), we describe the physical model for the jet, the main assumptions and the integration method; in Section 3 we deal with the treatment of the evolution equation for the relativistic particle distribution and of the shock acceleration; the simulations results are discussed and the conclusions are given in Section 5.

2   The jet physical model

We consider a fluid, non-relativistic jet that propagates in a uniform medium and is in pressure equilibrium with the exteriors. The environment is permeated by a *passive* magnetic field, that is advected by the fluid and has no effect in the momentum conservation equation ($\beta_{\text{plasma}} \to \infty$). Under these conditions, the relevant equations are the standard hydrodynamic equations of mass, momentum and energy conservation for the thermal pressure p, the density ρ, the fluid velocity $\mathbf{v}$.

We then restrict our analysis to an infinite jet in cylindrical geometry (in the coordinates r, z); in this limit, the system of hydrodynamic equations can be complemented by the equations for the passive magnetic field, in the form:

$$\frac{\partial}{\partial t}(rA_\phi(r,z)) + \mathbf{v} \cdot \nabla(rA_\phi(r,z)) = 0 \tag{1}$$

$$\frac{\partial}{\partial t}\left(\frac{B_\phi(r,z)}{\rho r}\right) + \mathbf{v} \cdot \nabla\left(\frac{B_\phi(r,z)}{\rho r}\right) = 0, \tag{2}$$

where $A_\phi(r,z)$ is the only component of the vector potential ($rA_\phi(r,z)$ is usually called 'stream function'), as appropriate for the chosen geometry, and $B_\phi(r,z)$ is the toroidal field. One can notice that Eqs. (1,2) have in common the standard form of a 'tracer' equation:

$$\frac{\partial \mathcal{T}}{\partial t} + \mathbf{v} \cdot \nabla \mathcal{T} = 0 , \tag{3}$$

where $\mathcal{T} \equiv rA_\phi$ in Eq. (1) and $\mathcal{T} \equiv B_\phi/\rho r$ in Eq. (2).

As mentioned before, we consider an axially symmetric, cylindrical jet in a r, z coordinate system. The flow velocity is initially uniform along the z direction (V_z) and the jet is in pressure equilibrium with the ambient. For the initial velocity and density profiles in the r coordinate, and the perturbation to the transverse velocity $V_r(r,z)$ see Bodo *et al.* (1994).

The initial configuration for the magnetic field is assumed:

$$B_z = 1, \; B_r = 0, \; B_\phi = \frac{2r}{\cosh(r^m)}, \tag{4}$$

with $m = 8$.

In the calculations we measure lengths in units of the jet initial radius a, time in units of the radius sound crossing time $t_c \equiv a/c_s$ (c_s is the isothermal sound speed), and the magnetic field in units of the initial value B_0.

The system of hydrodynamic equations is solved numerically by means of a PPM (Piecewise Parabolic Method) hydrocode (Woodward and Colella 1984) over an integration domain of 256×256 grid points, with the jet radius spanning over 60 grid points, for a total domain of $0 \le z \le D$, $0 \le r \le R$ (here was $D = 10\pi$ and $R = 20$). The axis of the jet is coincident with the bottom boundary of the domain ($r = 0$), where symmetric (for p, ρ and V_z) or antisymmetric (for V_r) boundary conditions are given. At the $z = 0$ and $z = D$ boundaries we have set periodic conditions, and at the upper boundary ($r = R$) we have chosen a free outflow condition, by imposing for each variable null gradient ($d/dr = 0$).

3   Evolution of the relativistic particles

We consider a distribution of relativistic particles that is passively advected by the fluid. The relativistic particles are assumed to be injected, at $t = 0$, with a given, spatially uniform, distribution $N(E,0) = N_0 E^{-\gamma}$. A sample of $n = 10$ parcels of this distribution are then selected and followed, as lagrangian particles, as they travel with the fluid, undergo adiabatic expansion-compression effects, synchrotron losses and shock acceleration.

The temporal evolution equation for a distribution function $N(E,t)$ of relativistic particles, subject to adiabatic effects and synchrotron losses, can by written as (Kardashev 1962):

$$\frac{\partial N}{\partial t} = \frac{\partial}{\partial t}\left[\left(-\alpha(t)E + \beta(t)E^2\right)N\right] \ , \tag{5}$$

where

$$\alpha(t) = \frac{1}{3}\nabla \cdot \mathbf{v} \,, \tag{6}$$

takes into account adiabatic effects, and

$$\beta(t) = bB^2 \,, \tag{7}$$

includes synchrotron losses. Particle energy is expressed in units of the particle energy lost by synchrotron emission in a time unit and in the initial field $E_0 = 1/(bB_0^2 t_c)$.

Eq. (5) is solved numerically along the selected fluid parcels, *away from shocks* and at every time step, using a Lax-Wendroff scheme, and the coupling between Eq. (5) and the hydrodynamic equations is given by the expansion-compression term $\nabla \cdot \mathbf{v}$. The energy range for the integration is $E_{\min} = 10^{-4}$, $E_{\max} = 1$.

When one of the selected fluid particles, representing a parcel of the distribution, enters a shock, Fermi acceleration takes place. Under the assumptions that the acceleration time scale $t_{\rm acc}$ ($\sim \kappa/v^2$, with κ the spatial diffusion coefficient) is much smaller than both the synchrotron time t_{sync}, and the dynamical time for shock evolution ($\sim t_c$), we can apply the stationary diffusive shock acceleration (DSA) model (Bell 1978a,b; Drury 1983; Blandford and Ostriker 1980).

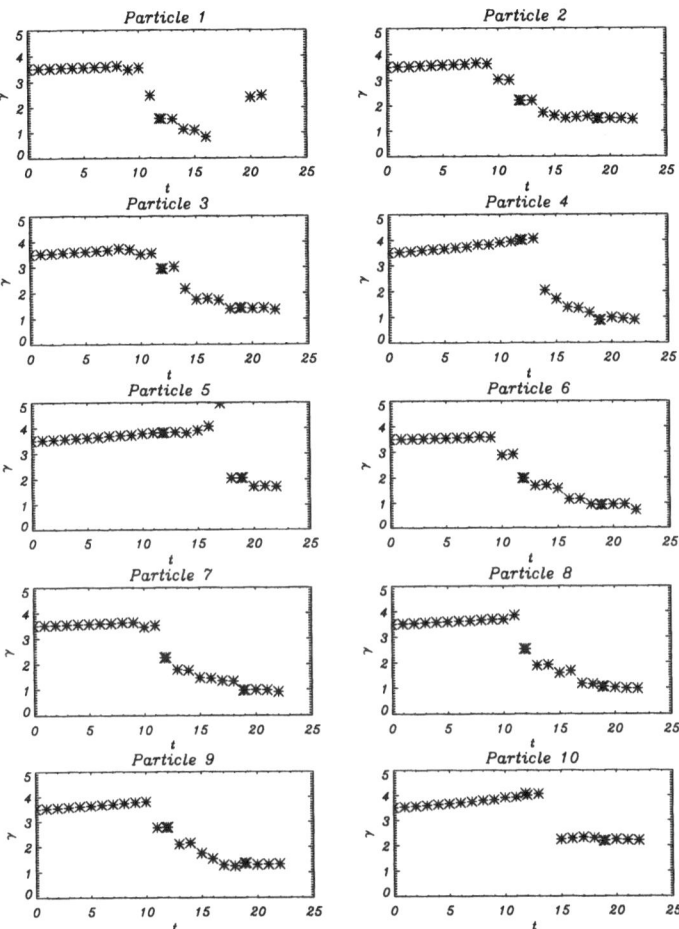

Fig. 1. Spectral index γ of the distribution functions interpolated by a power-law $\propto E^{-\gamma}$ vs time, for each particle, from $t = 0$ ($\gamma = 3.5$) to $t = 22$

4 Results and conclusions

The calculations for a more complete set of parameters will be discussed in a forthcoming paper, here we present the results for Mach number $\mathcal{M} = 5$, density ratio $\nu(\equiv \rho_\infty/\rho_{\text{jet}}) = 5$, and initial spectral index for the distribution function $\gamma = 3.5$. The initial positions for the 10 selected lagrangian particles are given in Table 1. Consistently with the assumed boundary conditions, particles leaving the domain on the $z = D$ are re-injected at $z = 0$, and those leaving the domain at the upper boundary are lost.

Table 1. Initial positions of the particles in the domain.

particle	z	r
1	12.21	0.325
2	//	0.5
3	//	0.66
4	//	0.825
5	//	0.99
6	22.03	0.24
7	//	0.41
8	//	0.575
9	//	0.74
10	//	0.91

The temporal evolution of the particle distribution function depends on the path of every fluid parcel considered, on the number of shocks crossed and their compression ratios, on the value of the magnetic field encountered. Again, a complete and detailed description of the lagrangian particle trajectories and of the evolution of the associated distribution functions will be given elsewhere, here we only show in Fig. 1 the temporal behavior of the spectral index γ. We note than the spectral index tends to reach a quasi asymptotical value within an interval $1 < \gamma < 2$ for most of the particles.

The relativistic particles undergo acceleration independently of their initial position inside the jet. Particles initially located nearer to the jet axis cross a higher number of shocks and the spectral index in their distribution functions typically reaches values ~ 1; particles initially located near the edges of the jet, instead, see few but stronger shocks and the resultant spectral index is ~ 2.

References

Achterberg, A. (1990): Astron. Astrophys. **231**, 251

Bell, A.L. (1978a): Mon. Not. R. astr. Soc. **182**, 147

Bell, A.L. (1978b): Mon. Not. R. astr. Soc. **182**, 443

Blandford, R.D., Ostriker, J.P. (1980): Astrophys. J. **237**, 793

Blandford, R.D., Eicheler, D. (1987): Physics Reports **154**, 1

Bodo, G., Massaglia, S., Ferrari, A., Trussoni, E. (1994): Astron. Astrophys. **283**, 655

Drury, L.O., (1983): Rep. Prog. Phys. **46**, 973

Ferrari, A., Melrose, D.B. (1997): Vistas in Astronomy **41**, 259

Kirk, J.G., Melrose, D.B., Priest, E.R. (1994): *Plasma Astrophysics*, Saas-Fee Advanced Course 24 (Lecture Notes 1994), Springer–Verlag, Berlin

Meisenheimer, K., Röser, H.-J., Schlötelburg, M. (1996): Astron. Astrophys. **307**, 61

Meisenheimer, K. (1996): *Jets from Stars and Galactic Nuclei*, Proceedings of a Workshop Held at Bad Honnef (3-7 July 1995), 57

Melrose, D.B., Pope, M.H. (1993): *Proc. ASA* **10**, 222

On Disks and Jet(s)
in the Defunct Quasar M 87

Max Camenzind

Landessternwarte Königstuhl
D 69117 Heidelberg, Germany

Abstract. M 87 is a prime candidate of an old quasar where the fuelling rate has been reduced drastically to $\simeq 10^{-3}\,\dot{M}_{Ed}$. Under this condition, accretion towards a rapidly rotating black hole occurs in an optically thin fashion. The accretion rate is limited to $10^{-3}\,\dot{M}_{Ed}$ by the jet luminosity. This low-accretion rate is in agreement with the fuelling provided by stellar winds in the central core of M 87. Due to the low angular momentum of the stellar component in this giant elliptical, stellar mass-loss accumulates in a ring-like structure on the parsec-scale which is observed by HST. Accretion towards the central black hole occurs from here over geometrically thin advection dominated flows (ADAFs) that also drag inwards magnetic flux from the parsec-scale.

This magnetic structure builds up a dipolar magnetosphere which immerses the central black hole. Disk plasma is injected into rotating field lines that connect to the horizon. We discuss the energetics and wind properties of the magnetically driven outflows. The observed kinetic luminosity is essentially Poynting-flux transformed into kinetic energy along collimated flux-tubes. The injection conditions near the horizon determine the asymptotic outflow velocities, Lorentz factors of 3 – 5 are quite natural. The resulting collimation radius is in agreement with VLBI data.

1 Introduction

The giant elliptical galaxy M 87 at the center of the Virgo cluster is well known for its spectacular one-sided jet. Much information is available for the stellar light distribution and the X-ray emission from this object. The observed radiative power of the central AGN in M 87 does not exceed $L_{\mathrm{rad}} \simeq 10^{42}$ erg s^{-1}, and the kinetic luminosity of the jet does probably not exceed $L_j \simeq 10^{44}$ erg s^{-1}. If the jet speed is nearly the speed of light (Biretta et al. (1999)), this would correspond to a mass-outflow of not more than $\dot{M}_j \simeq 10^{-3}\,M_\odot$ yr^{-1}. As will be shown, this can easily be supplied by the central stellar core in M 87.

Extragalactic jets are intimately related with accretion processes onto supermassive black holes residing in the centers of elliptical galaxies. The sparse stellar density, in particular in giant ellipticals (Kormendy & Richstone (1995); Faber *et al.* (1997)), obviously favours the formation of collimated outflows from the innermost regions. In this respect, M 87 represents a prime candidate of an old quasar, having a central mass M_H of $3 \times 10^9\,M_\odot$

and showing the presence of a rotating gas disk on the parsec-scale lying apparently normal to the jet direction (Ford *et al.* (1994); Harms *et al.* (1994)). Similar structures have been found by HST observations in other nearby radio galaxies. In addition, low-luminosity radio activity is indeed very common in nearby ellipticals (Wrobel (1991); Slee *et al.* (1994)). This has to be expected if black holes are remnants in at least core ellipticals (Faber *et al.* (1997)).

In distinction to these low-luminosity sources, quasars at higher redshifts dispose of much larger gas reservoirs on the parsec-scale and have therefore much higher fuelling rates converging towards Eddington accretion rates (Camenzind (1997)). This is the biggest distinction between real quasars and radio galaxies found nearby. The accretion rate in M 87 is only about $10^{-3} \dot{M}_{\mathrm{Ed}}$. According to recent developments in accretion theory (Abramowicz *et al.* (1996); Narayan *et al.* (1997); Peitz and Appl (1997)), standard accretion disk theory is no longer valid for these low accretion rates. The corresponding global solutions become hot and optically thin, at least within a distance of about 100 Schwarzschild radii. This has to be taken into account for modeling outflows in M 87.

Outflows from the immediate vicinity of a rotating black hole are driven by magnetohydrodynamic processes. Efficient jet production needs a black hole immersed into a dipolar magnetic structure. The very origin of such a structure is still unclear. Dynamo action near a rotating black hole is one possible mechanism (Khanna and Camenzind (1996); Nunez (1997)); advection of magnetic structures from the parsec-scale another (Khanna and Camenzind (1992); Kudoh and Kaburaki (1996)). Magnetohydrodynamics in a disk near rotating black holes differs in some important aspects from the non-relativistic formulation: the gravitomagnetic potential produced by the spin of the black hole couples into the equations of the magnetic flux and current (Camenzind (1998a)). This makes the arguments invalid used in the proof of Cowling's antidynamo theory (Khanna and Camenzind (1996); Nunez (1997)).

We discuss the formation of collimated outflows in terms of collimated rotating magnetospheres emerging from the immediate vicinity of the black hole. Together with the overall structure of relativistic accretion disks, this implies that jet matter is essentially formed by disk winds emerging from the innermost region of the disk. The core of the jets could in addition be filled up with electron-positron pairs generated in a gap near the horizon of the black hole (Beskin *et al.* (1992); Hirotani and Okamoto (1997)).

These are probably the most general models for the overall structure of MHD jets: the envelope of the jet is formed by collimated disk winds, the core is filled up by pair plasma leaving a tiny region along the axis which only contains poloidal magnetic field. In the case of M 87, all radii scale with the gravitational radius $M_H = GM_H/c^2 = 30\,\mathrm{AU}$ for a mass of $3 \times 10^9\,M_\odot$. The light cylinder of the rotating magnetosphere is about 100 AU and the jet outflow extends to at least 1000 AU. This conforms with constraints from VLBI observations, showing that the jet is collimated within 0.05 light years.

2   The central region of M 87

Stars in M 87 are measured to have a linear velocity dispersion of $\sigma_* \simeq$ 300 km/s. This defines the radius of influence for gravity of the central object out to a scale given by

$$R_H = \frac{GM_H}{3\sigma_*^2} = 3 \times 10^5 \, M_H \simeq 50 \, \text{pc} \, \frac{M_H}{3 \times 10^9 \, M_\odot}. \tag{1}$$

This is about a factor 10 smaller than the break radius of 7.6 arcsec (585 pc) observed in the stellar light distribution, which defines the core radius R_c of this giant elliptical (Byun et al. (1996)). In fact, the dispersion is observed to increase within 100 pc from the center, corresponding to about the one-arcsecond scale. All matter within 50 parsecs from the center moves essentially only under the influence of the gravitational force of the central object, in particular the stars and the gaseous disk detected by HST (Macchetto et al. (1997)).

2.1   The mechanical equilibrium for gas in the core

X-ray observations tell us that the core region is filled up with hot gas of a density $n_G \simeq 0.5 \, \text{cm}^{-3}$ (the volume of the core $V_c \simeq 2 \times 10^{64} \, cm^3 \simeq 5 \times 10^8 \, pc^3$). This gas has a sound velocity $c_S \simeq 500 \, \text{km/s}$, comparable to the velocity dispersion of the stars. This amounts to a total mass $M_G \simeq 10^6 \, M_\odot$ of hot gas confined by the stellar core of M 87. This is not much gas when compared to the gas masses observed e.g. in quasars such as 3C 273, where $M_G \simeq 10^{10} \, M_\odot$. This gas results most probably from the stellar wind injection with $\dot{M}_* \simeq 10^{-13} \, M_\odot \, yr^{-1}$ per solar-type star. Since the stellar core has a mass of about $10^{11} \, M_\odot$, we expect a gas injection rate of $10^{-2} \, M_\odot \, yr^{-1}$. Due to the deep potential of the core, this gas is not lost from the core. but will be accreted. Without accretion, the total mass accumulated in the core would be $M_G \simeq 10^8 \, M_\odot$, which is about a factor 100 higher than the presently observed gas mass. If in addition a cooling flow occurs, then a much higher amount of gas will be deposited in the core.

The black hole accretes gas within the accretion radius $R_{ac} \simeq h_{i_l}$ that is similar to the radius of influence for gas and stellar motion. In previous days, the gas distribution has been modeled by a simple spherical hydrostatic equilibrium. This is, however, not correct if the gas has some angular momentum. Since the stars in the core of M 87 have some angular momentum, this will also influence the gas distribution. According to the observed rotation curve inside 10″, the rotation is only about 10 km/s, which amounts to a maximal angular momentum for the stars of

$$j_* = v_{\rm rot} R_b \simeq 2 \times 10^{27}\,{\rm cm}^2\,{\rm s}^{-1} \tag{2}$$

at a typical distance $R \simeq 500\,{\rm pc}$. Accretion can therefore not proceed in a spherical fashion, the gas settles into a ring-like structure given by the centrifugal radius in terms of the specific angular momentum of the gas $j_G \simeq j_*$,

$$R_c = \frac{j_G^2}{GM_H} \simeq 3.0\,{\rm pc} \left(\frac{j_G}{2 \times 10^{27}\,{\rm cm}^2\,{\rm s}^{-1}} \right)^2 \frac{3 \times 10^9\,{\rm M}_\odot}{M_H}, \tag{3}$$

corresponding roughly to $0\!''\!05$ (*i.e.* the resolution limit of HST). Viscosity in the ring will spread the gas distribution to much larger radii. This ring could be the origin of the emission line gas detected by HST in the core of M 87 (Crane *et al.* (1993), Ford *et al.* (1994)). Similar gaseous disks have been found in the nuclei of a number of other galaxies. The rotation curve of this inner arcsecond of ionized gas has recently been determined to a distance as close as 5 pc (Macchetto *et al.* (1997)). These observations show that beyond a scale of $0\!''\!4$ the observed rotation curve considerably deviates from a Keplerian law. This might be due to a non-Keplerian disk distribution. *The gas distribution on the parsec-scale of M 87 is not a standard accretion disk, but probably more a ring-like structure due to the accumulation of stellar winds at the centrifugal radius.*

The natural unit of angular momentum near a black hole is given by

$$j_H = GM_H/c = 1.3 \times 10^{25}\,{\rm cm}^2\,{\rm s}^{-1}\,\frac{M_H}{3 \times 10^9\,{\rm M}_\odot} \tag{4}$$

for a black hole mass of M 87. This is at least a factor 200 lower than the angular momentum j_G observed in the HST ring. If the angular momentum distribution is roughly Keplerian on the sub-parsec-scale, gas has to fall down for about 40000 gravitational radii before crossing the horizon. This indeed corresponds roughly to the scales observed in the core of M 87.

Due to a non-vanishing angular momentum, the gas is not in hydrostatic equilibrium within the core, but in a mechanical equilibrium given by

$$\frac{1}{\rho}\,dP_G = -d\Phi, \tag{5}$$

where Φ is the effective potential determined by the gravitational potential of the point source and the stellar distribution, as well as the centrifugal potential (Camenzind (1995))

$$\Phi = -\frac{GM_H}{\sqrt{R^2 + z^2}} - \frac{GM_*}{\sqrt{R^2 + z^2 + R_b^2}} + \frac{j_G^2}{2R^2}. \tag{6}$$

For an isothermal equation of state, $P_G = \rho c_S^2$, this can easily be integrated to yield the density distribution of the gas

$$n_G(R, z) = n_0\,\exp\left(-\frac{\Phi(R, z)}{c_S^2} \right), \tag{7}$$

with $n_0 \simeq 0.2$ cm^{-3} as the density where the total potential vanishes, $\Phi = 0$ (the last bound surface). We have assumed a Plummer potential for the stellar potential. This could be replaced by any other suitable potential law for the stars (see *e.g.* Zhao (1996)). When all radii are expressed in terms of the core radius R_c, the potential is given as

$$\Phi = -\frac{GM_*}{R_c} \left(\frac{M_H/M_*}{\sqrt{R^2 + z^2}} + \frac{1}{\sqrt{1 + R^2 + z^2}} - \frac{L_c^2}{2R^2} \right). \tag{8}$$

$L_c = j_G/\sqrt{GM_*R_c}$ is the dimensionless angular momentum. In M 87, we find $L_c \simeq 0.01$, $R_c \simeq 600$ pc and $M_H/M_* \simeq 0.01$.

In this approach, surfaces of equal density coincide with the equipotential surfaces $\Phi = const$. These are toroidal surfaces centered around the centrifugal radius (Camenzind (1995)). Along the axis, a funnel is evacuated due to the existence of a non-vanishing angular momentum. For the values of the parameters of M 87, the funnel has a width of about one parsec.

2.2   Accretion towards the gravitational center

The observed parsec-scale disk represents the fuel presently available for feeding the central monster. Angular momentum is probably extracted by means of magnetic fields which are present on the equipartition level. In this way, a kind of accretion disk is formed on the sub-parsec level.

Standard accretion disks on the sub-parsec-scale would be mostly dominated by gas pressure. The transition between gas pressure dominance in the outer part and radiation pressure support in the inner part occurs at a radius

$$R_{\rm RG} \simeq 5700 \, M_H \, \alpha_T^{2/21} \, \dot{m}^{16/21} \, M_9^{2/21} \simeq 60 \, M_H \tag{9}$$

for accretion rates $\dot{m} \simeq 10^{-3}$. This transition is practically independent of the turbulence parameter α_T. This middle region has an optical depth independent of radius and mass

$$\tau_{ff} \simeq 2000 \, \alpha_T^{-4/5} \, \dot{M}_{-2}^{1/5} , \tag{10}$$

and depends strongly on the turbulence. The disk has a central temperature

$$T_c(R) \simeq 10^4 \, K \, \alpha_T^{-1/5} \, M_9^{-1/2} \left(\frac{R}{100 \, M_H} \right)^{-3/4} . \tag{11}$$

The effective temperature of the disk is, however, $T_{\rm eff} \simeq 5000$ K at this radius, and it drops as $R^{-3/4}$. In this part, the disk would be marginally optically thick with opacity mainly given by scattering processes. The standard accretion disk could exist in M 87 down to a radius of $R_t \simeq 50$ gravitational radii for accretion rates $\dot{M} \simeq 10^{-3} \, M_{ED}$. The total luminosity from this standard disk would be given by the transition radius R_t

$$L_D = \frac{1}{2} \frac{GM_H}{c^2 R_t} \dot{M}c^2 \simeq 5 \times 10^{42} \, \mathrm{erg \, s^{-1}} \frac{100 \, M_H}{R_t} \dot{m}_{-3} \, . \tag{12}$$

This is astonishingly close to the observed luminosity of the central point source (Tsvetanov (1998)). The above estimate also shows that a standard disk cannot exist much below a radius of 100 gravitational radii. If the observed point source is non-thermal in origin, the limits are even tighter, the transition radius must be bigger or the accretion rate lower.

Another interesting radius is given by the scale where self-gravity becomes important. This follows from the Toomre parameter

$$Q = \frac{2\pi G \Sigma}{G M_H H/R^3} \simeq \frac{4\pi \rho_c}{M_H/R^3} \tag{13}$$

determined by the surface mass density Σ (or central density ρ_c). This gives a transition radius for $Q(R_{GP}) = 1$

$$R_{GP} \simeq 2200 \, M_H \, \alpha^{2/9} \, \dot{m}^{4/9} \, M_9^{-2/9} \simeq 300 \, M_H \, . \tag{14}$$

This would indicate that the HST disk ($R \gg R_{GP}$) is probably unstable against spiral modes.

Due to the above temperature structure, accretion disks around supermassive black holes are expected to have a partial ionization zone, as in Galactic binaries, and therefore to be subject of a similar thermal instability. This zone also forms at a distance of a hundred gravitational radii from the center (Lin and Shields (1986); Clarke (1989)). Depending on the viscosity, the instability can develop in a very narrow unstable zone and propagate over the entire disk resulting in large-amplitude outbursts on time-scale of the order of 10^5 years (Siemiginowska et al. 1996).

In standard disk models one finds a local equilibrium relationship between a steady state accretion rate $\dot{M}$ (or effective temperature) and the surface density Σ for a given mass, viscosity parameter α and distance r from the center. This relation has a characteristic S-shape for an optically thick, geometrically thin disk. Three characteristic regions on the S-curve describe different physical conditions. The lower branch is thermally stable, cool, and neutral hydrogen dominates the chemical composition; molecules strongly contribute to the opacity. A disk on the upper branch is also thermally stable, but is hot, and hydrogen is fully ionized; bound-free transitions in heavy metals, free-free transitions and electron scattering determine the opacity. The middle branch corresponds to a partially ionized disk that is thermally unstable due to the rapid increase of opacity with temperature. The instability strip is located at an effective temperature of $T_A \simeq 4000 \, \mathrm{K}$, and its radius r_A can be derived from the local values of Σ_A and T_A (Siemiginowska et al. 1996)

$$r_A \simeq 10^3 \, M_H \, M_8^{-0.6} \, \dot{M}^{0.4} \, \alpha_{-1}^{-0.05} \, . \tag{15}$$

For conditions encountered in the core of M 87, this would also correspond to about 50 gravitational radii. Surface density and local accretion rate occur on the viscous time-scale $t_{\text{visc}} \simeq r/v^r$, or

$$t_{\text{visc}} \simeq 2 \times 10^5 \, yr \, \alpha_{-1}^{-0.8} \, M_8^{0.25} \, \dot{M}^{-0.3} \, r_{16}^{1.25} \, , \tag{16}$$

where $r_{16} = r/10^{16}$ cm. Thus the time-scale on which global accretion events would occur is of the order of a few hundred thousand years for a mass of $10^9 \, M_\odot$.

M 87 could have started a recent outburst a few thousand years ago. The last outburst would have created the outer jet extending now to a few 100 kpc and having stored a total energy of $\simeq 10^{57}$ erg in their radio lobes (Klein (1999)). With an age of a few 10^5 years, this corresponds to a total kinetic luminosity similar to the kinetic luminosity of the present jet.

3   The central Black Hole, its disk and jet

3.1   A modern view on the gravity of rotating black holes

Accretion and outflows inside the transition radius R_t need a fully relativistic description. A physical approach for black holes physics is based on the 3+1 split of the metric (Thorne *et al.* (1986)), which contains five functions,

$$ds^2 = -\alpha^2 \, dt^2 + \tilde{\omega}^2 \, (d\phi - \omega \, dt)^2 + \exp(\mu_1) \, dr^2 + \exp(\mu_2) \, d\theta^2 \, . \tag{17}$$

α is the redshift factor which is asymptotically one. The horizon is located at $\alpha(r_H) = 0$, $r_H = M_H + \sqrt{M_H^2 - a_H^2}$, where M_H denotes the gravitational radius, $M_H = GM/c^2$, and a_H the specific angular momentum of the black hole, limited by $|a_H| < M_H$. The function $\omega \propto a_H$ is the gravitomagnetic potential that is essential for the description of physical processes near rotating black holes. It vanishes in the case of the Schwarzschild black hole.

Rotating black holes dispose of free rotational energy

$$E_{\text{rot}} = M_H c^2 \left(1 - \sqrt{\frac{1}{2} \left(1 + \sqrt{1 - (a_H/M_H)^2} \right)} \right) \tag{18}$$

which can be tapped by magnetic processes. If this energy can be dissipated on a cosmological time-scale t_{diss}, this provides a luminosity due to rotational energy-loss

$$L_{\text{rot}} \simeq \frac{E_{\text{rot}}}{2t_{\text{diss}}} \simeq 2.5 \times 10^{45} \, \text{erg s}^{-1} \, \frac{M_H}{3 \times 10^9 \, M_\odot} \, \frac{10^{10} \, yr}{t_{\text{diss}}} \, . \tag{19}$$

This is at least a factor 10 higher than the presently observed jet luminosity and can, therefore, easily account for the energetisation of the jet activity over a cosmological time-scale. The efficiency for the energy conversion is a question of the magnetospheric physics near black holes. This problem is in, a way, similar to the pulsar problem where also rotational energy of a neutron star is dissipated into kinetic energy of outflowing pair winds.

3.2   Present understanding of inner accretion disk

Much of this magnetospheric physics depends on the inner structure of accretion disks around rotating black holes. The final accretion towards the black hole cannot be described in a Newtonian language. Global disk solutions have to be constructed which link the region of a hundred gravitational radii with the horizon of the black hole (Fig. 1). This accretion can only occur if a sufficient angular momentum transport exists, either by anomalous viscosity or by means of magnetic fields.

For accretion rates lower than a critical one, $\dot{m} \simeq 0.3\alpha_T^2$, optically thick disks are no longer a useful concept (Narayan and Yi (1995)). Global relativistic disks have been constructed by Peitz and Appl (1997). Depending on the viscosity law, stationary solutions can be found with transonic accretion and a sonic radius near the marginally stable orbit. The temperature is near free-fall temperature and the disks become geometrically thick beyond a radius of about 20 gravitational radii. The disk height scales roughly as $H \propto r \ln r$. This produces a wide funnel around rotating black holes which is favourable for precollimation of outflows (see Fig. 1). The detailed structure of this disk may depend on cooling and viscosity, but the overall behaviour of the disk height seems to be universal for advection dominated accretion flows (ADAFs).

The spectra emitted by such disks have been recently compared by Reynolds et al. (1997) with observations, however for a Newtonian ADAF solution which is not a global solution of the black hole accretion problem. The overall spectral features will however not depend strongly on the detailed structure functions of the disk, except for emission from the innermost part which is missed in Newtonian models. Cyclotron emission from thermal electrons in the magnetic fields advected by the accretion produces a peak in the energy distribution in the radio regime around 100 GHz. This emission is somewhat promiscuated with self-absorbed synchrotron emission from the jet. A second prominent peak occurs from bremsstrahlung of the hot electrons near an energy of 100 keV. These peaks will however not exceed 10^{40} erg s^{-1} and are barely observable with present-day instruments. The spectra presented by Reynolds et al. miss the high optical point measured with HST (the so-called non-thermal central optical peak Harms et al. (1994)). This could be synchrotron emission from the innermost part of the jet which has a magnetic field strength of about 10 Gauss, requiring energies in the relativistic electrons of 10 GeV, which is in agreement with even higher energies observed from the jet on a scale of hundred parsecs (Meisenheimer (1996)). The essential point is that the radio data limit the accretion rate to $\dot{m} \leq 10^{-3}$ (Reynolds et al. (1996)), provided magnetic fields in the disk will not exceed the equipartition values.

The pressure in Newtonian ADAFs is

$$P_d = \frac{\dot{M}c}{4\pi r^2} \sqrt{\frac{M_H}{r}} \frac{r}{H} \, cgs \simeq 1.0 \times 10^7 \, \alpha^{-1} \, \frac{\dot{m}}{M_{H,9}} r^{-5/2} \,, \tag{20}$$

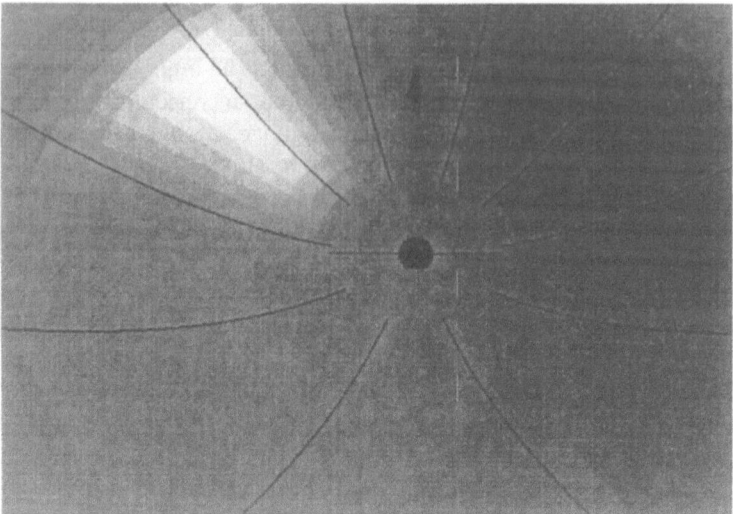

Fig. 1. The three essential elements of modern black hole models for quasars: gravity is completely determined by a rotating supermassive black hole with its accretion disk perpendicular to the the angular momentum of the black hole; hot plasma is peeled off the surface of the disk and flows away along collimated magnetic fields (marked as lines). Accretion disks connect down to the horizon of the black hole and do not end at the marginal stable orbit.

where radii are measured in gravitational units GM_H/c^2. Fields in equipartition with this pressure

$$B_e = \sqrt{8\pi P_d} = 15\,\text{kG}\,\alpha_T^{-1/2}\,\sqrt{\frac{\dot{m}}{M_{H,9}}}\,r^{-5/4}, \tag{21}$$

quite in analogy to the equipartition field strengths in standard accretion disks. Near the black hole in M 87 the disk can support field strengths up to one thousand Gauss, if $\dot{m} \simeq 10^{-3}$ and $\alpha_T \simeq 0.1$.

3.3   Magnetic fields near rotating black holes

The above estimates for the magnetic fields are quite uncertain. Magnetic fields evolve in the disk according to Maxwell's equations which are hyperbolic. For long term evolution, one can derive a parabolic diffusion type equation for the transport of the poloidal magnetic flux (Khanna and Camenzind (1996))

$$\gamma\frac{\partial\Psi}{\partial t} + (\alpha\gamma\mathbf{v}_p \cdot \nabla)\Psi - \eta\,\tilde{\omega}^2\,\nabla\cdot\left[\frac{\alpha}{\tilde{\omega}^2}\,\nabla\Psi\right] = -\eta\,\tilde{\omega}^2\,\mathbf{E}_p\cdot\nabla\omega. \tag{22}$$

η is the magnetic diffusivity in the disk given by turbulent processes, and the electric field $\mathbf{E}_p$ follows from Ohm's law. Differential rotation of absolute space generates a dynamo term that could amplify poloidal fields (Khanna and Camenzind (1996), Nunez (1997)). The advective velocity $\alpha\gamma\mathbf{v}_p$ transports flux along the disk. This term is finite at the horizon, since $\alpha\gamma$ remains finite there. The second term represents diffusion of magnetic flux against advection with the Grad-Shafranov operator.

In the stationary approximation for thin disks with half-angular thickness Θ_D and variable

$$\xi = \frac{\theta - \pi/2}{\Theta_D} \tag{23}$$

this equation reads as

$$\frac{\partial \Psi}{\partial r} + D(r,\theta)\frac{\partial^2 \Psi}{\partial \xi^2} = \frac{\eta}{\alpha\gamma u^r}\,\mathbf{E}_p \cdot \nabla\omega, \tag{24}$$

where, $u^r = \tau v^r > 0$, and

$$D(r,\theta) = \frac{\eta\,\alpha\exp(-2\mu_2)}{(\alpha\gamma u^r)\,\Theta_D^2} = \frac{\alpha\eta}{(\alpha\gamma u^r)\,r^2\,\Theta_D^2} \tag{25}$$

is the positive diffusion coefficient (Khanna and Camenzind (1992); Kudoh and Kaburaki (1996)). Near the horizon, diffusion is naturally damped by the redshift factor α, since $\alpha\gamma u^r$ remains finite there. This has the effect that the fields are radially advected towards the horizon — near the horizon, the magnetosphere of a black hole is quite naturally spherically symmetric.

The source term for Ψ decays very rapidly when moving away from the horizon, since $\nabla\omega \propto 1/r^4$. In addition, the poloidal electric field stays mainly perpendicular to the flux surfaces so that $\mathbf{E}_p \cdot \nabla\omega$ is probably a tiny effect near the horizon. Under this condition, the solution of the diffusion type equation can be written (Camenzind (1998b))

$$\Psi(r,\xi) = \frac{\Psi_0}{2\sqrt{\pi\,\Gamma}}\,\exp\left(-\frac{\xi^2}{4\Gamma}\right) \tag{26}$$

with the covariance function

$$\Gamma(r) = \int_{r_0}^{r} D(r')\,dr'. \tag{27}$$

r_0 is the outer radius where diffusion starts from. Due to the damping by the redshift factor and damping by supersonic flow near the horizon, the covariance function $\Gamma(r)$ stays nearly constant in the region of a few horizon radii. As a consequence, the flux surfaces close radially towards the horizon.

Fig. 2. A global view of the magnetosphere around a black hole on the scale 10 – 100 gravitational radii. An ADAF closes to the horizon with a scale height $H \propto r^2$. Such advection dominated disks become geometrically thick on scales between 20 and 50 gravitational radii and form ideal wide funnels for the injection of disk winds into the magnetosphere of the black hole. Formation of a hot corona is probably favoured by Kelvin-Helmholtz instabilities near the surface of the disk which is in pressure equilibrium with the magnetosphere in the inner parts. Field strengths in this region are of the order of a few hundred Gauss in M 87.

3.4   Structure of rotating wind magnetospheres

This field structure in the disk extends towards the polar region and has to be in pressure equilibrium with the disk. This part of the magnetic structure around black holes forms a magnetosphere that is in good approximation axisymmetric. Its field structure is described in terms of the poloidal flux function $\Psi(r, \theta)$

$$\Psi(r, \theta) = \frac{1}{2\pi} \int_A \mathbf{B}_p \cdot d\mathbf{A}, \tag{28}$$

and the axial current flowing downward along the central axis

$$I(r, \theta) = -\int_A \alpha \mathbf{j}_p \cdot d\mathbf{A} = -\frac{c}{2} \alpha \tilde{\omega} B_T. \tag{29}$$

A denotes an upward directed surface around the central axis. The contour surfaces of Ψ are the magnetic surfaces which are generated by rotating field lines. The contour surfaces of I are the surfaces where the poloidal currents flow.

Due to stationarity and axisymmetry, the electromagnetic fields assume a simple form

$$\mathbf{B} = \frac{1}{\tilde{\omega}} \nabla \Psi \times \mathbf{e}_\phi - \frac{2I}{\tilde{\omega} \alpha c} \mathbf{e}_\phi \tag{30}$$

$$\mathbf{E} = -\frac{\Omega^F(\Psi) - \omega}{\alpha c} \nabla \Psi \quad, \quad \mathbf{E}_T = 0. \tag{31}$$

Infinite conductivity requires $\mathbf{B} \cdot \mathbf{E} = 0$, and thus the integration constant $\Omega^F(\Psi)$ represents the angular velocity of field lines.

The structure of a rotating magnetosphere can now be derived from the well-known Grad-Shafranov equation for rigidly rotating magnetospheres (Beskin *et al.* (1992); Camenzind (1998b))

$$\boxed{\nabla \cdot \left(\frac{D_K}{\alpha \tilde{\omega}^2} \nabla \Psi \right) + 4\pi \alpha \frac{\partial U_K}{\partial \Psi} = 0} \tag{32}$$

with the potential U_K given by

$$U_K = \frac{I^2}{2\pi \alpha^2 \tilde{\omega}^2} + P + \frac{M^2 B_p^2}{8\pi \alpha^2}. \tag{33}$$

M^2 is the square of the Alfvén Mach number, discussed below, and $D_K = \alpha^2 - x^2(1 - \omega/\Omega^F)^2 - M^2$ the Alfvén function defined on Kerr-space. Here, we use the scaled cylindrical radius $x \equiv \tilde{\omega}\Omega^F/c$. The potential energy is due to the current interaction, the pressure and the poloidal motion. This action generalizes various special forms discussed in the literature. The potential energy depends in general also on the motion of the plasma, given in terms of the Mach numbers along flux surfaces. The full solution is therefore a complicated non-linear problem, which has not yet been addressed in the literature. Neglecting this last term and the plasma pressure term, one is left with the force-free approximation.

Force-free magnetospheres are good approximations for high magnetisation $\sigma_* \gg 1$, or $M^2 < x^2$ globally (Blandford and Znajek (1977); Okamoto (1992); Fendt (1997))

$$\tilde{\omega}^2 \nabla \cdot \left(\frac{D_K}{\alpha \tilde{\omega}^2} \nabla \Psi \right) + \frac{2}{\alpha} \frac{dI^2}{d\Psi} = 0, \tag{34}$$

with $D_K = \alpha^2 - x^2(1 - \omega/\Omega_*)^2$, $\Omega_* \equiv \Omega_F$ globally. Global solutions for this equation have been recently calculated by Fendt (Fendt (1997)). Under

certain conditions one can find solutions which are spherical near the horizon and asymptotically collimate into cylinders (Fig. 3). These solutions are
suitable for carrying collimated outflows. The entire magnetosphere consists
of a family of nested surfaces $\Psi = const$ which collimate to cylinders at large
vertical distances (see Fig. 2). This demonstrates that the surfaces are indeed
radial near the stellar surface and cylindrical at large distances.

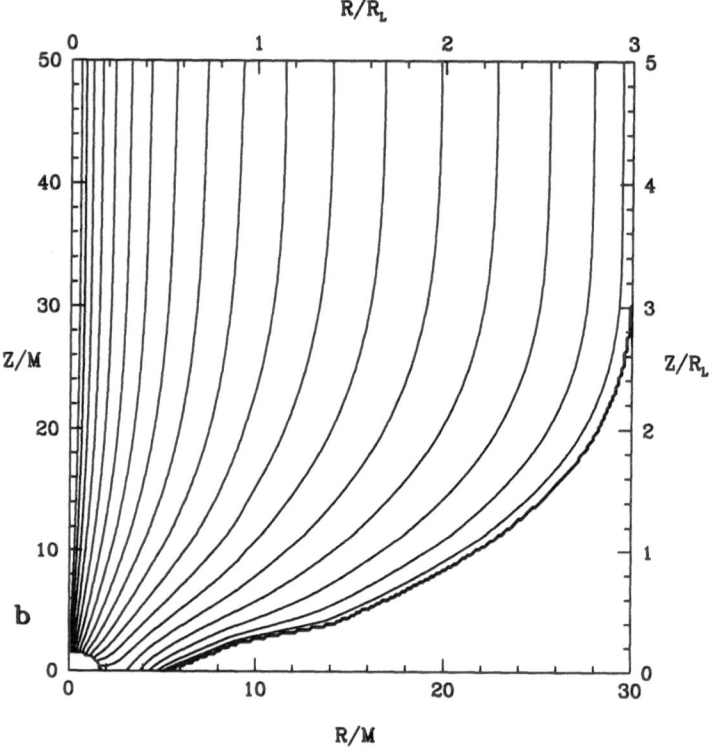

Fig. 3. Magnetic surfaces in the meridional plane for self-collimated wind solutions
around a rotating black hole (Fendt (1997)). The magnetic field lines are radial near
the horizon, but collimated into a cylindrical structure beyond the light cylinder
radius R_L.

3.5   Energetics of black hole driven winds

When a rotating body carries an axisymmetric magnetosphere, the rotation
induces electric fields $\mathbf{E}_\perp$ over the unipolar induction such that the magnetic
surfaces become equipotential surfaces. These electric fields together with the
toroidal magnetic fields $\mathbf{B}_T$ give rise to a Poynting flux

$$\mathbf{P}_p = \frac{c}{4\pi} \mathbf{E}_\perp \times \mathbf{B}_T . \tag{35}$$

This can be written in the form, $I_* = -(cRB_\phi/2)_*$

$$\mathbf{P}_p = \frac{\Omega_* I_*}{2\pi c} \mathbf{B}_p . \tag{36}$$

When integrating this Poynting flux over the entire surface of the central object, we end up with a magnetic luminosity

$$L_{\mathrm{mag}} = \frac{1}{c} \Omega_* I_* \Psi_* , \tag{37}$$

depending on the rotational state of the object, the total magnetic flux Ψ_* carried by the magnetosphere and the total current I_* driven by the rotation of the object. This magnetic luminosity is a new source of energy which is usually neglected in the energy budgets. It is however an important source of energy for strongly magnetized and rapidly rotating objects.

In the case of supermassive black holes sitting in the center of bright elliptical galaxies, the field rotation is related to the angular velocity Ω_H of the horizon which is a function of the angular momentum parameter $a_H < 1$ and the radius r_H of the horizon

$$\Omega_* < \Omega_H = \frac{1}{2} \frac{a_H c}{r_H} \simeq 10^{-4} \, \mathrm{rad \, s}^{-1} \frac{10^9 \, M_\odot}{M_H} . \tag{38}$$

This rotation rate together with the typical values for the magnetic flux Ψ_H covering a black hole and the typical current driven through the magnetosphere (Okamoto (1992))

$$I_* \simeq \frac{1}{2} (\Omega_H - \Omega_*) \Psi_H \simeq 10^{17} \, A \, \Psi_{H,31} \frac{3 \times 10^9 \, M_\odot}{M_H} \tag{39}$$

provides the following estimate for the magnetic luminosity of the rotating black hole in M 87

$$L_{\mathrm{mag}} \simeq 10^{43} \, \mathrm{erg \, s}^{-1} \frac{3 \times 10^9 \, M_\odot}{M_H} \frac{\Psi_H}{10^{31} \, \mathrm{G \, cm}^2} \frac{I_*}{10^{17} \, A} . \tag{40}$$

A more reliable estimate for the current I_* has to be given in terms of a detailed accretion theory. This has to be compared with the kinetic luminosity of the jet in M 87

$$L_j = (\gamma_j - 1) \dot{M}_j c^2 \simeq 10^{43} \, \mathrm{erg \, s}^{-1} \frac{\dot{M}_j}{0.001 \, M_\odot \, \mathrm{yr}^{-1}} \frac{\gamma_j}{3} . \tag{41}$$

The mass outflow is typically a few percents of the mass inflow, or the accretion rate that determines the overall bolometric luminosity of quasars. *Therefore, we see that the magnetic luminosity of accretion disks around supermassive black holes can easily explain the kinetic luminosity in jets.*

3.6   Plasma injection and outflow speeds

Various mechanisms have been discussed in the past for the injection of plasma into the magnetosphere near a black hole. While most investigators dream in terms of pair plasma (Beskin *et al.* (1992); Blandford and Levinson (1995); Hirotani and Okamoto (1997)), I will discuss a possible mechanism that can inject normal disk plasma into the magnetosphere of the black hole. For the configuration discussed above, the edge of the disk is in pressure equilibrium with the outer magnetosphere. Since the magnetic field lines are rotating with angular velocity Ω^F that is different from the rotation of the plasma in the disk, Kelvin-Helmholtz instabilities are excited near the surface of the disk which inject normal plasma into a coronal region above the disk. The fate of this plasma depends on the position along the disk. Near the horizon, gravity is dominant and the plasma accretes towards the horizon, far away from the horizon, the centrifugal force dominates and the coronal plasma flows away as an MHD disk wind. Accretion and ejection can only occur in the region between the inner (ILC) and outer light surface (OLC). These light surfaces are spanned by the vector field $\partial_t + \Omega_* \partial_\phi$ and therefore given by the vanishing norm of this vector field

$$\alpha_L^2 - R_L^2 (\Omega_* - \omega_L)^2 / c^2 = 0. \tag{42}$$

In the following we assume that the entire magnetosphere rigidly rotates with angular velocity Ω_*.

For a Kerr geometry, this equation has two solutions. The outer light surface is formed by the rapid rotation in the same manner as in pulsar models. It is essentially a cylinder with radius $R_L = c/\Omega_*$, however slightly deformed near the equatorial plane. This light cylinder is fairly compact for a rapidly rotating object in M 87, $a_H \simeq M_H$,

$$R_L = \frac{c}{\Omega_*} = \frac{c}{\Omega_H} \frac{\Omega_H}{\Omega_*} = 2 r_H \frac{M_H}{a_H} \frac{\Omega_H}{\Omega_*} \simeq 400 \, \text{AU}. \tag{43}$$

The existence of the inner light surface is a consequence of the frame dragging effect. For slowly rotating field lines, the inner light surface would move towards the static limit, for corotation with the horizon, $\Omega_* = \Omega_H$, it would fall towards the horizon. For the realistic situation $\Omega_* < \Omega_H$, the inner light surface is located somewhere within the ergosphere. Plasma injected into the magnetosphere will accrete inside the corotation radius

$$r_c = M_H (1 - a_H \Omega_*)^{1/3} \left(\frac{R_L}{M_H} \right)^{2/3}. \tag{44}$$

But outside the corotation radius plasma will stream away from the black hole and form a disk wind. This radius is always in-between the two light surfaces.

It is known that the wind equation for polytropic plasma flows on a black hole background can be expressed in terms of various dimensionless quantities

(Camenzind (1996), Camenzind (1998c)). One of the basic parameters is the pressure parameter p_* defined at injection

$$p_* = \frac{P_*}{mc^2 n_*} = \frac{kT_*}{m_p c^2} \qquad (45)$$

which is for normal disk plasmas $p_* \ll 1$, except for pair plasmas. Near the black hole we find typically $p_* \simeq 10^{-3}$ in hot accretion. This parameter has to be compared with the dimensionless Poynting flux in the energy equation

$$\frac{E}{mc^2} = \frac{\mu_*}{mc^2}(\alpha_* \gamma_* + \omega_* u_\phi) + \frac{I_* \Omega_*}{2\pi m_p \eta}. \qquad (46)$$

Using the expression for the current I_* at injection in the limit for injection $M_*^2 \ll 1$, $x_*^2 \ll \epsilon$, $\epsilon \equiv \Omega_* L/E$, we get the expansion for the energy

$$\frac{E}{mc^2} = \left[1 + \frac{\Gamma}{\Gamma - 1} p_* \left(\frac{M_*^2}{M^2}\right)^{\Gamma - 1}\right](\alpha_* \gamma_* + \omega_* u_\phi) + \frac{\epsilon}{\epsilon - 1}. \qquad (47)$$

For $\epsilon \to 1$, the total energy is completely dominated by Poynting flux, and not by pressure. We are therefore in the cold wind limit, except for hot pair outflows.

The next important parameter in the wind equation is the *dimensionless magnetisation parameter* σ_* defined as

$$\boxed{\sigma_*(\Psi) \equiv \frac{(B_{p*} R_*^2)(\Psi)c}{4\pi \mu \eta(\Psi) R_L^2(\Psi)} = \frac{V_A^2}{cV_*}\left(\frac{R_*}{R_L}\right)^2.} \qquad (48)$$

V_A is the classical Alfvén velocity at the injection radius $R = R_*$, $V_* \simeq c_S$ the injection velocity which is always near the local sound speed c_S. This altogether leads to the wind equation for the Mach number M, or the poloidal velocity u_p,

$$\boxed{x^2 \frac{F_S(x; M^2, \epsilon)}{D_S^2(x; M^2)}\left(\frac{E}{\mu}\right)^2 = \alpha^2 x^4 + \sigma_*^2 \Phi_\Psi^{-2} M^4.} \qquad (49)$$

This fundamental equation clearly shows that the solutions of the wind equation, when formulated for the Mach number M, depends on the following parameters

- the parameter ϵ, which defines the position of the Alfvén point, $\epsilon < 1$,
- the dimensionless energy $\bar{E} = E/mc^2$,
- the magnetisation parameter σ_*,
- the flux tube function Φ_Ψ
- the central pressure p_* and the initial Mach number M_*.

These are essentially 5 parameters for each flux surface $\Psi = const$. It can be shown that the relativistic wind equation has the same critical points as the non-relativistic one: the slow magnetosonic point, the Alfvén point $D_S(R_A) = 0$ and the fast magnetosonic point. It is very important in this respect that the light cylinder is not a critical point.

The requirement that a wind solution passes through all three critical points fixes therefore three of the five parameters. We may consider σ_* and p_* as free parameters which are fixed by injection physics. Relativistic outflows occur for high magnetisation, $\sigma_* > 1$, for low magnetisation, $\sigma_* \ll 1$ only non-relativistic speeds are produced (Fendt and Camenzind (1996)).

For the disk corona in M 87 we estimate a magnetic field $B_* \simeq 100$ Gauss and a typical density $n_c \simeq 10^8 \, cm^{-3}$, following from the above mass-loss rates, and providing $V_A \simeq c$. The corona in M 87 is probably extremely hot, since the underlying accretion plasma already has near free-fall temperature. Then the sound speed $c_S \simeq 0.1c$. This corresponds to a magnetisation

$$\sigma_*^{(M\,87)} \simeq 10 \left(\frac{R_*}{R_L} \right)^2 \simeq 3 - 5 \tag{50}$$

according to the estimates given for the compactness of the light surface. *As a result, we find for disk wind outflows a natural bulk Lorentz factor $\Gamma_j^\infty \simeq \sigma_* \simeq 3 - 5$ for disk properties in M 87.*

4   Conclusions

M 87 is a prime object for studying the connection between accretion and ejection of plasma. Due to its low accretion rate, which however is still much higher than the accretion rate in the Galactic center, standard scenarios developed for high accretion rates cannot be applied. One should be aware of the fact that the observed HST disk is at least 10000 gravitational radii away from the horizon of the putative black hole. It is also by far unclear, whether accretion inside the HST disk occurs continuously or in a kind of burst mode with a cycle time of about a few hundred thousand years.

One thing is however quite clear; the disk surrounding the horizon cannot be a standard disk due to the observed low luminosity of the very center of M 87. From the theoretical point of view we expect the existence of a geometrically thick, but optically thin disk within about one hundred gravitational radii. How this disk goes over into the larger structure is also unclear. This inner disk is responsible for the creation of the outflow that is collimated into the parsec-scale jet. Acceleration and collimation of this disk outflow is most probably achieved over an axisymmetric magnetic structure anchored in the disk. Newtonian simulations of this scenario have to be critically analysed, since relativity is important near the horizon of a rapidly rotating compact object. All these simulations provide a too narrow collimation compared with observations due to the lack of relativistic effects. Stationary model calculations provide evidence for relativistic outflows with bulk Lorentz factors in the range of 3 – 5 in agreement with observed proper motion measurements.

Acknowledgements: Projects on jets and AGN were funded by our Sonderforschungsbereich 328 in Heidelberg. Special thanks go to the organizers of this Ringberg meeting which was very stimulating for me.

References

Abramowicz, M.A. *et al.* 1996, ApJ 471, 762

Beskin, V.S., Istomin, Ya.N., Par'ev, V.I. 1992, Sov. Astron. 36(6), 642

Biretta, J., 1998, these proc.

Blandford, R.D. 1994, in Cosmical Magnetism, ed. D. Lynden-Bell, Kluwer (Dordrecht), p. 171

Blandford, R.D., Levinson, A. 1995, ApJ 441, 79

Blandford, R.D., Znajek, R.L. 1977, MNRAS 179, 433

Buyn, Y.-I. *et al.*, 1996, AJ 111, 1889

Camenzind, M. 1993, in Lecture Notes in Phys. **421**, 109

Camenzind, M. 1996, in *Rev. Mod. Astron.* **8**, ed. G. Klare, p. 201

Camenzind, M. 1996, in *Solar and Astrophysical Magnetohydrodynamic Flows*, ed. K. Tsinganos, Kluwer (Dordrecht), p. 699

Camenzind, M., 1997, *Les noyaux actifs de galaxie*, Lecture Notes in Phys. **m46**, Springer-Verlag (Heidelberg)

Camenzind, M. 1998a, in *Relativistic Astrophysics*, eds. H. Riffert, H. Ruder, H.-P. Nollert, F.W. Hehl, Vieweg (Braunschweig), p. 82

Camenzind, M. 1998b, in prep.

Camenzind, M. 1998c, in *Astrophysical Jets : Open Problems*, eds. S. Massaglia, G. Bodo, Gordon and Breach (Amsterdam), p. 3

Clarke, C.J., 1989, MNRAS 235, 881

Crane, P. *et al.*, 1993, AJ 106, 1371

Dopita, M.A., Karathar, A.P., *et al.* 1997, ApJ 490, 207

Faber, S. *et al.*, 1997, AJ 114, 1771

Fendt, C., 1997, A&A 319, 1025

Fendt, C., Camenzind, M., 1996, A&A 313, 591

Ford, H.C., Harms, R.J., Tsvetanov, Z.I. *et al.* 1994, ApJ 435, L27

Harms, R.J. *et al.*, 1994, ApJ 435, L35

Hirotani, K., Okamoto, I. 1997, ApJ, in press

Khanna, R., Camenzind, M. 1992, A&A 263, 401

Khanna, R., Camenzind, M. 1996, A&A 307, 665

Klein, U., 1999, these proceedings

Kormendy, J., Richstone, D., 1995, ARA&A 33, 581

Kudoh, T., Kaburaki, O., 1997, ApJ 460, 199

Lin, D.N.C., Shields, G.A., 1986, ApJ 305, 28

Macchetto, F.D., Marconi, A., Axon, D.J. *et al.* 1997, ApJ 489, 579

Macdonald, D.A. 1984, MNRAS 211, 313

Meisenheimer, K., 1996, in *Jets from Stars and Galactic Nuclei*, ed. W. Kundt, Lecture Notes in Phys. **471**, p.

Narayan, R., Yi, I. 1995, ApJ 444, 231

Narayan, R., Kato, S., Honma, F., 1997, ApJ 476, 49

Nunez, M., 1997, Phys. Rev. Lett. 79, 796

Okamoto, I. 1992, MNRAS 254, 192

Peitz, J., Appl, S. 1997, MNRAS 286, 681

Reynolds, C.S. *et al.*, 1996, MNRAS 283, 111

Siemiginowska, A., Czerny, B., Kostyunin, V., 1996, ApJ 458, 507

Slee, O.B., Sadler, E.M., Reynolds, J.E., Ekers, R.D. 1994, MNRAS 269, 928

Thorne, K.S., Price, R.H., Macdonald, D.A., 1986, *Black Hole — The Membrane Paradigm*, Yale Univ. Press, New Haven

Tsvetanov, Z.I. *et al.*, 1998, ApJ 493, L83

Wrobel, J.M. 1991, AJ 101, 127

Zhao, HongShen 1996, MNRAS 278, 149

Motion and Structure of Relativistic Jets

Philip E. Hardee

Department of Physics & Astronomy
The University of Alabama
Tuscaloosa, AL, 35487
USA

Abstract. The time-dependent structures predicted to arise on relativistic jets are presented and compared to structures appearing in jet simulations. Apparent motions of these structures and the relation to jet flow speeds is discussed.

1 Introduction

There is compelling evidence for the existence of relativistic jets in extragalactic objects (Cawthorne 1991). Among the extragalactic radio sources observed superluminal motions indicate jet flow speeds as large as 99.9% of lightspeed, *e.g.*, 3C 345 (Zensus, Cohen & Unwin 1995). It is possible that all jets associated with AGNs (Bicknell 1994, 1995) are initially relativistic and observed motions of features in the M 87 jet (Biretta, Zhou & Owen 1995, Biretta, these proceedings) provide strong evidence for near lightspeed flow in this jet. Many relativistic jets including the M 87 jet exhibit complex structures and motions. In this proceedings article we present the types of time-dependent structures expected to appear in relativistic flows and the relationship between motions of these structures and the underlying flow.

2 Jet Structures

A general approach to analyzing time-dependent jet structures is to linearize the fluid equations along with an equation of state where the density, velocity, and pressure are written as $\rho = \rho_0 + \rho_1$, $\mathbf{v} = \mathbf{u} + \mathbf{v}_1$ and $P = P_0 + P_1$, and subscript 1 refers to a perturbation to the equilibrium quantity. In cylindrical geometry a random perturbation of ρ_1, $\mathbf{v}_1$ and P_1 can be considered to consist of Fourier components of the form $f_1(r, \phi, z) = f_1(r) \exp[i(kz \pm n\phi - \omega t)]$ where the flow is along the z-axis, r is in the radial direction, and ϕ is the azimuthal angle. The propagation and growth or damping of the Fourier components are calculated from a dispersion relation, *cf.* Hardee (1987), where values of n = 0, 1, 2, etc. describe pinch, helical, elliptical, etc., normal mode distortions to the jet. Each normal mode consists of a *surface* wave and multiple *body* wave distortions. Inside the jet, displacements of jet fluid from an initial position (r_0, ϕ_0, z_0) associated with a particular solution to the dispersion relation can be written in the form $\boldsymbol{\xi}(r_0, \phi_0, z_0) =$

$A(r_0)e^{i\Delta(r_0)} \exp[i(kz_s \pm n\phi_s - \omega t)]$ where (z_s, ϕ_s) is the axial and azimuthal position of the displacement at the jet surface. Fluid displacements are modified in amplitude by a factor $A(r)$ relative to those at the jet surface and shifted in phase by $\Delta(r)$. The pinch mode ($n = 0$) *surface* wave has a negligible growth rate on supersonic jets, and is not expected to be important. With the exception of the helical mode ($n = 1$), *surface* wave distortions decrease in amplitude in the jet interior whereas *body* wave distortions maintain significant amplitudes in the jet interior. Surface wave distortions at longer wavelengths reveal only modest phase shifts between surface and interior distortions whereas body wave distortions reveal relatively large phase shifts in the jet interior. Jet distortions associated with surface and body wave modes are illustrated in Figure 1 (*cf.* Hardee, Clarke & Rosen 1997, Hardee *et al.* 1998)

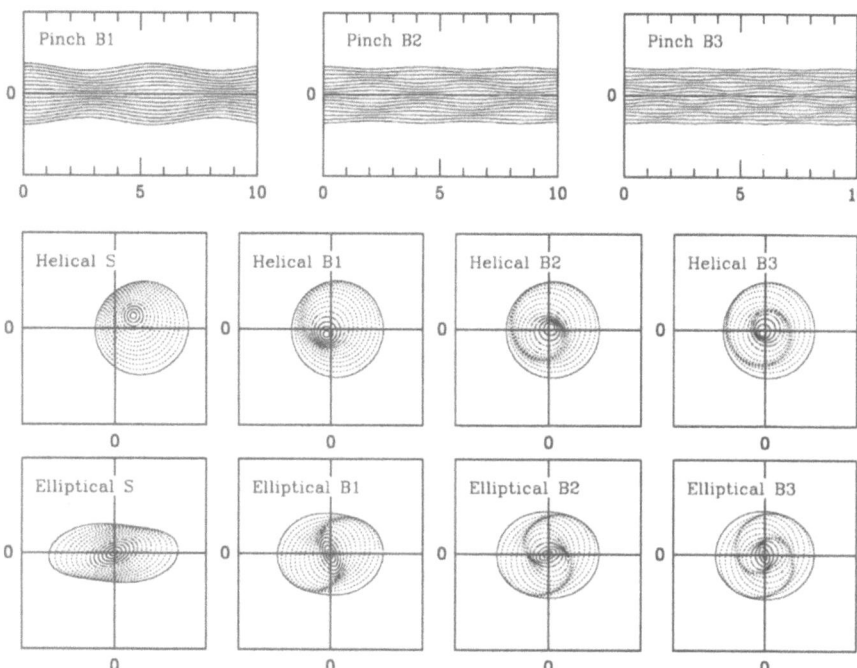

Fig. 1. Axial section displacements for pinch body modes (B1, B2, B3) and cross section displacements for helical and elliptical surface (S) and body modes (B1, B2, B3).

The motion of structures associated with jet distortions depends on the frequency or wavelength and the normal mode or modes excited by perturbations to the jet. In the low frequency limit the *surface* wave solution to higher order normal modes, $n > 0$, is given by

$$v_w/u \equiv \omega/ku \approx \frac{\gamma^2\eta_{rel}}{1+\gamma^2\eta_{rel}} + i\frac{\gamma\eta_{rel}^{1/2}}{1+\gamma^2\eta_{rel}} \ , \tag{1}$$

In equation (1), $\gamma \equiv (1 - u^2/c^2)^{-1/2}$, $\eta_{rel} = h_{jt}/h_{ex} = (a_{ex}/a_{jt})^2$, the enthalpy $h = [\rho + \Gamma P/(\Gamma - 1)c^2]$, the sound speed

$$a = [\frac{\Gamma P}{\rho + \frac{\Gamma}{\Gamma-1}\frac{P}{c^2}}]^{1/2} \ ,$$

where Γ is the adiabatic index and quantities are determined in the rest frame of the jet (jt) or external (ex) cocoon medium – *note not the undisturbed ambient medium*. The positive imaginary part of eq. (1) indicates wave growth.

In the low frequency limit the real part of the *body* wave solutions is given by

$$kR \approx k^{min}R \equiv \frac{(n + 2m - 1/2)\pi/2}{\gamma[M_{jt}^2 - 1]^{1/2}} \ , \tag{2}$$

where $m \geq 1$ is the body wave mode number, R is the jet radius and $M_{jt} \equiv u/a_{jt}$. At wavelengths $\lambda < \lambda^{max} \lesssim 2\pi/k^{min}$ the *body* wave solutions are growing and $v_w/u \equiv \omega/ku > 0$ but less than v_w/u given by eq. (5) below.

With the exception of the pinch mode surface wave which has a broad plateau in the growth rate, all surface and body wave modes have a relatively sharp maximum in the growth rate when the jet is supersonic. In the supersonic limit a resonance, *i.e.*, maximum in the growth rate, is achieved when (*cf.* Payne & Cohn 1985, Hardee 1987)

$$\omega R/a_{ex} \approx \omega^* R/a_{ex} \equiv (n + 2m + 1/2)\pi/2 \ , \tag{3}$$

$$\lambda/R \approx \lambda^*/R \equiv \frac{2\pi}{\omega^* R/u} \frac{\gamma[M_{jt}^2 - u^2/c^2]^{1/2}}{[M_{ex}^2 - u^2/c^2]^{1/2} + \gamma[M_{jt}^2 - u^2/c^2]^{1/2}} \ , \tag{4}$$

$$v_w/u \approx v_w^*/u \equiv \frac{\gamma[M_{jt}^2 - u^2/c^2]^{1/2}}{[M_{ex}^2 - u^2/c^2]^{1/2} + \gamma[M_{jt}^2 - u^2/c^2]^{1/2}} \ . \tag{5}$$

The spatial growth rate is approximately given by

$$k_I \approx k_I^* \equiv -(2\gamma M_{jt}R)^{-1}\ln[4\frac{\omega^* R}{a_{ex}}] \ , \tag{6}$$

where the minus sign on the imaginary part of the wavenumber, k_I, indicates spatial growth. The expressions for resonant frequency, wavelength, wave speed, and spatial growth rate provide reasonable estimates when $M_{jt} \gg 1$, $M_{ex} \equiv u/a_{ex} \gg 1$, $\omega R/a_{ex} \gg 1$ and $(u^2/c^2)(\omega/ku) \ll 1$.

In the high frequency limit the real part of the solutions to the dispersion relation for the all surface and body wave modes is given by

$$v_w/u \approx \frac{1 \pm a_{jt}/u}{1 \pm a_{jt}u/c^2} \ , \tag{7}$$

and in this limit the growth rate is negligible. As can be seen from equations (1), (5) & (7), the propagation speed of waves and the accompanying jet distortions can be different depending on the frequency, wavelength and mode.

3   Numerical Simulations

Figure 2 contains schlieren-type images that show the gradient of the pressure from relativistic axisymmetric propagating jet simulations with $\gamma = (1 - u^2/c^2)^{-1/2} \approx$ (A) 1, (B) 2.5 & (C) 5.5, $a_{jt}/c \approx 0.05, 0.28$ & 0.52, and $a_{ex}/c \approx 0.1, 0.3$ & 0.55, respectively (cf. Duncan & Hughes 1994, Hardee et al. 1998).

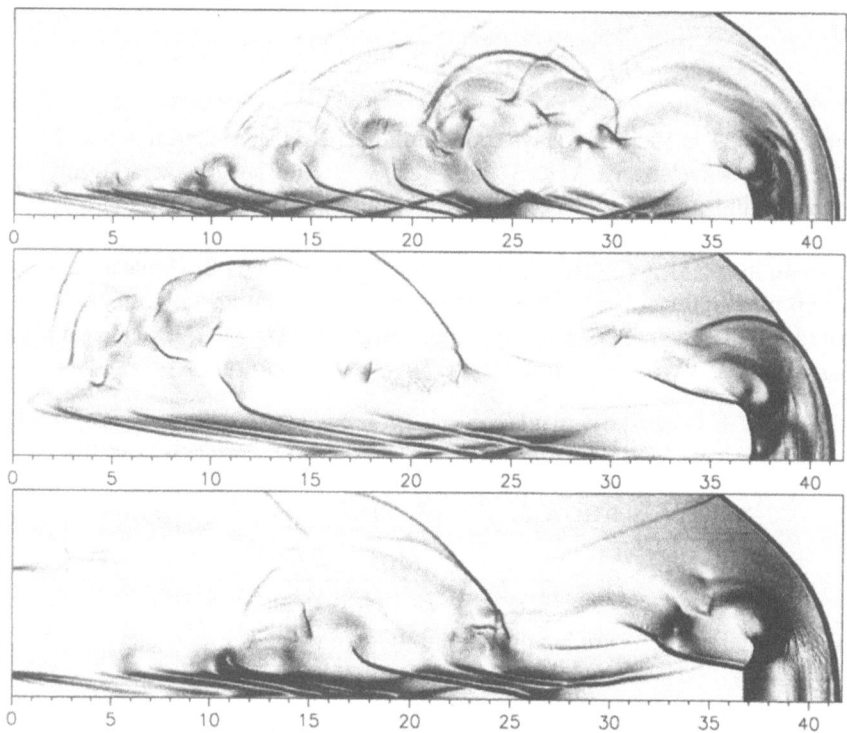

Fig. 2. Schlieren-type images of the pressure gradient in relativistic jet simulations (A) top, (B) middle, (C) bottom.

Strong pressure waves are driven into the jets by large scale vortices in the surrounding cocoon with subsequent reflection of the cocoon driven pressure waves from the jet axis forming a diamond pattern within the jet. The initial perturbation moves with the axial speed of the cocoon vortices which can be backwards or at speeds up to the head advance speed. In all three simulations the initial perturbation couples to the pinch body mode B_3 ($n = 0, m = 3$) at the longest unstable wavelength, λ^{max}, computed from the dispersion relation (Hardee *et al.* 1998). Resulting jet structures move no faster than the observed head advance speed $v_h/c \approx$ (A) 0.06, (B) 0.31 & (C) 0.55. These values are $\approx$ 78% of the 1D estimate given by $v_h = [\gamma\eta_{rel}^{1/2}/(1 + \gamma\eta_{rel}^{1/2})]u$, where $\eta_{rel} = h_{jt}/h_{am}$ (Mart´i *et al.* 1997). In the enthalpy, h, $\rho_{jt} = 1$, $\rho_{am} = 10$, $P_{jt} = P_{am}$, and $\Gamma = 5/3$ are normalized rest frame mass densities, pressures and adiabatic factors in the jet (jt) and undisturbed ambient (am) medium in all three simulations.

In the relativistic simulations we see the effects of perturbations that are driven by large scale cocoon vortices exaggerated by the assumption of axisymmetry. On the other hand, Figure 3 contains a time series of line-of-sight integrations corresponding to total synchrotron intensity images from a fully 3D simulation with perturbation due to a precessional motion at the jet origin (*cf.* Hardee, Clarke & Rosen 1997). In this non-relativistic simulation the poloidally magnetized jet is 4 times denser and in total pressure balance with an unmagnetized external medium (cocoon). The relevant Mach numbers are $M_{ex} = 4$ and $M_{jt}^{ms} \equiv u/(a_{jt}^2 + V_A^2)^{1/2} = 7.7$, where the magnetosonic Mach number takes the place of the sonic Mach number in unmagnetized flows.

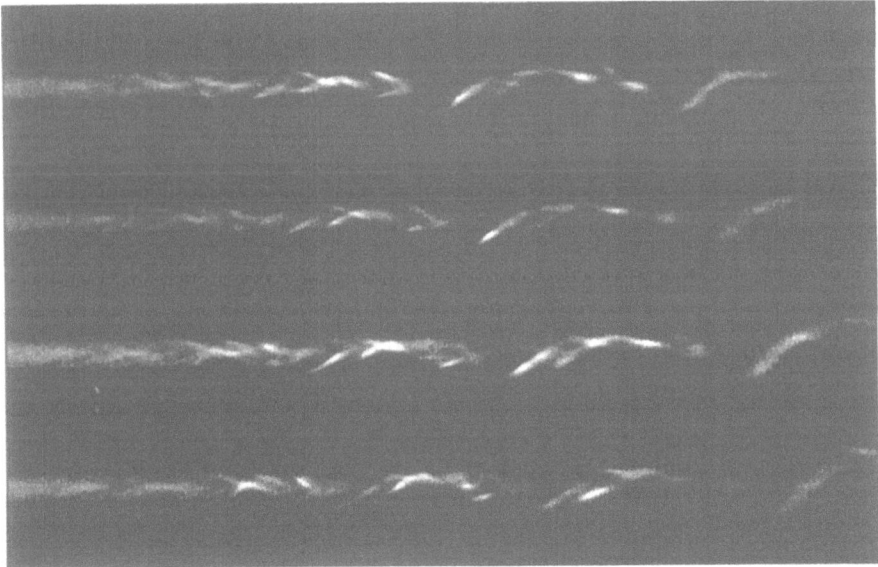

Fig. 3. Four line-of-sight synchrotron intensity images showing motions of twisted jet structures.

In general, emission structures appear as helically twisted filaments on different scales that can be associated with helical and elliptical surface and body normal modes of the jet. Wavelengths and motions are consistent with resonant excitation of the normal modes and the average motion of $v_w/u \approx 0.61$ is approximately given by eq. (5) with $M_{jt} = M_{jt}^{ms}$ and $u/c \approx 0$. While the average motion is very constant there can be deviations in the motion by as much as a factor of two from this average in local regions where emission features brighten or fade and the apparent structure changes.

4   Implications

The numerical simulations show that jets may contain structures produced by "local" perturbations, *e.g.*, cocoon vortices, that can move slowly or even backwards, or may contain structures induced by perturbations at the jet's origin. Perturbations induced near to the jet's origin will move at or near to the resonant wave speed at distances far removed from the source of perturbation. Cocoon vortices will not play as large a role on non-axisymmetric jets as in axisymmetric numerical simulations but still should be capable of inducing some local stationary or slowly moving jet structures. However, in all cases the "average" motion of structures is less than the jet's flow speed.

Apparent motions of features in the M 87 jet (Biretta, these proceedings) range from essentially stationary to as large as 5c with an apparent "average" motion of $1/2 - 2/3$ lightspeed on the kiloparsec jet. For viewing angles between $20° - 45°$ to the line of sight this is also approximately the range for the true "average" speed of jet structures. The twisted nature of emission structures on the M 87 jet appear similar to the twisted normal mode structures in the 3D numerical simulation and thus we identify the "average" speed with the resonant wave speed given by eq. (5). Higher or lower apparent speeds can result from transient interaction between different surface and body modes moving at slightly different speeds as is seen in the fully 3D numerical simulation. Transient interaction between slowly moving "locally" produced and faster moving resonant structure is also possible. Additionally, slower moving sheath and faster moving spine structures can be produced if there is significant velocity shear across the jet.

I thank my collaborators D. Clarke, C. Duncan, P. Hughes and A. Rosen, and acknowledge support from the National Science Foundation through grant AST-9318397 to the University of Alabama. Simulations where performed at the Ohio Supercomputer Center and at the Pittsburgh Supercomputing Center.

References

Bicknell, G.V.: 1994, ApJ, 422, 542
Bicknell, G.V.: 1995, ApJS, 101, 29
Biretta, J.A., Zhou, F., & Owen, F.N.: 1995, ApJ, 447, 582
Cawthorne, T.V.: 1991, *Beams and Jets in Astrophysics*, ed. P.A. Hughes (Cambridge: CUP), 187
Duncan, G.C., & Hughes, P.A.: 1994, ApJ, 436, L119
Hardee, P.E.: 1987, ApJ, 313, 607
Hardee, P.E., Clarke, D.A., & Rosen, A.: 1997, ApJ, 485, 533

Hardee, P.E., Rosen, A., Hughes, P.A., & Duncan, C.: 1998 ApJ, 500, 599

Mart´i, J.Mª., Müller, E., Font, J.A., Ibáñez, J.Mª., & Marquina, A.: 1997, ApJ, 479, 151

Payne, D.G., & Cohn, H.: 1985, ApJ, 291, 655

Zensus, J.A., Cohen, M.H., & Unwin, S.C.: 1995, ApJ, 443, 35

The Nuclear Disk in M 87: A Review

Holland Ford[1,2] and Zlatan Tsvetanov[1]

[1] Johns Hopkins University
Homewood Campus
Baltimore, MD 21218
USA

[2] Space Telescope Science Institute
3700 San Martin Drive
Baltimore, MD 21218
USA

Abstract. The disk in the center of M 87 is a prototype for gas orbiting a massive central object. Three sets of HST+COSTAR FOS and FOC observations provide strong evidence that the nuclear disk in M 87 is in Keplerian rotation around a black hole with a mass of $(2 - 3) \times 10^9 \, M_\odot$. A deep (6 orbits), high resolution Hα+[N II] PC2 HST image shows a trailing, three arm spiral superposed on the underlying nuclear disk. Several of the extended filaments appear to connect directly to the disk. The filaments extending to the NW appear to be twisted, as in NGC 4258. Earlier arguments that the NW filaments are flowing from the nucleus are supported by the presence of blue shifted non-Keplerian components within 20 pc of the nucleus. The gas in the blue and red shifted non-Keplerian components has negative energy and will fall back into the nucleus. The morphological and kinematical observations can be explained by assuming that the filaments originate in a bidirectional wind emanating from the disk. Such a wind will carry away angular momentum, enabling gas in the disk to move toward the black hole.

Small ($r \sim 1''$; $r \sim 100 - 200$ pc), well-defined dusty (D-type) and ionized (I-type) "nuclear" disks are common in elliptical galaxies. We suggest that the size of the black hole's radius of influence R_{BH} relative to the radius of the nuclear disk R_{disk} determines whether the disk will be a D-type or I-type. I-type disks (M 87 and M 81) occur when $R_{BH} \geq R_{disk}$. Differential Keplerian rotation throughout the disk may then generate turbulence and shocks that ionize the gas. D-type disks (NGC 4261 and NGC 6251) occur when $R_{BH} \ll R_{disk}$. The regions of a disk that are exterior to R_{BH} will rotate at approximately constant angular velocity in the galaxy's stellar potential. In the absence of differential rotation, shocks will be suppressed, and the gas will remain cold and dusty. Intermediate D/I types (3C 264) may occur when R_{BH} is a significant fraction of the disk's radius. Comparison of R_{BH} with the sizes of the ionized regions in M 87, NGC 4261, and NGC 6251 supports these suggestions.

1 Introduction

The discovery of a small, well-defined dusty disk in the center of NGC 4261 (Jaffe *et al.* 1994 [J94]; 1996) with a major axis nearly perpendicular to the galaxy's large scale radio axis inspired the Faint Object Spectrograph

Investigation Definition Team to search for a similar disk in M 87. Hα on-band/off-band images taken with the newly installed WFPC2 (Ford *et al.* 1994) revealed a small ($r \sim 1''$; 70 pc at 15 Mpc) disk-like structure of ionized gas whose apparent minor axis was within $\sim 15°$ degrees of the position angle of the jet. The hypothesis that the gas is a rotating disk was confirmed when $0\overset{''}{.}26$ aperture FOS observations made with the newly installed COSTAR showed Keplerian motion around a central mass of $2.4 \pm 0.7 \times 10^9 M_\odot$ (Harms *et al.* 1994; H94). The large mass-to-light ratio $(M/L)_V \sim 500$ of the central mass led H94 to conclude that the disk is rotating around a massive black hole. Subsequent observations with the FOS $0\overset{''}{.}086$ aperture confirmed the disk's Keplerian motion, but showed that large non-circular motions are also present (Ford *et al.* 1996a,b; F96).

In this paper we discuss the size, morphology, and alignment of M 87's nuclear disk in the broader context of nuclear disks in AGN. We review the kinematical observations, and suggest a connection between the non-circular motions and the large scale morphology of the ionized filaments. Finally, we present data which supports the hypothesis that the size of the ionized disk depends on the size of the central black hole's radius of influence. Throughout this paper we assume that the distance to M 87 is 15 Mpc.

2 Nuclear Disks in AGN: M 87

The disk-like structure in the center of M 87 is not unique. In a recent review (Ford *et al.* 1997; F97) we showed that small ($r \sim 1''$; $r \sim 100 - 200$ pc) well-defined gaseous disks, which we call nuclear disks, are common in the centers of elliptical galaxies. The minor axes of the disks are closely aligned with the directions of the large-scale radio jets, suggesting that the direction of the jet is ultimately determined by the angular momentum in the nuclear disk. The disks are most commonly dusty with unresolved or partially resolved HII in their centers, rather than completely ionized as in M 87. We will call these D-type disks. NGC 4261 is the prototype of the D-type nuclear disks, whereas M 87 is the prototype of ionized nuclear disks, which we call I-type. Figure 1 gives an example of two D-type disks (NGC 4261 and NGC 6251) and two I-type disks (M 87 and M 81). This is a striking similarity between the Hα "disks" in M 81 and the disk in M 87. However, we do not yet have kinematical observations of M 81 which show that in fact the disk-like structure is rotating.

J94 and Ferrarese, Ford & Jaffe (1996; FFJ96) discussed the morphological and kinematical evidence for warping in the NGC 4261 disk. FOS $0\overset{''}{.}086$ aperture observations show that the disk is in Keplerian motion around a central mass $(4.9 \pm 1.0) \times 10^8 M_\odot$ (FFJ96). The mass-to-light ratio is $(M/L)_V \sim 2100 M_\odot/L_\odot$ within the inner 14.5 pc. The large M/L and the fact that NGC 4261 is a radio galaxy strongly point to the dark mass resid-

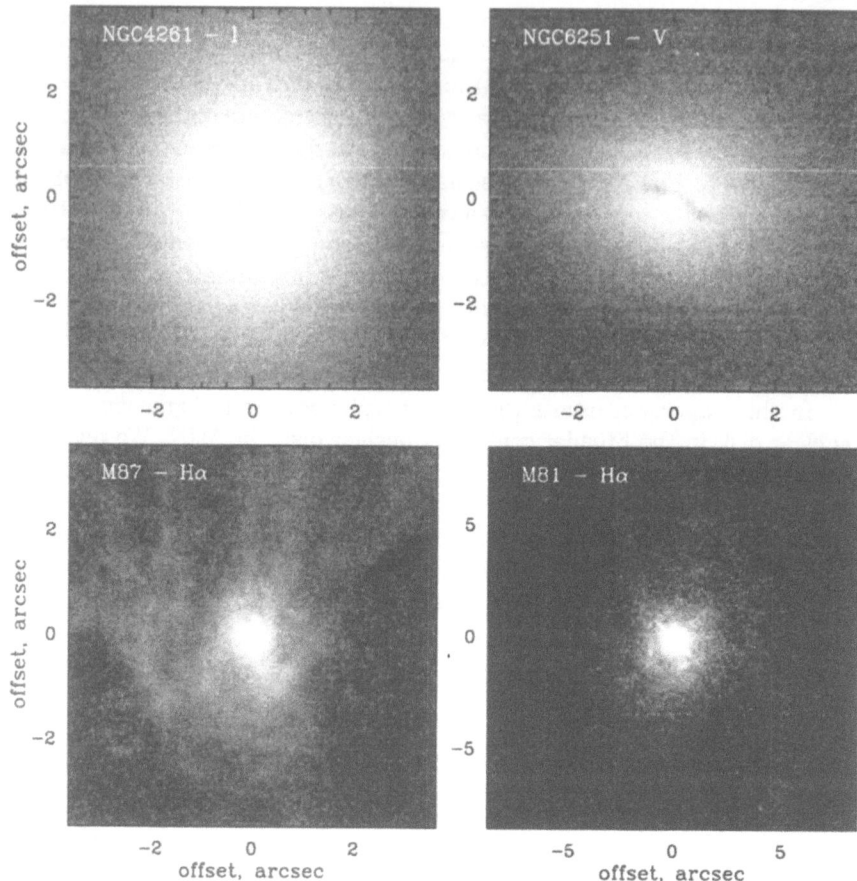

Fig. 1. Examples of nuclear disks in galaxies with active nuclei. Two D-type nuclear disks are shown in the upper panels, and two I-type disks in the lower panels.

ing in a black hole. The angle between the kinematical minor axis and the radio axis is 21°.

The dusty disk in NGC 6251 also is obviously warped. The fit of the FOS 0''.086 aperture observations to a Keplerian model gives a central mass of $(7.5 \pm 2.2) \times 10^8 M_\odot$ inside 0''.086 (= 43 pc) and a central mass-to-light ratio $(M/L)_V \sim 680 M_\odot/L_\odot$ (Ferrarese, Ford, & Jaffe, 1998; FFJ98), leading to the conclusion that the disk is rotating around a massive black hole. The kinematical major axis is rotated $\sim 63°$ from the major axis of the dusty disk. The angle between the kinematical minor axis and the projection of the radio jet is $\sim 40°$. If the jet is perpendicular to the accretion disk, the large scale warping of the disk persists down to the accretion disk.

To gain deeper insights into the morphology of M 87's Hα disk, and the relationship of the disk to the extended "filaments" (Ford and Butcher, 1979; Sparks, Ford, and Kinney 1993; SFK), we used 6 HST orbits to take deep, dithered, high resolution PC2 F658N observations of M 87. Figure 2 shows the on-band Hα sum of 6 orbits from which a model of M 87's light distribution has been subtracted. The Hα images are discussed in detail by Tsvetanov *et al.* (1998a) in this volume. Here we summarize the principal results. Out to a radius of ∼ 1″ the observed emission-line distribution is well represented by a trailing three armed spiral superposed on an elliptical (in projection) power-law disk. Between 20 and 85 pc the position angle of the major axis is approximately constant at ∼ 10°, and the ellipticity varies smoothly from 0.2 to 0.4. The PA of the jet is 290.5°; the disk minor axis and jet are aligned to ∼ 10°. The position angle of the line of nodes and the mean inclination are close to the values used by H94 and F96 to measure the central mass.

2.1 Outflow From the Disk

The filaments in Figure 2 that extend ∼ 17″ (1200 pc) to the NW at position angle ∼315° appear to be composed of three distinct twisted strands. The appearance of these filaments is similar to the filaments in NGC 4258, which are morphologically (Ford *et al.* 1986) and kinematically (Cecil, Wilson, & Tully 1992) twisted. The NW filaments at position angles ∼315° and ∼343° appear to connect directly to the disk. Based on the presence of absorption from dust in the filaments at ∼315°, SFK argued that these filaments are most likely tipped toward us, as is the jet. SFK also found that these filaments are blue shifted with respect to systemic velocity. Consequently, they concluded that the filaments are flowing away from the nucleus, rather than falling into the nucleus. There is now direct evidence for an outflow. Tsvetanov *et al.* (1999b; T99) in these Proceedings found UV absorption lines blue shifted by 150 km s^{-1} with respect to M 87's systemic velocity. And, as discussed in Section 4, FOS emission line spectra show clear evidence for large non-circular motions associated with the disk.

We suggest that the filaments which appear to connect to the disk originate in a wind blowing from the disk. Because the disk is rotating, the filaments will carry away angular momentum, allowing gas to flow through the disk toward the black hole in the center. The apparent twisting of the filaments may originate in angular momentum carried by the wind. These ideas can be tested by using high spatial resolution spectroscopy to search for kinematical twisting in the filaments.

3 Kinematics and the Central Mass

Radial velocities in M 87's nuclear disk have been measured with the Hubble Space Telescope three times. The first measurements were made with the

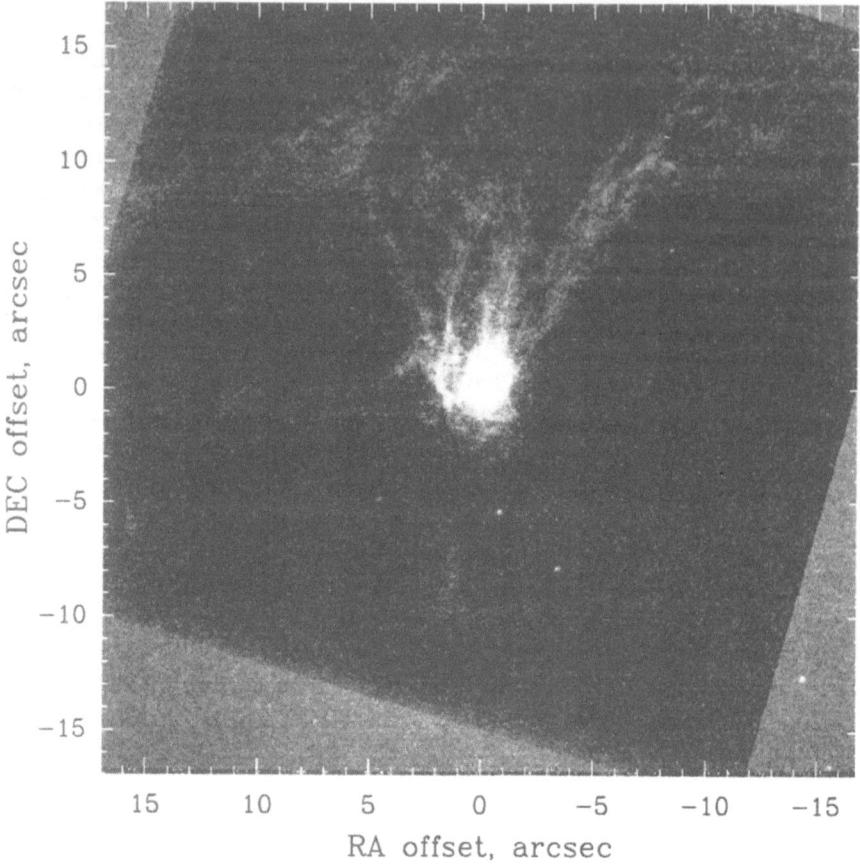

Fig. 2. The sum of 6 HST orbits of F658N PC2 imaging of M 87. The starlight has been removed by subtracting a model galaxy.

FOS by H94 shortly after the installation of COSTAR. Subsequently the FOS team used FOS observations with $0''\!\!.086$ and $0''\!\!.26$ apertures to measure velocities along the disk's minor axis and within 6 pc of the nucleus along the major axis. These observations revealed large non-circular motions at positions along the minor axis and at some positions close to the nucleus along the major axis. However, these non-Keplerian velocities appear to be superposed on Keplerian rotation around a massive black hole (F96). Finally, Macchetto *et al.* (1997; M97) used COSTAR and the FOC $0''\!\!.06 \times 13''\!\!.5$ slit to measure [O II] $\lambda\lambda3727+29$ velocities across the disk. Although their difficult target acquisition missed the nucleus, their data is a good fit to Keplerian rotation around a mass of $3.2 \pm 0.9 \times 10^9\,M_\odot$.

Table 1 summarizes the results from the three sets of observations. The F96 analysis includes the H94 data, so the two mass determinations are not independent. The central mass derived from the FOS and FOC observations agree to within their respective error bars. The conclusion that the disk is in rapid rotation around a central mass of $(2\text{--}3) \times 10^9 \, M_\odot$ appears to be firm.

The gas in the disk is very turbulent; $600 \, \text{km s}^{-1}$ is the characteristic FWHM. The energy in this turbulence has not been included in the measurements of the central mass made by H94, F96, and M97. At positions 9a,b and 11a,b (cf. Section 4) the circular velocity in the disk is $\sim 1200 \, \text{km s}^{-1}$ for a mass of $2 \times 10^9 \, M_\odot$. We take $3\sigma^2/v_{\text{circ}} \sim 0.13$ as a measure of how much the mass has been underestimated. Neglecting the turbulence does not appear to have much effect on the estimated mass. This reflects the fact that in spite of the turbulence, the disk does appear to be in Keplerian rotation.

Table 1. HST Spectroscopic Observations of the Nuclear Disk in M 87

Paper	Instrument	No. of Positions	Mass $(M_\odot)$	Radius (pc)	$(M/L)_V$ $(M/L)_\odot$
H94	FOS+COSTAR	$5 \times 0\!''\!.26$	$2.4 \pm 0.7 \times 10^9$	18	~ 540 (starlight)
F96	FOS+COSTAR	$6 \times 0\!''\!.086$ $2 \times 0\!''\!.26$	$2.0 \pm 1.0 \times 10^9$	~ 6	~ 3100 (starlight)
M97	FOC+COSTAR	$3 \times 0\!''\!.06 \times 13\!''\!.5$	$3.2 \pm 0.9 \times 10^9$	~ 6	~ 110 (total light)

F96's mass-to-light ratio $(M/L)_V \sim 3100 \, (M/L)_\odot$ excludes the non-thermal light from the unresolved point source (cf. T99), whereas the value reported by M97 is based on the total light. An M/L that is orders of magnitude larger than the M/L found in star clusters of similar color is compelling, but not conclusive, evidence that the dark mass resides in a black hole. Maoz (1995, 1998) has investigated the maximum lifetimes of central dark clusters composed of compact objects such as white dwarfs, neutron stars, and stellar mass black holes. He finds that present observations impose maximum ages of central dark clusters that are much less than the age of the galaxies in only two galaxies, the Milky Way and NGC 4258. Unless the dense objects in the hypothetical nuclear clusters have masses $\leq 0.03 M_\odot$ (e.g., low-mass black holes or elementary particles), the respective maximum ages of dark clusters are $\sim 10^8 \, \text{yr}$ and $\sim 2 \times 10^8 \, \text{yr}$. Because these lifetimes are much less than the ages of galaxies, dark clusters are highly improbable sources for the mass. Consequently, the Milky Way and NGC 4258 are the strongest candidates for hosting massive black holes.

In spite of the theoretical possibility that a dark cluster may provide M 87's dark mass, we think there are many reasons for concluding that the dark mass resides in a black hole. These arguments, which are largely encapsulated within the "AGN Paradigm", include large energy release in the small volumes implied by variability, and the production of relativistic jets from the presumed accretion disks around a massive black hole. Given the high probability that massive black holes exist in two AGN, we think it likely that the large dark masses routinely being found in AGN such as M 87, NGC 4261, and NGC 6251 (FFJ98) also reside in massive black holes.

4 Non-Keplerian Motions

Non-Keplerian motions are readily evident when non-systemic velocities are observed at positions along the minor axis of the disk, or when the line profiles show the unmistakable presence of two or more emission line components. Because of the difficulty of separating individual velocity systems in the Hα+[N II] blend, we used the [O III] λ5007 profiles to identify the non-Keplerian components. Figure 3 is a schematic that shows the positions on the major and minor axes where we find non-circular motions. Figure 4 shows the spectra of Hβ through [O III] at position 7 and 8 on the minor axis.

At position 7 on the minor axis, which is on the same side as the jet, the systemic component of the disk is present at $\sim 1300\,\mathrm{km\,s^{-1}}$ and $\sim 1193\,\mathrm{km}$ $\mathrm{s^{-1}}$ in [O III] λ5007 and Hβ. A blue shifted component is clearly present in both [O III] λ5007 and Hβ at respectively $675\,\mathrm{km\,s^{-1}}$ and $544\,\mathrm{km\,s^{-1}}$. Note that the blue shifted component is stronger than the systemic component in [O III], whereas the reverse is true in Hβ. [O III] λ4959 has the same profile as [O III] λ5007, although the weaker systemic component is present only as an inflection on the red side of the line profile.

At position 8 on the minor axis opposite the jet, the expected systemic component at $v \sim 1325\,\mathrm{km\,s^{-1}}$ is present, as well as a red shifted component at $v \sim 1775\,\mathrm{km\,s^{-1}}$ and a *blue* shifted component at $v \sim 884\,\mathrm{km\,s^{-1}}$. The line [O III] λ4959 has the same profile as λ5007, so we are confident that all three components are present. Hβ is very broad, undoubtedly due to the presence of multiple components. As at position 7, [O III] and Hβ have different profiles, showing that the excitation varies between the different components.

Table 2. Non-Keplerian Velocities Within 0″.26 of the M 87 Nucleus

Position	x (″)	y (″)	v_{obs} km s^{-1}	v_k km s^{-1}	r_0 (pc)	v_w/v_{circ}
9a	0.0250	0.0750	766	1984	5.5	0.48
8	0.2220	−0.0770	1736	1232	20.1	0.92
	0.2220	−0.0770	884	1232	20.1	0.69
7	−0.2440	0.1030	675	1287	22.7	1.15
11a	−0.0355	−0.0845	851	576	6.4	0.42
	−0.0355	−0.0845	1424	576	6.4	0.18
11b	−0.0345	−0.1045	1503	629	7.6	0.29

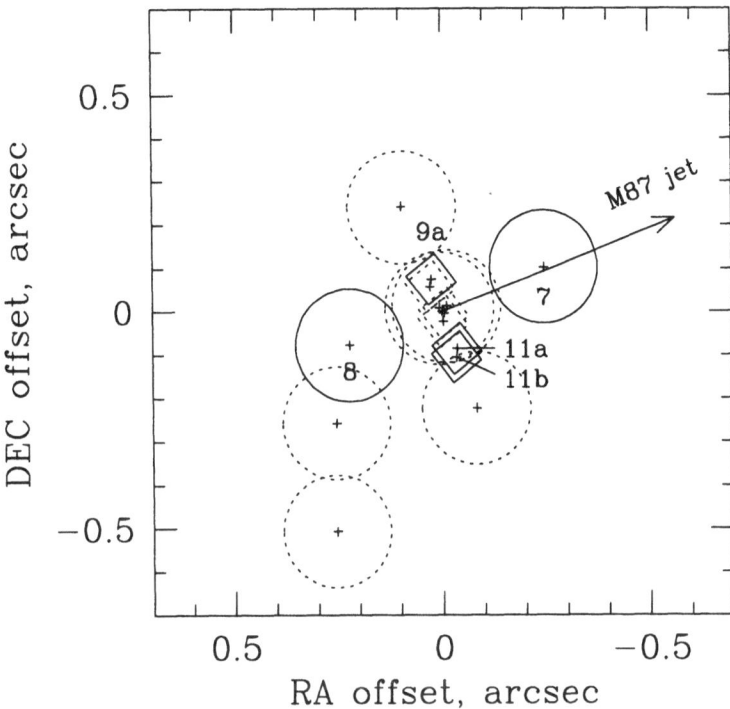

Fig. 3. A schematic showing the positions where FOS measurements were made. The position of the nucleus is at the base of the arrow representing the jet. Solid lines show the positions where non Keplerian velocities were found.

The observed non-Keplerian components are summarized in Table 2. The first column gives the position designations assigned after the observations were made (*e.g.* Position 11a was the first visit to Position 11, and Position 11b was the second visit). Columns 2 and 3 give the locations of the apertures relative to the nucleus derived from the FOS peak-up data (x and y are positive to the E and N). Column 4 gives the observed velocities of the non-Keplerian components. Column 5 gives the *predicted* line-of-sight velocities from the model disk (F96), calculated using the equations in H94. Column 6 lists the radius in the disk at the position of the aperture. Finally, column 7 gives the "wind" velocity, $v_w = (v_{obs} - 1250)/cos(i)$, divided by the circular velocity v_{circ} at the position r_0, assuming that the wind is perpendicular to the disk.

The blue shifted velocity component ($675\,\mathrm{km\,s^{-1}}$) at position 7 along the minor axis on the side of the jet could be due to gas entrained by the jet. There is a corresponding red shifted component ($1736\,\mathrm{km\,s^{-1}}$) at position 8 on the minor axis opposite the jet. This could be gas entrained by the unseen counter jet. However, there also is a *blue* shifted component ($884\,\mathrm{km\,s^{-1}}$) at

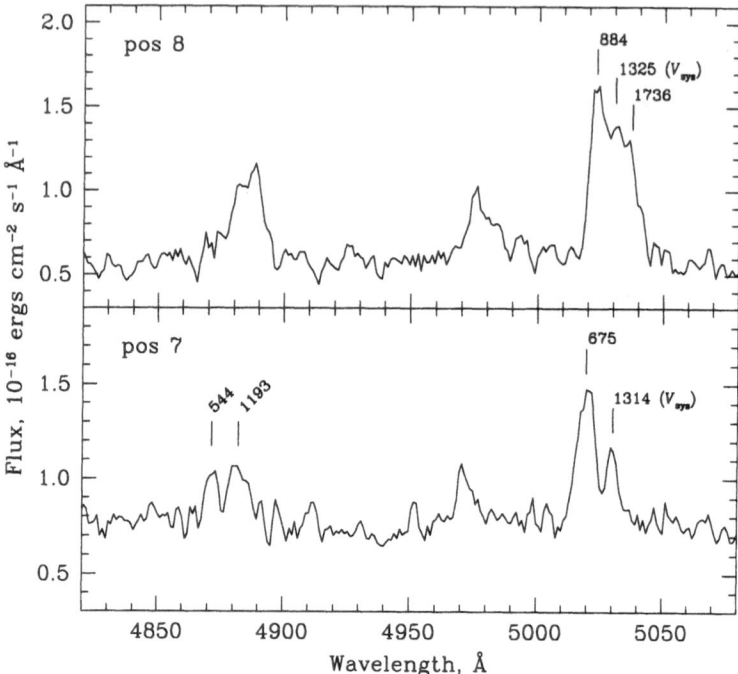

Fig. 4. The upper panel shows the nuclear spectrum from Hβ through [O III] λ5007 at position 8 on the minor axis opposite the jet. The systemic component from the disk is flanked by blue shifted and red shifted components. The lower panel shows the spectrum at position 7 on the minor axis on the same side as the jet. The blue shifted component is obvious in both [O III] λ5007 and Hβ.

position 8. Further more, there are both red and blue shifted components at positions 9a, 11a, and 11b along the major axis on both sides of the nucleus. We think the most likely explanation for these components is that a "wind" is coming off the disk. As previously noted, SFK argued that the filaments extending to the NW are an outflow from the nucleus. Because these filaments appear to connect to the disk, it is natural to suppose that they originate in a "wind" that is carrying away angular momentum. Although we cannot directly associate the non-Keplerian velocity components with these filaments, we think they lend support to this suggestion.

Escape from the black hole's sphere of influence requires $v_w \geq \sqrt{2}\, v_{circ}$. Unless the gas is far from the disk, Table 2 shows that the gas is bound to the black hole, and will fall back in. This may account for the apparent "loop" structure which extends $\sim 8''$ to the NNW of the nucleus (see Fig. 2). The fact that the gas is bound to the nucleus should not be surprising. Unless there is a large unobserved reservoir of energy in the nuclear disk, such as a strong magnetic field, it will be impossible to drive off gas at velocities exceeding escape velocity.

5   Disk Type and the Black Hole's Radius of Influence

NGC 4261 and NGC 6251 are examples of well defined, dusty (D-type) disks that have small, partially resolved regions of ionized gas in their centers. Conversely, the disk in M 87 is entirely ionized (I-type). The face-on disk in 3C 264, shown in Figure 5, is intermediate between NGC 4261 and M 87. Disks such as this one, wherein the ionized region is more than $\sim 20\%$ of the radius of the dusty region, will be referred to as D/I-type.

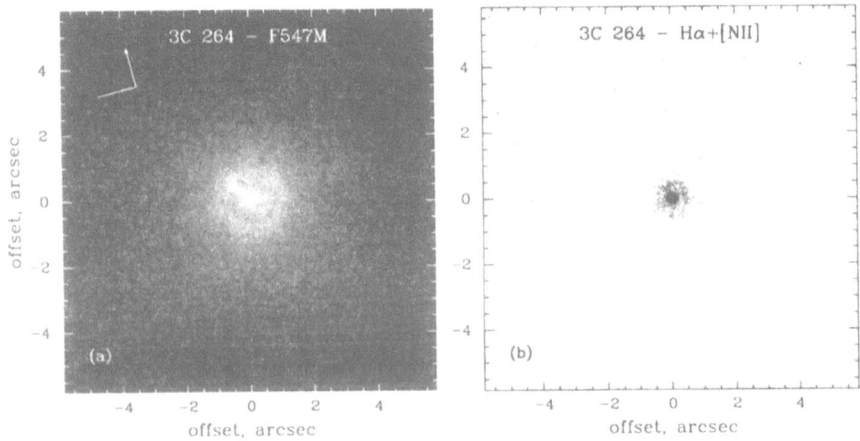

Fig. 5. The left panel shows a PC2 F547M continuum image of 3C 264. The bright linear feature in the F547M continuum image of 3C 264 that projects to just beyond the edge of the disk is an optical jet (*cf.* Baum *et al.* 1997). The right hand panel shows an Hα+[N II] image obtained by subtracting the F547M image from a F673N image. A small disk of ionized gas is present in the center of the dusty disk.

What determines the type of nuclear disk? We hypothesize that the type of disk is determined by the size of the disk relative to the black hole's radius of influence. If the dispersion of stars in the bulge of the parent galaxy is σ_0, the black hole's radius of influence will be

$$R_{BH} \sim GM_{BH}/\sigma_0^2 = 43\ M_8/(\sigma_{100})^2\ \text{pc} \qquad (1)$$

where $\sigma_{100} = \sigma_0/100\ \text{km s}^{-1}$, and $M_8 = M_{BH}/10^8 M_\odot$. We suggest that when $R_{\text{disk}} \leq R_{BH}$, differential Keplerian rotation generates turbulence and shocks throughout the disk that ionize the gas. H94 found that the gas in the disk is very turbulent with a characteristic FWHM $\sim 500\,\text{km s}^{-1}$. Dopita *et al.*

(1997) and Dopita (1998) find that the gas in the disk is ionized by fast shocks rather than by nuclear photoionization. Shock ionization is consistent with the presence of turbulence and the spiral features in the disk.

The regions of a disk that are exterior to the black hole's radius of influence will rotate at an approximately constant angular velocity in the galaxy's stellar potential. In the absence of differential rotation, shocks will be suppressed, and the gas will remain cold and dusty.

Table 3 lists the three galaxies to date that have nuclear disks and a measured black hole mass. The radius R_{BH} was calculated with the assumption that all three galaxies have a nuclear stellar velocity dispersion $\sigma_0 = 300$ km s^{-1}. The fact that the disk in M 87 is I-type, whereas the disks in NGC 4261 and NGC 6251 are D-type, is consistent with our hypothesis.

Table 3. Disk Radius and the Black Hole's Radius of Influence

Galaxy	M_{BH} $M_\odot$	R_{BH} (pc)	R_{Disk} (pc)	Disk Type
M 87	$(2-3) \times 10^9$	100–140	~100	I
N 6251	7.5×10^8	36	330	D
N 4261	4.9×10^8	19	130	D

6   Summary

The disk in the center of M 87 is not unique. Small ($r \sim 1''; r \sim 100 - 200$ pc), well-defined dusty (D-type) and ionized (I-type) "nuclear" disks are common in elliptical galaxies (F97). The minor axes of these disks are closely aligned with the directions of the large scale radio jets, suggesting that the disk's angular momentum determines the direction of the radio jets.

HST observations of M 87 with the FOS+COSTAR (H94 and F96) and the FOC+COSTAR (M97) provide strong evidence that the I-type nuclear disk in M 87 is in Keplerian rotation around a central mass of $(2-3) \times 10^9$ M$_\odot$. The stellar $(M/L)_V$ at radii $r \leq 6$ pc is at least $3100 (M/L)_\odot$, a value orders of magnitude larger than found in star clusters. The high M/L combined with i) arguments encapsulated in the "AGN Paradigm," and ii) the nearly inescapable fact that there are massive black holes in the centers of the Milky Way and NGC 4258, strongly suggest there is a massive black hole in M 87.

Deep, high resolution HST images show a three arm spiral superposed on the underlying nuclear disk . The filaments extending to the NW appear to be twisted, as are the filaments in NGC 4258. Several of the filaments appear to connect directly to the disk. SFK's arguments that the NW filaments are flowing out from the nucleus are supported by the presence of blue shifted non-Keplerian components seen in absorption and in emission. The non-Keplerian components do not have enough kinetic energy to escape the black hole's potential, and will fall back into the nucleus. The morphological and kinematical observations can be explained by assuming that the filaments originate in a bidirectional wind that is blowing off the disk. If this is true, the wind removes angular momentum and allows the gas in the disk to move toward the black hole.

The radius of the central black hole's radius of influence relative to the size of the nuclear disk determines whether the disk will be a dusty D-type or an ionized I-type disk. D-type disks (NGC 4261 and NGC 6251) occur when the black hole's radius of influence is much smaller than the disk's radius. I-type disks (M 87 and M 81) occur when the radius of influence is equal to or greater than the radius of the nuclear disk. Intermediate D/I types (3C 264) occur when the radius of influence is a significant fraction of the disk's radius.

References

Baum, S.A., O'Dea, C.P., Giovannini, G., Biretta, J., Cotton, W.B., De Koff, S., Feretti, L., Golombek, D., Lara, L., Macchetto, F. D., Miley, G. K., Sparks, W.B., Venturi, T., Komissarov, S.S. (1997): HST and MERLIN Observations of 3C 264 — A Laboratory for Jet Physics and Unified Schemes, *ApJ*, **483**, 178

Cecil, G., Wilson, A.S., & Tully, R.B. (1992): The Braided Jets in the Spiral Galaxy NGC 4258, *ApJ*, **390**, 365

Dopita, M.A., Koratkar, A.P., Allen, M.G., Tsvetanov, Z.I., Ford, H.C., Bicknell, G.V., & Sutherland, R.S. (1997): The LINER Nucleus of M 87: A Shock-excited Dissipative Accretion Disk, *A.J.*, **490**, 202

Dopita, M.A. (1999): The LINER Nucleus of M 87: A Shock-excited Dissipative Accretion Disk, this workshop

Ferrarese, L., Ford, H. & Jaffe, W. (1996): Evidence for a MBH in the Active Galaxy NGC 4261 from HST Images and Spectra, *ApJ*, **470**, 444 – 459 (FFJ96)

Ferrarese, L., Ford, H.C., & Jaffe, W. (1998): Evidence for a $(7.5 \pm 2.2) \times 10^8 M_\odot$ Black Hole in the Active Galaxy NGC 6251 from Hubble Space Telescope WFPC2 and FOS Observations, in preparation (FFJ98)

Ford, H.C. & Butcher, H. (1979): The System of Filaments in M 87 - Evidence for Matter Falling into an Active Nucleus, *ApJS*, **41**, 147

Ford, H.C., Dahari, O., Jacoby, G.H., Crane, P.C., and Ciardullo, R. (1986): Bubbles and Braided Jets in Galaxies with Compact Radio Nuclei, *ApJ*, **311**, L7

Ford, H., Harms, R., Tsvetanov, Z., Hartig, G., Dressel, L., Kriss, G., Davidsen, A., Bohlin, R., Margon, B. (1994): Narrow Band *HST* Images of M 87: Evidence for a Disk of Ionized Gas Around a MBH, *ApJ*, **435**, L27–L30

Ford, H.C., Ferrarese, L., Hartig, G., Jaffe, W., Tsvetanov, Z., and van den Bosch, F. 1996, "Hubble Space Telescope Observations of the Centers of Elliptical Galaxies," in *Proc. Nobel Symposium No. 98, Barred Galaxies and Circumnuclear Activity*, ed. A. Sandqvist and P.O. Lindblad (Heidelberg, Germany: Springer-Verlag), 293 (F96)

Ford, H., Tsvetanov, Z., Hartig, G., Kriss, G., Harms, R., and Dressel, L. (1996): HST FOS COSTAR Small Aperture Spectroscopy of the Disk of Ionized Gas in

M 87, in *Science with the Hubble Space Telescope — II*, ed. P. Benvenuti, F.D. Macchetto, & E.J. Schreier (Space Telescope Science Institute, 1996), 192 – 194 (F96)

Ford, H.C., Tsvetanov, Z., Ferrarese, L., Kriss, G., Jaffe, W., Harms, R., and Dressel, L. (1997): Gaseous Disks in the Nuclei of Elliptical Galaxies in *Accretion Phenomena and Related Outflows, IAU colloquium 163*, ed. D. T. Wickramasinghe, G. V. Bicknell, and L. Ferrario (ASP Conference Series) (F97)

Ford, H.C., Tsvetanov, Z.I., Ferrarese, L., & Jaffe, W. (1998): HST Detections of Massive Black Holes in the Centers of Galaxies, in the Proc. IAU Symp. No. 184, *The Central Regions of the Galaxy and Galaxies*, ed. Yoshiaki Sofue (Kluwer Academic Publishers, Dordrecht), in press.

Harms, R., Ford, H., Tsvetanov, Z., Hartig, G., Dressel, L., Kriss, G., Bohlin, R., Davidsen, A., Margon, B., Kochhar, A. (1994): *HST* FOS Spectroscopy of M 87: Evidence for a Disk of Ionized Gas Around a MBH, *ApJ*, **435**, L35 – L38 (H94)

Jaffe, W. Ford, H.C., O'Connell, R.W., van den Bosch, F., & Ferrarese, L. (1994): Hubble Space Telescope Photometry of the Central Regions of Virgo Cluster Elliptical Galaxies I. Observations, *AJ*, **108**, 1567 (J94)

Jaffe, W., Ford, H.C., Ferrarese, L., van den Bosch, F., & O'Connell, R.W.O. (1996): The Nuclear Disk of NGC 4261: HST Images and WHT Spectra, *ApJ* **460**, 214

Macchetto, F., Marconi, A., Axon, D.J., Capetti, A., Sparks, W., & Crane, P. (1997): The Supermassive Black Hole of M 87 and the Kinematics of its Associated Gaseous Disk, *ApJ*, **489**, 579 (M97)

Maoz, E. (1995): A Stringent Constraint on Alternatives to a Massive Black Hole at the Center of NGC 4258, *ApJ*, **447**, L91

Maoz, E. (1998): Stellar Dynamical Constraints On Alternatives To a Black Hole, in the Proc. IAU Symp. No. 184, *The Central Regions of the Galaxy and Galaxies*, ed. Yoshiaki Sofue (Kluwer Academic Publishers, Dordrecht), in press.

Sparks, W. B, Ford, H. C., & Kinney, A. (1993): The Dusty Emission Filaments of M 87,*ApJ*, **413**, 531 (SFK)

Tsvetanov, Z. I., Allen, M. G., Ford, H. C. *et al.* (1999a): Morphology of the Nuclear Gaseous Disk in M 87, these proceedings

Tsvetanov, Z. I., Hartig., G. F., Ford, H.C., Kriss, G. A., Dopita, M. A., Dressel, L. L., & Harms, R. J. (1999b): The Nuclear Spectrum of M 87, these proceedings

The Supermassive Black Hole of M 87 and the Kinematics of Its Associated Gaseous Disk

F. Duccio Macchetto[1,2]

[1] Space Telescope Science Institute
3700 San Martin Drive
Baltimore, MD 21218
USA
[2] Affiliated to the Astrophysics Division
Space Science Department of ESA
Space Telescope Science Institute

Abstract. We have carried out the *first* HST long-slit observations of a gaseous disk around a candidate super-massive black hole. The results of this study on the kinematics of the gaseous disk in M 87 are a considerable improvement in both spatial resolution and accuracy over previous observations and requires a projected mass of $M_{BH}(\sin i)^2 = (2.0 \pm 0.5) \times 10^9 \, M_\odot$ ($M_{BH} = 3.2 \times 10^9 \, M_\odot$ for a disk inclination $i = 52°$) concentrated within a sphere whose radius is less than $0\rlap{.}''05$ (3.5 pc) to explain the observed rotation curve. The kinematics of the ionized gas is well described by a thin disk in Keplerian motion. A lower limit to the mass-to-light ratio of this region is $M/L_V \simeq 110$, significantly strengthening the claim that this mass is due to the presence of a central black hole in M 87.

1 Introduction

One of the cornerstones of the contemporary view of the physics of active galactic nuclei is that their energy output is generated by accretion of material on to a massive black hole (*e.g.* Blandford 1991, Antonucci 1993 and references therein). Considerable controversy has surrounded attempts to verify the existence of black holes in nearby giant elliptical galaxies using ground-based stellar dynamical studies (Sargent *et al.* 1978, Young *et al.* 1978, Duncan & Wheeler 1980, Binney and Mamon 1982). To-date the best available data remains inconclusive, largely because of the difficulty of detecting the high-velocity wings on the absorption lines, which are the hallmark of a black hole (van der Marel 1994, Dressler & Richstone 1990, Jarvis & Pelletier 1991, Kormendy & Richstone 1995). One of the major goals of HST has been to establish or refute the existence of black holes in active galaxies by probing the dynamics of AGN at much smaller radii than can be achieved from the ground. Already persuasive evidence has been found that apparently quiescent nearby galaxies contain black holes (van der Marel 1997a, Kormendy *et al.* 1996) but, locally, there are few candidate galaxies which might harbour a super-massive black hole large enough to have sustained Quasar type activity earlier in their evolution (Rees 1997). M 87 is the nearest and brightest elliptical believed to harbour such a super-massive black hole.

M 87, the dominant giant elliptical galaxy at the center of the Virgo cluster, was the first galaxy for which tenable stellar dynamical and photometric evidence was advanced for the presence of a supermassive black hole (Sargent *et al.* 1978, Young *et al.* 1978). Subsequent stellar dynamical models showed however that the projected stellar density and the observed rise in velocity dispersion did not necessarily imply the presence of a black hole, but could still be explained by the presence of a stellar core with an anisotropic velocity tensor(Duncan & Wheeler 1980, Young 1980, Binney and Mamon 1982).

An important step in the quest to verify the existence of a massive black hole in M 87 has been the discovery from HST emission line (Crane *et al.* 1993, Ford *et al.* 1994) and continuum (Macchetto 1996) imagery of a circumnuclear disk of ionized gas which is oriented approximately perpendicularly to the synchrotron jet (Sparks *et al.* 1996). Similar gaseous disks have also been found in the nuclei of a number of other massive galaxies (Jaffe *et al.* 1993, Ferrarese *et al.* 1996). Because of surface brightness limitations on stellar dynamical studies at HST resolutions, the kinematics of such disks are in practice likely to be the only way to determine if a central black hole exists in all but the very nearest galaxies.

Previous HST spectroscopic observations (Harms *et al.* 1994, Ford *et al.* 1996) of the gas disk at several discrete locations on opposite sides of the nucleus showed a velocity difference of $> 1000\,\text{km s}^{-1}$. On the assumption that these motions arise in a thin rotating Keplerian disk, this lead to a central dark mass of M 87 in the range 1 to $3.5 \times 10^9\,\text{M}_\odot$. An important shortcoming of this estimate is the assumption about the inclination of the disk with respect to the line of sight which cannot be properly determined neither by using the few velocities derived from the FOS observations nor from the WFPC2 imaging data. Harms *et al.* (1994) assumed a disk inclination of $42° \pm 5°$, as derived by an ellipse fitting of the Hα+[N II] image, and this resulted in a misleadingly accurate mass value. Actually, such value for the inclination was determined at distances between $0\rlap{.}''3$ and $0\rlap{.}''8$ from the nucleus, because in the inner regions the disk emission was subsumed by the bright central point-source. Indeed Ford *et al.* (1996) quoted a larger uncertainty in their mass determination but it is not clear from that short paper whether they have fully taken into account all of the possible values of the disk inclination which are consistent with the data. Implicit in this measurement of the mass of the central object is also the assumption that the gas motions in the innermost regions reflect Keplerian rotation and not the effects of non-gravitational forces such as interactions with the jet which are known to dominate the gas motions in the inner regions of many AGN (Whittle *et al.* 1988, Axon *et al.* 1989). Establishing the detailed kinematics of the disk is therefore critical. Currently their result still provides the most convincing observational evidence in favour of the black hole model.

In order to determine whether the motions observed in M 87 are due to Keplerian rotation, we have re-investigated the velocity field at high spatial resolution with HST. Our results demonstrate that both the observed rotation curve and line profiles are consistent with a thin-disk in Keplerian motion and allow us to determine an improved estimate for the mass of the black hole. On the assumption that the distance to M 87 is 15 Mpc, $0\rlap{.}''1 \simeq 7$ pc.

2   Observations and Data Reduction

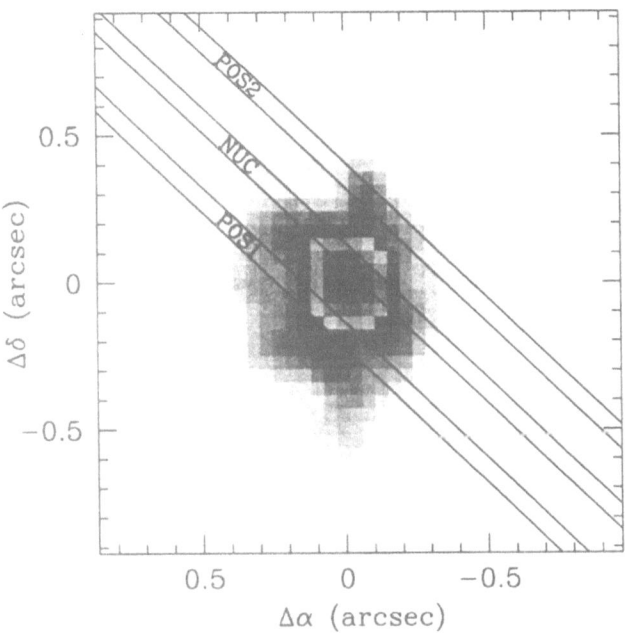

Fig. 1. The slit positions discussed in the text are marked with solid lines on an Hα+[N II] image from the HST archive. North is up and east is left.

The circumnuclear disk of M 87 was observed on 1996 July 25th, with the COSTAR corrected Faint Object Camera f/48 long-slit spectrograph on board the Hubble Space Telescope. An F305LP filter, was used to isolate the first order spectrum which covers the 3650–5470 Å region. All the spectra have a pixel size of $0\rlap{.}''0287$ in the spatial direction and cover the spectral lines of [O II]λλ3726,3729 Å, Hβλ4861 Å, and [O III]λ4959,5007 Å at a resolution of 1.78 Å per pixel and were taken in the 1024 × 512 non-zoomed format of the Faint Object Camera. An interactive acquisition image (F140W filter) was first obtained to accurately locate the nucleus. The slit, with dimensions of 13″5×0″063 at a position angle of 47°, was then stepped across the nucleus at three positions separated by 0″2. The integration times for the inner spectrum was 7761 seconds while those of the two outer spectra where 2261 seconds. Comparing the continuum flux observed in the spectra with the luminosity distribution derived from a F342W FOC, f/96 archival image we could estimate that the inner slit position was located to within 0″07±0″01 of the nucleus. The slit positions used are shown in Fig. 1 overlaid on an Hα+[N II] WFPC2 image of the gaseous disk of M 87 obtained from the HST data archive.

The distortion induced by the optics and the magnetic focusing of the detector was corrected by using the equally spaced grid of reseau marks, etched onto the photocathode of the detector (Nota, Jedrzejewski & Hack 1995). The distortion induced by the spectroscopic mirror and the grating were corrected using spectra of the planetary nebula NGC 6543 and of the globular cluster 47 Tucanae for the slit and dispersion directions respectively which are then characterized by maximum uncertainties of 0.45 pix and 0.28 pix (*i.e.* 0.8 Å and 8 mas) when comparing measurements all across the detector. After the background subtraction, the only line with enough signal to be suitable for velocity measurements was [O II]λ3726,3729 Å, which was then fitted (row by row) using the task LONGSLIT in the TWODSPEC FIGARO package (Wilkins & Axon 1992). In all cases the line profile is well represented by a Gaussian function. Full details of the observation strategy and data reduction and results are given in Macchetto *et al.* 1997.

3   Results

The measured variations of radial velocity and continuum flux for the central slit are shown in Fig. 2. It is immediately apparent that the observed velocity field does not show the very steep rise in velocity one would expect to be the signature of a massive black hole. Instead over the central ±0″2 the velocity varies approximately linearly with position, with an amplitude of ±600 km s^{-1}, while at larger radii it flattens, and eventually starts to turn over. As shown in the following section, this apparent flattening is a consequence of both the finite distance between the slit and the nucleus and the smearing of the instrumental spatial PSF.

The comparison between the new FOC data and the archival FOS observations is shown in Fig. 3 where we include only those FOS apertures which overlapped our slit at NUC. Within their substantially larger uncertainties, the previous velocity measurements (Harms *et al.* 1994, Ford *et al.* 1996) are in reasonable accord with our results. The comparison shows clearly the increase in spatial resolution and sampling of the FOC observations with respect to the FOS ones. Moreover, our data provide considerably more reliable relative spatial positions, since all the data points were obtained simultaneously on the detector. On the contrary, the limitations of the FOS target

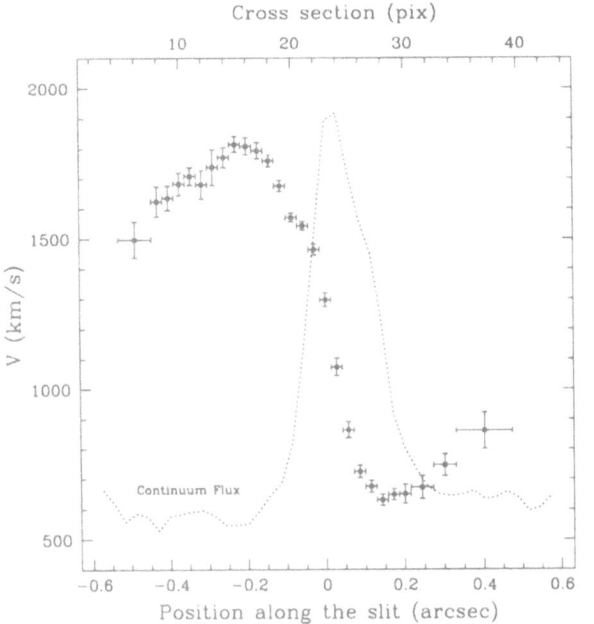

Fig. 2. Observed rotation curve from the [O II]$\lambda\lambda$3726,3729 Å doublet for the spectrum with the smallest impact parameter to the nucleus of M 87 compared with the continuum flux distribution along the slit. The uncertainties in the measured velocities and positions are 65 km s^{-1}, at 3727 Å, and 0.28 pixels (8 mas) respectively.

acquisition procedure can produce errors as large as 0.''1 (Van der Marel *et al.* 1997b) and such uncertainty in the position is crucial in the inner 0.''2 when comparing observations taken at different epochs.

4 Discussion

To understand the implications of our results for the mass distribution in the center of M 87 we built models of the gas kinematics both under the assumption of the existence of a central black hole, and an extended mass distribution following a Plummer Potential.

In the hypothesis of a thin disk rotating in circular orbits, neglecting the slit width, instrumental PSF and luminosity distribution of the line, the observed velocity is given by:

$$V = V_{sys} - \Phi(R)^{0.5}\frac{X}{R}\sin i \tag{1}$$

where

$$X = -b\sin\theta + s\cos\theta \quad Y = b\cos\theta + s\sin\theta$$

and

$$R = \left(X^2 + Y^2\right)^{0.5} \ .$$

b is the distance of the slit from the nucleus, s is the coordinate along the slit and R represents the "true" distance of each point on the slit from the nucleus ($s = 0$ when $R = b$). V_{sys} is the systemic velocity, i is the inclination of the disk with respect to the line of sight ($i = 90°$ for the edge-on case), θ is the angle between the slit and the line of nodes. $\Phi(R)$ is the gravitational potential and results

$$\Phi(R) = \frac{GM_{BH}}{R} \quad \text{Keplerian}$$

$$\Phi(R) = \frac{GM}{(R^2 + R_C^2)^{0.5}} \quad \text{Plummer}$$

where M_{BH} and R_C are the black hole mass and core radius, respectively.

The effects of the f/48 spatial PSF and the finite slit size are taken into account by averaging the velocity field using the luminosity distribution and PSF as weights. The model rotation curve is thus given by:

$$V_{ps}(S) = \frac{\int_{S-\Delta S}^{S+\Delta S} ds \int_{B-h}^{B+h} db \int \int_{-\infty}^{+\infty} db' ds' V(s', b') I(s', b') P}{\int_{S-\Delta S}^{S+\Delta S} ds \int_{B-h}^{B+h} db \int \int_{-\infty}^{+\infty} db' ds' I(s', b') P} \tag{2}$$

where $V(s', b')$ is the Keplerian velocity derived in (1), $I(s', b')$ is the intrinsic luminosity distribution of the line, $P = P(s' - s, b' - b)$ is the spatial PSF of the f/48 relay along the slit direction. B is the impact parameter (measured at the center of the slit) and $2h$ is the slit size, S is the position along the slit at which the velocity is computed and $2\Delta S$ is the pixel size of the f/48 relay. For the PSF we have assumed a Gaussian with 0.''08 FWHM $i.e.$

$$P = \frac{1}{\sqrt{2\pi\sigma^2}} \exp\left(-\frac{1}{2}\frac{(s'-s)^2}{\sigma^2} - \frac{1}{2}\frac{(b'-b)^2}{\sigma^2}\right) \ . \tag{3}$$

Because the intrinsic surface brightness distribution of the emission line disk interior to the HST PSF is not known, we modeled this with two extreme parameterized forms, either as a power law or as an exponential. We then fitted the model rotation curves using a χ^2 minimization. The free parameters of the fit, in the case of the black hole model, were the inclination of the disk with respect to the line of sight, the angle between the disk and the line of nodes (positive angles indicate that the line of nodes has a larger position angle than the slit), the position along the slit of the point closest to the nucleus, the impact parameter of the slit, the systemic velocity and the mass of the black hole. The uncertainties are dominated by the poorly determined inclination of the gas disk within the HST PSF. Taking into account the possible ranges of variation of the model free parameters, which were derived using a large Monte Carlo generated grid of solutions, we find a best estimate of the projected mass of the black hole of $M_{BH}(\sin i)^2 = (2.0\pm0.5)\times10^9\,M_\odot$. Three representative model fits are shown in Fig. 3. The solid, dashed and dotted lines represent three black hole model fits which are compatible with the observations and differ in the value of the angle between the slit and the line of nodes of the gaseous disk ($\theta \simeq -9°$, $7°$ and $1°$, respectively). The best fit model, represented by the solid line in the upper panel has a disk inclination, $i = 51°(39°; 65°)$, angle of line of nodes, $\theta = -9°(-15°; 13°)$, impact parameter, $b = 0\rlap{.}''08(0\rlap{.}''06; 0\rlap{.}''08)$ and a systemic velocity, $V_{sys} = 1290\,\mathrm{km\,s^{-1}}$ (1080; 1355), leading to a mass estimate for the black hole of $M_{BH} = (3.2\pm0.9)\times10^9\,M_\odot$. The values in parentheses represent the allowed errors on the values of the parameters. As shown by this analysis, the main source of uncertainty on the *deprojected* mass is the value of the disk inclination which more than doubles the total error.

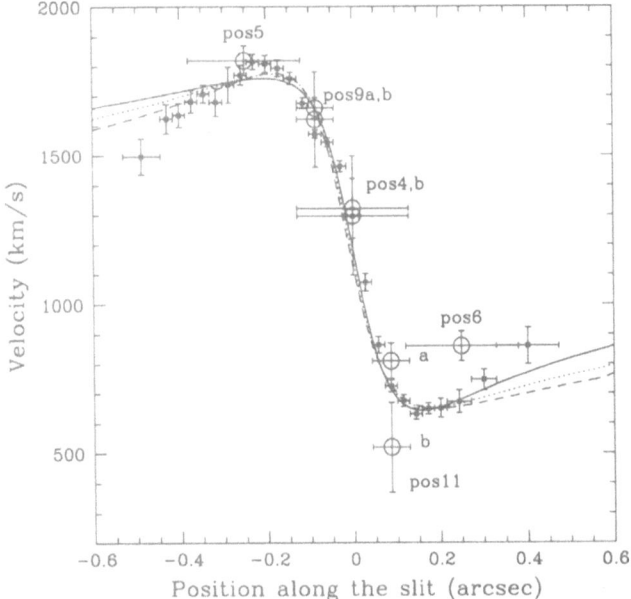

Fig. 3. Comparison between the FOC rotation curve and the archival FOS observations. The labels "pos#" refer to the notation used by Ford *et al.* (1996). "a" and "b" indicate positions which were nominally coincident but resulted slightly shifted due to acquisition uncertainties. The error-bars in position represent the aperture diameters, those in velocity represent the dispersion of the measurements on different lines. The solid, dotted and dashed lines are model fits which are discussed in the text.

In the case of an extended mass (Plummer) distribution the observations are reproduced only if at least 60% of the total mass is confined within a sphere of $0\!\!''\!05$ ($\simeq 3.5\,\mathrm{pc}$). Since the maximum spatial extent is determined by the PSF FWHM ($0\!\!''\!08$) this possible mass distribution is therefore indistinguishable from a point mass.

We have demonstrated that the data are fully compatible with a thin disk in Keplerian rotation around a central mass condensation, at least within the inner $0\!\!''\!2$–$0\!\!''\!3$. At larger distances, the signal-to-noise in our data decreases and, consequently, the error bars increase. The few points that appear to deviate from the Keplerian rotation curve, do not provide enough information or weight to warrant the fit of more complex models, such as thick disks or sub-Keplerian disks with outflows.

Can we rule out the alternative explanation that the observed rotation is due to an extended mass? A strong argument against this interpretation is that a mass of $(3.2 \pm 0.9) \times 10^9\,M_\odot$ distributed within $3.5\,\mathrm{pc}$ yields a mean density of $\approx 2 \times 10^7\,M_\odot\,\mathrm{pc}^{-3}$, higher than the highest density encountered in the collapsed cores of galactic globular clusters. Moreover, the V band flux in a region $16 \times 16\,\mathrm{pc}^2$ of the nucleus yields a mass-to-light ratio $M/L_V \simeq 110\,M_\odot/L_{V\odot}$ ($L_{V\odot}=0.113\,L_\odot$) which is far higher than those predicted for stellar clusters by evolution synthesis Bruzual (1995). The mass condensation in the nucleus of M 87 can therefore not be a cluster of normally evolved stars. Even more exotic possibilities (*e.g.* massive clusters of neutron stars, other dark objects etc.) have been discussed by van der Marel *et al.* (1997a) in the case of M 32 and found to be equally improbable. Therefore the most natural explanation for the central mass condensation of M 87 is that it is due to the presence of a black hole.

The data and the analysis presented by Harms *et al.* (1994) and Ford *et al.* (1996) indicated the presence of a velocity shear which could be consistent which the presence of a black hole. Here we have demonstrated the existence of a Keplerian rotation curve in the disk and we have also excluded the possibility that it might be due to a mass concentration more extended than our spatial resolution.

5   Conclusions

We have presented the *first* HST long-slit spectrum of a gaseous disk around a candidate supermassive black hole. We have obtained a rotation curve in the [O II]$\lambda\lambda$3736,3729 which extends up to $\simeq 1''$ from the nucleus. We have modeled the rotation curve in the case of a thin disk in circular orbits around a supermassive black hole and we have derived a projected mass of $M_{\mathrm{BH}}(\sin i)^2 = (2.0 \pm 0.5) \times 10^9\,M_\odot$ ($M_{\mathrm{BH}} = 3.2 \times 10^9\,M_\odot$ for a disk inclination, $i = 52°$) concentrated within a sphere whose radius is less than $0\!\!''\!05$ ($3.5\,\mathrm{pc}$). We have shown that the observed characteristic of the data are well explained under the working hypothesis of circular Keplerian orbits and that there is no substantial contribution from a mass distribution more extended than our spatial resolution. Given the inferred mass-to-light ratio $M/L_V \simeq 110\,M_\odot/L_{V\odot}$ the most natural and likely explanation is that of a supermassive black hole. In conclusion, with respect to previous determinations, we have improved the accuracy of the mass estimate, demonstrated its reliability by verifying the assumption of the thin disk in circular, Keplerian rotation and excluded the possibility of a mass distribution more extended than the spatial PSF.

To make further progress there are a number of possibilities the easiest of which is to make a more comprehensive and higher signal-to-noise 2D velocity map of the disk to better constrain its parameters. The biggest limitation of the present data is that, even by observing with HST at close to its optimal resolution at visible wavelengths, some of the important features of the disk kinematics are subsumed by the central PSF. Until a larger space based telescope becomes available, the best we can do is to study the gas disk in Lyα and gain the Rayleigh advantage in resolution by moving to the UV. This may be the only way to proceed because of the difficulty of detecting the high velocity wings which characterize the stellar absorption lines in the presence of a supermassive black hole.

Acknowledgements

This work was carried out with a team of friends, whose names appear in the refereed publications (Marconi *et al.* 1997, Macchetto *et al.* 1997). They are in alphabetical order, D. Axon, A. Capetti, P. Crane, A. Marconi and W. B. Sparks.

References

Antonucci, R. (1993): *ARA&A*, **31**, 473

Axon, D. J., Unger, S. W., Pedlar, A., Meurs, E. J. A, Whittle, D. M. and Ward, M. J. (1989): *Nature*, **341**, 631

Binney, J. and Mamon, S. (1982): *MNRAS*, **200**, 361

Blandford, R. D. (1991): *Physics of AGN*, Proceedings of Heidelberg Conference, Springer-Verlag, eds. W. J. Duschl and S. J. Wagner, p. 3

Bruzual, G. A. (1995): *From Stars to Galaxies: the Impact of Stellar Physics on Galaxy Evolution*, Proceedings of Crete Conference, eds. C. Leitherer, U. Fritze-von Alvensleben and J. Huchra, ASP Conf. Series, vol. 98, p. 14

Capetti, A., Axon, D. J., Macchetto, F., Sparks, W. B. and Boksenberg, A. (1996): *ApJ*, **469**, 554

Crane, P., Stiavelli, M., King, I. R., Deharveng, J. M., Albrecht, R., Barbieri, C., Blades, J. C., Boksemberg, A., Disney, M. J., Jakobsen, P. (1993): *AJ*, **106**, 1371

Dressler, A., Richstone, D. O. (1990): *ApJ*, **348**, 120

Duncan, M. J. and Wheeler, J. C. (1980), *ApJ*, **237**, L27

Ferrarese, L., Ford, H. C., Jaffe, W. (1996): *AJ*, **470**, 444

Ford, H. C., Harms, R. J., Tsvetanov, Z. I., Hartig, G. F., Dressel, L. L., Kriss, G. A., Bohlin, R. C., Davidsen, A. F., Margon, B., Kochhar, A. K. (1994): *ApJ*, **435**, L27

Ford, H. C., Tsvetanov, Z. I., Hartig, G. F., Kriss, G. A., Harms, R. J., Dressel, L. L. (1996): *Science with the HST–II*, eds. P. Benvenuti, F. Macchetto and E. Schreier, p. 192

Harms, R. J., Ford, H. C., Tsvetanov, Z. I., Hartig, G. F., Dressel, L. L., Kriss, G. A., Bohlin, R. C., Davidsen, A. F., Margon, B., Kochhar, A. K. (1994): *ApJL*, **435**, L35

Kormendy, J. and Richstone, D. (1995): *ARA&A*, **33**, 581

Kormendy, J., *et al.* (1996): *ApJ*, **459**, L57

Jaffe, W., Ford, H. C., Ferrarese, L., van den Bosch, F. C., O'Connell, R. W. (1993): *Nature*, **364**, 214

Jarvis, M., Peletier, R. F. (1991): *A&A*, **247**, 315

Macchetto, F. (1996): *Science with the HST–II*, Paris, eds. P. Benvenuti, F. Macchetto and E. Schreier, p. 59

Macchetto, F. D., Marconi, A., Axon, D. J., Capetti, A., Sparks, W. B., Crane, P. (1997): *ApJ*, **489**, 579

Marconi, A., Axon, D. J., Macchetto, F. D., Capetti, A., Sparks, W. B., Crame, P. (1997): *MNRAS*, **289**, L21

Nota, A., Jedrzejewski, R., Hack, W. (1995): *Faint Object Camera Instrument Handbook Version 6.0*, Space Telescope Science Institute

Rees, M. J. (1997): in *Unsolved problems in astrophysics*, Princeton University Press, eds. J. N. Bachall and J. P. Ostriker, 181

Sargent, W. L. W., Young, P. J., Boksenberg, A., Shortridge, K., Lynds, C. R., Hartwick, F. D. A. (1978): *ApJ*, **221**, 731

Sparks, W. B., Biretta, J. A. and Macchetto, F. (1996): *ApJ*, **473**, 254

van der Marel, R. P. (1994): *MNRAS*, **270**, 271

van der Marel, R. P., de Zeeuw, P. T., Rix, H. W., Quinlan, G. D. (1997a): *Nature*, **385**, 610

van der Marel, R. P., de Zeeuw, P. T., Rix, H. W. (1997b): *ApJ* submitted

Whittle, D. M., Ward, M. J., Meurs, E. J. A., Pedlar, A., Unger, S. W. and Axon, D. J. (1988): *ApJ*, **326**, 125

Wilkins, T. W. and Axon, D. J. (1992) in *Astronomical data analysis software and systems I*, Ast. Soc. Pac. Conf. Ser. 25, p. 427

Young, P. J., Westphal, J. A., Kristian, J., Wilson, C. P., Landauer, F. T. (1978): *ApJ*, **221**, 721

Young, P. J. (1980): *ApJ*, **242**, 1232

Morphology of the Nuclear Disk in M 87

Z.I. Tsvetanov[1], M.G. Allen[1,2], H.C. Ford[1,3], and R.J. Harms[4]

[1] Johns Hopkins University, Baltimore, MD 21218, USA
[2] Mount Stromlo and Siding Spring Observatories, ACT 2611, Australia
[3] Space Telescope Science Institute, Baltimore, MD 21218, USA
[4] RJH Scientific, 5904 Richmond Highway, Alexandria, VA 22303, USA

Abstract. A deep, fully sampled diffraction limited (FWHM $\sim$ 70 mas) narrow-band image of the central region in M 87 was obtained with the Wide Filed and Planetary Camera 2 of the *Hubble Space Telescope* using the dithering technique. The Hα+[N II] continuum subtracted image reveals a wealth of details in the gaseous disk structure described earlier by Ford *et al.* (1994). The disk morphology is dominated by a well defined three-arm spiral pattern. In addition, the major spiral arms contain a large number of small "arclets" covering a range of sizes ($0''.1$–$0''.3$ = 10–30 pc). The overall surface brightness profile inside a radius $\sim 1''.5$ (100 pc) is well represented by a power-law $I(\mu) \sim \mu^{-1.75}$, but when the central $\sim$ 40 pc are excluded it can be equally well fit by an exponential disk. The major axis position angle remains constant at about PA$_{\rm disk} \sim 6°$ for the innermost $\sim 1''$, implying the disk is oriented nearly perpendicular to the synchrotron jet (PA$_{\rm jet} \sim 291°$). At larger radial distances the isophotes twist, reflecting the gas distribution in the filaments connecting to the disk outskirts. The ellipticity within the same radial range is $e = 0.2 - 0.4$, which implies an inclination angle of $i \sim 35°$. The sense of rotation combined with the dust obscuration pattern indicate that the spiral arms are trailing.

1 Introduction

The disk of ionized gas in the nucleus of M 87 is currently the best example of a family of similar small ($r \sim 100$ pc) gaseous disks found to be common in the centers of elliptical galaxies with active nuclei (for a review see Ford *et al.* 1998). Several *HST* kinematical studies have shown that in M 87 the gas is in Keplerian rotation, orbiting a massive black hole with a mass $M_{\rm BH} \sim 2-3 \times 10^9 M_\odot$ (Harms *et al.* 1994; hereafter H94, Ford *et al.* 1996a,b; and Macchetto *et al.* 1997, hereafter M97). The few other galaxies studied kinematically so far (NGC 4261 — Ferrarese *et al.* 1996, NGC 6521 — Ferrrarese *et al.* 1998, NGC 4374 — Bower *et al.* 1998) have further shown that nuclear gaseous disks offer an excellent tool for measuring the central black hole mass.

Recent studies have revealed other important characteristics of the nuclear disk in M 87. Its apparent minor axis (F96, M97) is closely aligned with the synchrotron jet ($\Delta\theta \sim 10° - 15°$) suggesting a causal relationship between the disk and the jet. The system of filaments in the center of M 87 (Sparks, Ford & Kinney 1993; SFK) may also be causally connected to the disk. For example, the filaments extending $\sim17''$ (1200 pc) to the NW at PA $\sim 315°$

are blue shifted with respect to systemic velocity and show dust absorption implying they are on the near side of M 87 as is the jet. These two findings led SFK to conclude that these filaments are streamers of gas flowing away from the center of M 87 rather then falling into it. The images in F94 (see also Ford & Tsvetanov, this volume, FT98) show an apparent connection between at least some of the larger scale filaments and the ionized nuclear disk.

Direct spectroscopic evidence for an outflow was found recently. Several UV and optical absorption lines from neutral and very mildly ionized gas were measured in the FOS spectrum of the nucleus (Tsvetanov et al. 1998; T98). These lines are broad (FWHM $\sim 400\,\mathrm{km\ s^{-1}}$) and blue shifted by $\sim 150\,\mathrm{km}$ $\mathrm{s^{-1}}$ with respect to M 87's systemic velocity implying both an outflow and turbulence. In addition, non-circular velocity components — both blue and red shifted — were found at several locations in the disk (F96, FT98), and observed emission lines are much broader than the expected broadening due to the Keplerian motion across the FOS aperture. All these properties are best understood if a bi-directional wind from the disk were present. This wind may be an important mechanism for removing angular momentum from the disk to allow accretion through the disk onto the central black hole.

Whatever the physical conditions in the disk it is important to map its morphology in detail. The first *HST* images (F94) have hinted that a spiral pattern could be present, but the signal-to-noise was too low for a definitive conclusion. In this paper we present deep, fully sampled diffraction limited narrow band images of the nuclear region in M 87. We use these images to characterize the ellipticity, brightness distribution, and morphology of the disk. In this paper we adopt a distance to M 87 of 15 Mpc, corresponding to a scale of $1'' = 73\,\mathrm{pc}$.

2   Observations

M 87 was imaged though the WFPC2 narrow-band filter F658N (central wavelength/effective width = $6590/28\,\text{Å}$) which isolates the Hα+[N II] emission line complex at the systemic redshift. The nucleus was placed near the center of the planetary camera (PC) and the *HST* roll angle was 307°, orienting the M 87 synchrotron jet roughly along the diagonal of the PC field of view.

The PC has a scale of $45.54\,\mathrm{mas\ pixel^{-1}}$ (Holtzman et al. 1995) and undersamples the *HST* diffraction limited PSF at all wavelengths shorter of $\sim 1\mu$. To achieve full sampling we took images at four adjacent positions, offsetting the telescope by 5.5 PC pixels between each images — the so called dithering technique. The pattern used is illustrated in Fig. 1. At each of the four subpixel positions we took 3 images for cosmic ray rejection and cleaning of permanent defects. We re-calibrated the images using the best recommended calibration files.

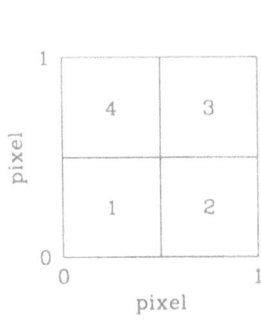
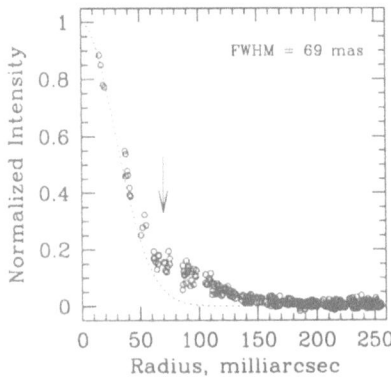

Fig. 1. *Left* panel illustrates the dithering pattern. Three images were taken at each of the positions numbered 1 to 4. The telescope offset was 5.5 PC1 pixels between positions. The *right* panel shows the radial profile of an unsaturated star ~6″ south of nucleus. The overplotted Gaussian has a FWHM equal to the *HST* diffraction limit at the wavelength of observation (6580 Å). The first Airy ring is marked by the arrow.

To build a fully sampled image we used the variable-pixel linear reconstruction algorithm (informally known as "drizzling") developed originally for the Hubble Deep Field (Fruchter & Hook 1997). The relative offsets of individual images were estimated from the positions of good signal-to-noise objects in the field and also through a 2-D cross correlation. Both techniques yielded similar results. No significant offsets were found for the three frames taken at each one of the subpixel positions. These were combined to remove the cosmic ray events. The four combined frames were then combined on an output grid of 25 mas pixel^{-1} using the drizzling algorithm. The 25 mas pixel size is a good match for the ~ 70 mas *HST* diffraction limit at Hα. This is illustrated by the radial profile of a star in the field shown in Fig. 1 where the first Airy ring is clearly seen.

Because of serious observing time restrictions, and also because the underlying galaxy profile is known to be smooth, no continuum images were obtained simultaneously with the on-line ones. To create a suitable off-band image for continuum subtraction we used a number of M 87 images from the *HST* public archive. These include a series of F814W images taken on November 11, 1995, which have small relative offsets that allow us to build a fully sampled image using the same drizzling technique. The offset pattern in not ideal (as in the case of F658N), but this is less critical because of the smooth shape of the continuum light distribution.

To form an image suitable for continuum subtraction at the wavelength of Hα we used a $(V - I)$ colour image to correct the F814W image for colour effects, which are particularly important at the positions of the dust bands. Finally, the continuum subtracted Hα+[N II] image was flux calibrated using the estimated system throughput at the redshifted position of Hα and the averaged measured [N II]/Hα line ratio.

3 Disk morphology

The central $1''$ – $2''$ of the continuum subtracted Hα+[N II] image (Fig. 2, left panel) is dominated by a clockwise winding 3-arm spiral pattern superposed on an underlying disk-like morphology. At larger radial distances the gaseous structure is less well organized, asymmetric, and gradually transition into the larger scale Hα filaments. Several of these filaments can be traced to connect to the disk but none appears to go directly into the nucleus.

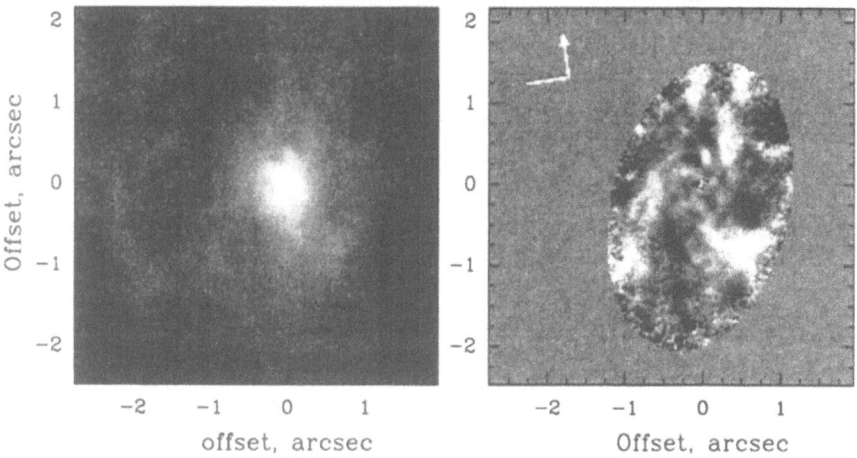

Fig. 2. The Hα+[N II] continuum subtracted image is shown in the *left* panel. The gray scale saturates the central few pixels, which are uncertain due to the imperfect colour matching and variability of the central point source. The *right* panel shows the ratio of the observed structure to the smooth disk model, as discussed in the text. The image is stretched linearly between ±70% relative to the model.

To estimate the parameters describing the surface brightness profile, ellipticity and orientation, we have performed an elliptical isophote analysis. We started by setting all parameters free, and gradually introduced constraints such as fixed center position, major axis position angle and/or ellipticity, in an attempt to explore the robustness of the solution. To our satisfaction the values for all major parameters of the fit remained stable regardless of the constraints used.

Figure 3 shows the surface brightness radial profile, major axis position angle, and ellipticity as a function of radial distance. Outside $r \sim 0''.5$ (40 pc), the surface brightness profile is well described by an exponential disk with

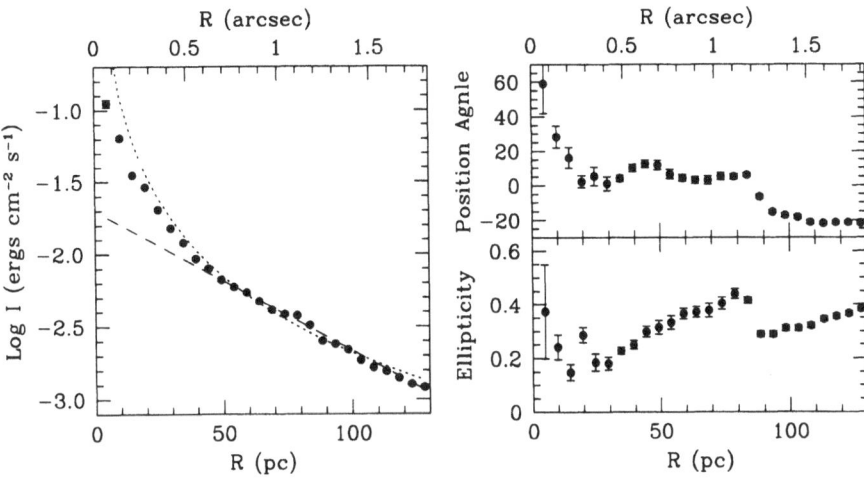

Fig. 3. *Left:* The surface brightness profile of the M 87 nuclear gaseous disk. The short dashed line is a power law $I_{\rm pl} \sim (r/50)^{-1.75}$, and the long dashed line is an exponential disk with $I_{\rm disk} \sim exp(-r/45)$, where r is in pc. *Right:* The radial dependence of the major axis position angle ($PA_{\rm maj}$) and ellipticity (e).

an effective radius of 45 pc. There is, however, a significant excess of light in the central $0\rlap{.}''5$, and a single power law $I_{\rm pl} \sim (r/50)^{-1.75}$ is a better representation of the overall profile. In the region 20–90 pc, the orientation of the major axis remains roughly constant at $PA_{\rm maj} \sim 6°$. At the same time the ellipticity increases smoothly from 0.2 to 0.4. The constancy of $PA_{\rm maj}$ is usually interpreted as a signature of a disk, while the ellipticity could be influenced by the spiral arms or there could be a warp. The central few points are less reliable because of possible colour mismatch of the on- and off-band images and the variability of the nuclear point source (T98).

The kinematic studies (F96, M97) have shown that the north part of the disk is receding and the south part is approaching and the estimated inclination is consistent with the one inferred from isophote analysis. The obscuration patches on the SE side (F94) indicate that the SE side is closer to us. Combining all this information indicates that the spiral arms are trailing.

This work is supported by NASA grant NAG-1640 to the HST FOS team.

References

Biretta, J. A., Stern, C. P., & Harris, D. E. 1991, AJ, 101, 1632 (BSH91)
Bower, G. *et al.* 1998, ApJ, 492, L111
Ferrrarese, L., Ford, H. C., & Jaffe, W. 1996, ApJ, 470, 444
Ferrrarese, L., Ford, H. C., & Jaffe, W. 1998, ApJ, in press
Ford, H. C. *et al.* 1994, ApJ, 435, L27 (F94)

Ford, H. C. *et al.* 1996, in Nobel Symp. 98, *Barred Galaxies and Circumnuclear Activity*, ed. A. Sandqvist & P. Lindblad (Heidelberg: Springer-Verlag), 293

Ford, H. C. *et al.* 1998, in IAU Symp. 184, *The Central Regions in the Galaxy and Galaxies*, in press

Fruchter, A. S., & Hook, R. 1997, http://www.stsci.edu/~fruchter/dither

Harms, R. J. *et al.* 1994, ApJ, 435, L35 (H94)

Holtzman, J. A. *et al.* 1995, PASP, 107, 156

Macchetto, F. *et al.* 1997, ApJ, 435, L35 (M97)

Sparks, W. B., Ford, H. C., & Kinney, A. 1993, ApJ, 413, 531 (SFK)

Tsvetanov, Z. I. *et al.* 1998, ApJ, 493, L83 (T98)

The Nuclear Spectrum of M 87

Z.I. Tsvetanov[1], G.F. Hartig[2], H.C. Ford[1,2], G.A. Kriss[1], M.A. Dopita[3], L.L. Dressel[4], and R.J. Harms[4]

[1] Johns Hopkins University, Baltimore, MD 21218, USA
[2] Space Telescope Science Institute, Baltimore, MD 21218, USA
[3] Mount Stromlo and Siding Spring Observatories, ACT 2611, Australia
[4] RJH Scientific, 5904 Richmond Highway, Alexandria, VA 22303, USA

Abstract. The nuclear spectrum of M 87 covering the Lyα-Hα wavelength range was obtained with the *HST* Faint Object Spectrograph (FOS) trough a $0''.21$ aperture. Contrary to some previous claims, a single power law ($F_\nu \sim \nu^{-\alpha}$) can not reproduce the observed continuum shape and at least a broken power law is required for a good fit ($\alpha = 1.75$ and 1.41 shortward and longward of the break at ~ 4500 Å). We detect a set of broad (FWHM ~ 400 km s^{-1}) absorption lines arising in the gas associated with M 87. These are only lines from neutral and very low ionization species blueshifted by ~ 150 km s^{-1} relative to the M 87 systemic velocity, indicating a net gas outflow and turbulence. The excitation sensitive emission line ratios suggest that shocks may be the dominant energy supplier.

The nuclear source in M 87 is significantly variable. From the FOS target acquisition data, we have established that the flux from the optical nucleus varies by a factor ~ 2 on time scales of ~ 2.5 months and by as much as 25% over 3 weeks, and remains unchanged ($\leq 2.5\%$) on time scales of ~ 1 day. These time scales limit the physical size of the emitting region to a few hundred gravitational radii. The variability, combined with other observed spectral properties, strongly suggest that M 87 is intrinsically of BL Lac type but is viewed at an angle too large to reveal the classical BL Lac properties.

1 Introduction

M 87 with its famous synchrotron jet is, perhaps, one of the most studied extragalactic objects in modern astronomical research. In this respect the spectrum of the nucleus is of particular importance for mapping the properties of the central source and a better understanding of its physics. Nevertheless, there is basically no good spectrum of the nucleus in the UV-optical-IR region. This is at least partially due to the difficulty of isolating from the ground the central point source from the bright host galaxy light. The best nuclear spectroscopy has concentrated on obtaining high resolution, high S/N ratio data for measuring the central black hole (BH) mass.

The situation changed dramatically with the installation of corrective optics (COSTAR) on *HST* in December 1993. Several critical observations have been made since then. These include the discovery of a nuclear gaseous disk (Ford *et al.* 1994) and vastly improved measurements of the BH mass (see Ford *et al.* and Macchetto *et al.* in these Proceedings) among others.

Here we present the first spectrum of the nucleus of M 87 obtained solely with the Faint Object Spectrograph on *HST* though a small 0″.21 aperture. At *HST* resolution the flux in this aperture is entirely dominated by the nuclear point source, and the contribution from the host galaxy can be neglected. The Lyα–Hα coverage allows us to study the continuum shape and map the absorption and emission line spectrum significantly better than any previous attempt. In accordance with several recent works we assume a distance to M 87 of 15 Mpc and a systemic velocity of 1280 km s^{-1}.

2   The nuclear spectrum

2.1   Continuum shape

The observed nuclear spectrum is dominated by a smooth featureless continuum with broad emission lines and a number of absorption lines imprinted on it (Fig. 1). We do not see any starlight features above the noise and estimate that the stellar contribution does not exceed the 10%–15% level at optical wavelengths.

Several earlier studies (*e.g.* Thomson *et al.* 1995) have strongly argued that the nuclear spectrum is synchrotron emission. Some authors have suggested that either optical-IR-radio (Zeilinger, Peletier & Stiavelli 1993) or optical-UV-X-ray (Biretta, Stern & Harris 1991) portions could be represented by a single power law. All these studies, however, relied on broad-band ground-based optical and IR photometry of the nuclear source.

In fitting the continuum shape we have adopted the strategy of finding the simplest possible model producing an acceptable fit to the data. The foreground galactic reddening is estimated at $E(B - V) = 0.017$ (Jacoby, Ciardullo & Ford 1990). In addition, the M 87 nuclear gaseous structure contains dust (see Sparks, Ford & Kinney 1993 and §2.2 below), and thus there will be some additional extinction if our line of sight to the nucleus passes through it.

The basic result from our fitting is that the observed continuum shape is inconsistent with a single power law regardless of the amount of extinction and the reddening law. There is additional curvature in the spectrum that can be accounted for by introducing a break in the power law (see Fig. 1). Of course, a smooth, curved model similar to the synchrotron self-Compton solutions employed by Stiavelli, Peletier & Carollo (1997) will also be a good representation of our data. In this respect, it is important to note that extrapolation of our best-fit broken power law misses both the radio and X-ray observations by a very significant margin.

2.2   Absorption and emission lines

We detect two clearly separated absorption line systems — one arising in the interstellar gas of our own galaxy and another one in that of M 87. Figure 2

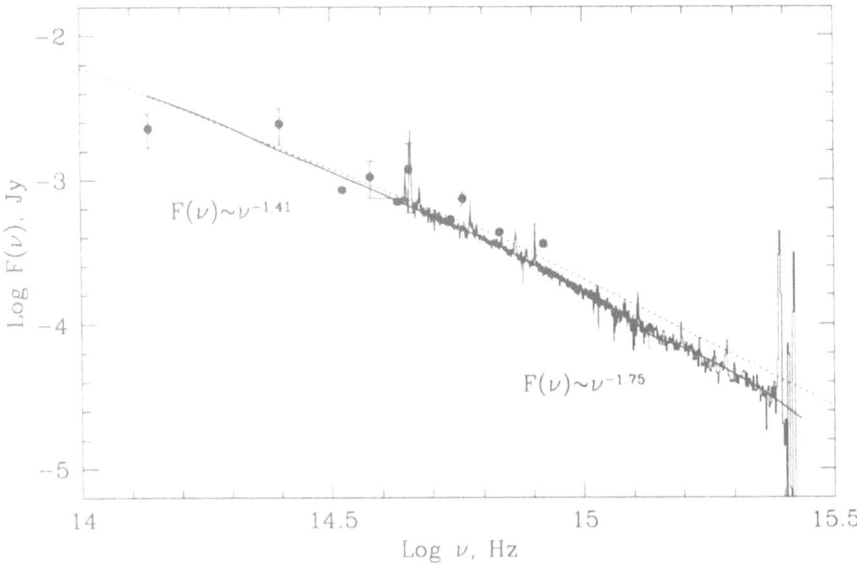

Fig. 1. The observed spectrum (light gray line) is plotted binned to the FOS resolution (1 diode) to reduce the noise. Our best fit — a broken power law with $E(B - V) = 0.039$ — is shown by a solid line. The two intrinsic (*i.e.* unreddened) power laws are shown as dashed lines connecting at the break of ~ 4500 Å. The filled circles are broad-band measurements from BSH91, Ze93 and St97. Our fit is extrapolated into the IR to show its position relative to the ground-based measurements.

shows the selected wavelength ranges with fitted Gaussian models overplotted onto the observed spectrum. We fitted all lines in each group simultaneously, imposing a minimum set of restrictions.

The averaged observed heliocentric velocity of the absorption lines arising in M 87 is $(1202 \pm 22)\,\mathrm{km\ s^{-1}}$, or $(1134 \pm 22)\,\mathrm{km\ s^{-1}}$ if corrected for the suspected wavelength zero point offset inferred from the comparison of positions of the Galactic and H I absorption features. This means that the gas responsible for the absorption in M 87 is blueshifted by $\sim 150\,\mathrm{km\ s^{-1}}$ relative to the systemic velocity of $1280\,\mathrm{km\ s^{-1}}$.

In M 87 we see absorption only from neutral and very mildly ionized gas — all ions have ionization potential $\chi < 10\,\mathrm{eV}$. In addition all M 87 absorption lines are broad with FWHM $\sim 400\,\mathrm{km\ s^{-1}}$. At the same time the Galactic absorption lines are unresolved by the FOS (FWHM $\sim 220\,\mathrm{km\ s^{-1}}$). We suggest that the M 87 absorption line properties can be understood in a simple model where our LOS to the nucleus passes through an outflow from the inclined ($\sim 30°$) and turbulent nuclear disk.

A significant number of emission lines are clearly visible in the entire wavelength region covered by our spectrum. These include both permitted

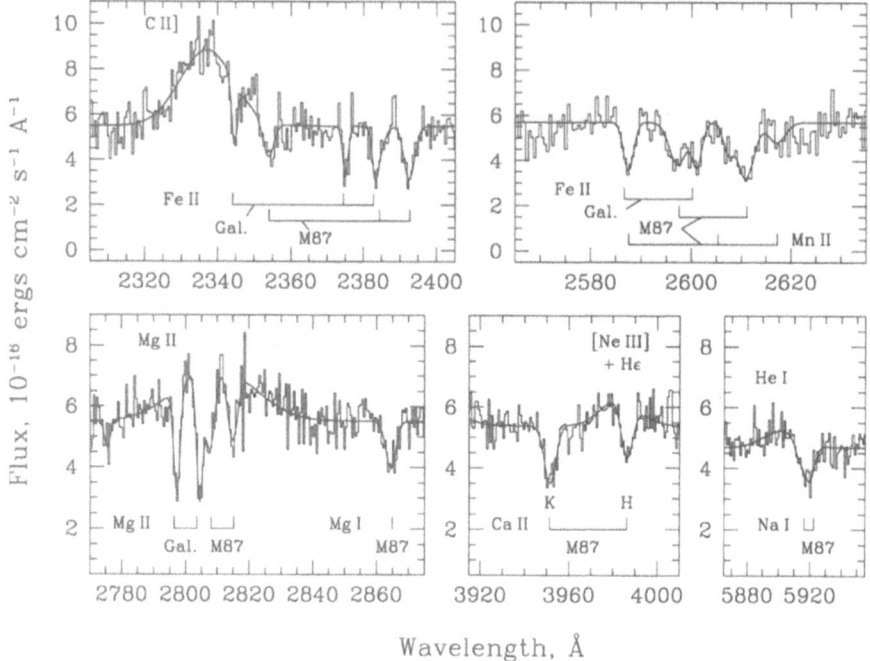

Fig. 2. Selected wavelength regions from the observed spectrum (histogram) and fitted Gaussian models (solid line). Absorption lines are labeled below the spectrum, and emission lines above it. Note that M 87 absorption lines are broader than Galactic ones (which remain unresolved at the FOS resolution).

and forbidden transitions of ions with a large range of ionization potentials — H I, He I, He II, O I, O II, O III, C II, C III, C IV, Ne III, S II, Mg II. With no exceptions all lines are very broad — the best isolated and high S/N ratio lines such as [O III] $\lambda 5007$ and Hβ have FWHM $\sim 2000\,\mathrm{km\,s^{-1}}$, and other line widths are consistent with this value. This is an expected result given the size of the FOS aperture ($0''\!.21 = 15\,\mathrm{pc}$) and the broadening due to the Keplerian rotation in the gaseous disk surrounding the central supermassive black hole (see Ford *et al.* and Macchetto *et al.* in these Proceedings).

Another important characteristic of the M 87 nuclear emission line spectrum is its LINER type. In fact, the spectrum of the nucleus is quite similar to the spectrum of the gaseous disk discussed in detail by Dopita *et al.* (these Proceedings). Here we note only that diagnostic diagrams involving critical UV emission line ratios clearly indicate shock heating is the dominant excitation mechanism.

3   Variability

A full presentation of the detected variability of the M 87 optical/UV nucleus is given in Tsvetanov *et al.* (1998). Measurements of the nuclear flux and the host galaxy at 6 epochs in the 1994 February – 1995 August time period are shown in Fig. 3. Here we summarize the basic results.

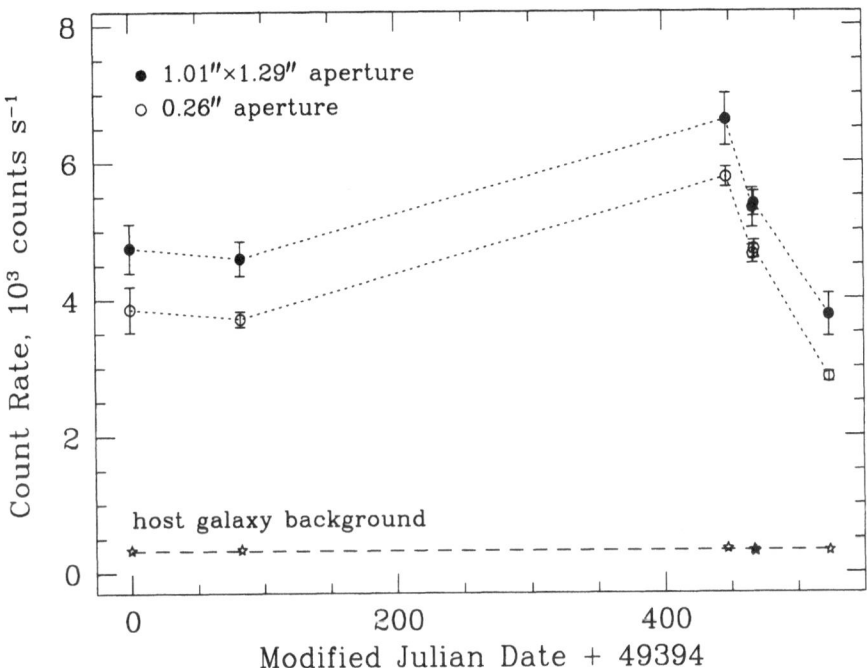

Fig. 3. Time dependence of the flux from the unresolved nucleus measured from the binary search data (filled circles) and from the peakup series (open circles) and the underlying galaxy (stars). The 1σ error bars are about 5% and 2.5% for the BS and PU data, respectively, and are smaller than the symbols for the galactic component. The extraction aperture is 1″01 × 1″29 for the BS and 0″26 for the PU data. The constant BS/PU flux ratio is characteristic of a point source for the aperture difference (Keyes *et al.* 1995).

Analysis of the target acquisition data yields the following results: (1) The nucleus remains unresolved at the *HST* resolution. The characteristic size of the FOS/COSTAR point-spread function imposes an upper limit of ~60 mas (= 5 pc) in diameter to the nuclear source size. (2) The flux from the nucleus changes by a factor ~2 over 2.5 months, ~25% over 3 weeks, and remains the same to within the errors (≤2.5%) during 1 day. (3) The continuum spectrum becomes bluer as it brightens while emission lines remain unchanged.

Some important conclusions can be drawn from the above results. (i) The one-month characteristic variability time scale and finite light travel time combined with the independently estimated gravitational radius of the BH in M 87 put an upper limit to the size of the emitting region of $l \leq 200R_g$. (ii) The changes by a factor ~ 2 indicate that at least half of the nuclear flux is variable. Yet, the total energy output in the variable component is only a small fraction of the Eddington limit, as is the total power released by the BH. (iii) The detected variability, combined with the observed continuum shape, relativistic boosting, and significant superluminal motions (Biretta *et al.* these Proceedings) imply strongly that M 87 is intrinsically of BL Lac type but is viewed from an angle too large to make its BL Lac properties dominate.

References

Biretta, J. A., Stern, C. P., & Harris, D. E. 1991, AJ, 101, 1632 (BSH91)

Ford, H. C. et al. 1994, ApJ, 435, L27

Jacoby, G. H., Ciardullo, R., & Ford, H. C. 1990, ApJ, 356, 332

Keyes, C. D. *et al.* 1995, *Faint Object Spectrograph Instrument Handbook*, version 6.0, (Baltimore: STScI), 27

Stiavelli, M., Peletier, R. F., & Carollo, C. M. 1997, MNRAS, 285, 181 (St97)

Sparks, W. B., Ford, H. C., & Kinney, A. 1993, ApJ, 413, 531 (SFK93)

Thomson, R. C. *et al.* 1995, MNRAS, 275, 921

Tsvetanov, Z. I. *et al.* 1998, ApJ, 493, in press

Zeilinger, W. W., Peletier, R. F., & Stiavelli, M. 1993, MNRAS, 261, 175 (Ze93)

The Inner Accretion Disk in M 87

C. S. Reynolds[1], T. Di Matteo[2], and A. C. Fabian[2]

[1] JILA
University of Colorado
Boulder CO 80309–0440
USA
[2] Institute of Astronomy
Madingley Road
Cambridge CB3 OHA
UK

Abstract. The observed AGN in M 87 is extremely quiescent compared to expectations based upon simple estimates for the mass accretion rate. We examine the possibility that the final stages of accretion are advection dominated by comparing current observational constraints on the flux from the M 87 core to the predictions of an advection dominated accretion flow (ADAF) model. We find that the observed self-absorbed radio core can be dominated by the ADAF, whereas the optical through to X-ray emission may be dominated by the jet. Speculation on the relevance of ADAFs to the FR-I/FR-II dichotomy is reported.

1 Why is M 87 so dim?

At the start of this workshop, Roger Blandford raised the question as to why the AGN in M 87 is so quiescent. In this contribution, we return to this important question and discuss the possible implications for our understanding of the central, and energetically dominant, regions of the accretion flow.

A simple energy budget allows us to place this issue on a quantitative footing. Summing the observed radiative flux (from radio to X-rays) from the non-thermal core of M 87 results in a broad-band (isotropic) radiative luminosity of $L_{rad} \sim 10^{42}$ erg s^{-1} (Biretta, Stern & Harris 1991). If some of this flux originates from the relativistic jet, it will be relativistically beamed and the *intrinsic* radiative luminosity of the core will be less (unless the Lorentz factor of the jet exceeds ~ 10, in which case we may lie outside of the beaming cone) . The kinetic luminosity of the AGN is a more difficult quantity to constrain. Models of the global energetics of the radio halo yields a total kinetic luminosity of $L_{kin} = 10^{43} - 10^{44}$ erg s^{-1} (Bicknell & Begelman 1996; Reynolds *et al.* 1996), at least an order of magnitude greater than the radiative luminosity.

To demonstrate the quiescence of this source, we now estimate the accretion rate. X-ray observations with *ROSAT* show that M 87 possesses an extensive hot interstellar medium (HIM). These X-ray data show that, at the centre of M 87, the sound speed of the HIM is $c_s \sim 500$ km s^{-1} and the gas density is at least $n_e \sim 0.5\,\mathrm{cm}^{-3}$ (C. B. Peres, private communication). Naively, the mass accretion rate onto the central black hole may be estimated from spherical Bondi accretion theory to be $\dot{M} \sim 0.2\,M_\odot\,\mathrm{yr}^{-1}$. If this mass is accreted onto the central black hole via a 'standard' radiatively-efficient accretion disk (Shakura & Sunyaev 1973) with 10 per cent efficiency, the power output of the AGN should be $L_{\mathrm{tot}} \sim 10^{45}\,\mathrm{erg\,s^{-1}}$. This is at least an order of magnitude greater than the inferred kinetic luminosity of the AGN, and three orders of magnitude greater than the observed radiative luminosity. If we believe all giant elliptical galaxies to possess massive nuclear black holes, this is a generic problem in all non-quasar elliptical galaxies (Begelman 1986; Fabian & Canizares 1988).

Assuming that our estimation of the actual source power is correct to within an order of magnitude, there are clearly two possible resolutions to this problem. Firstly, the spherical Bondi rate may be a poor estimator of the accretion rate onto the central black hole. For example, the accretion flow may be rotationally supported on large ($\sim 100\,\mathrm{pc}$) scales, or much of the accreting material may be expelled in the form of a wind before it ever falls close to the central hole (*e.g.* see contribution by James Binney in this volume). We note that any such process would have to be very efficient at starving the central hole of accreting material. The second possibility, and the one that we shall focus on in this contribution, is that the central hole may accrete in a very inefficient manner. We shall briefly introduce the subject of Advection Dominated Accretion Flows (ADAFs) and present an ADAF model for M 87. After describing some important caveats to the ADAF concept, some predictions and speculation shall be presented.

2   Advection Dominated Accretion Flows (ADAFs)

For sufficiently small accretion rates, a class of accretion disk solutions may exist that lie on a different branch from the 'standard' geometrically-thin and radiatively-efficient disk solutions (Ichimaru 1977; Rees *et al.* 1982; Narayan & Yi 1995; Abramowicz *et al.* 1995). These small-$\dot{M}$ solutions are optically-thin and hot — within 100 Schwarzschild radii or so of the central hole, the electron temperature is $T_{\mathrm{elec}} \sim 10^9\,\mathrm{K}$, and the ions are thermally decoupled and at the virial temperature ($\sim 10^{12}\,\mathrm{K}$ in the strong gravity regime). Ion pressure puffs the disk up into a geometrically-thick structure. Most importantly, most of the energy is in the ions which are assumed to be only weakly coupled to the electrons. Since only the electrons can radiate effectively, much of the energy of the accretion flow is advected in the flow across the event horizon as entropy of the ion gas. The overall radiative-efficiency of the disk can be as low as $\eta \sim 10^{-3}$ (where $L = \eta \dot{M} c^2$). Nowadays, these solutions are often called Advection Dominated Accretion Flow (ADAF) models.

The quiescence problem in M 87 is solved if one postulates that the final stages of accretion occur in an ADAF mode (Fabian & Rees 1995). Figure 1

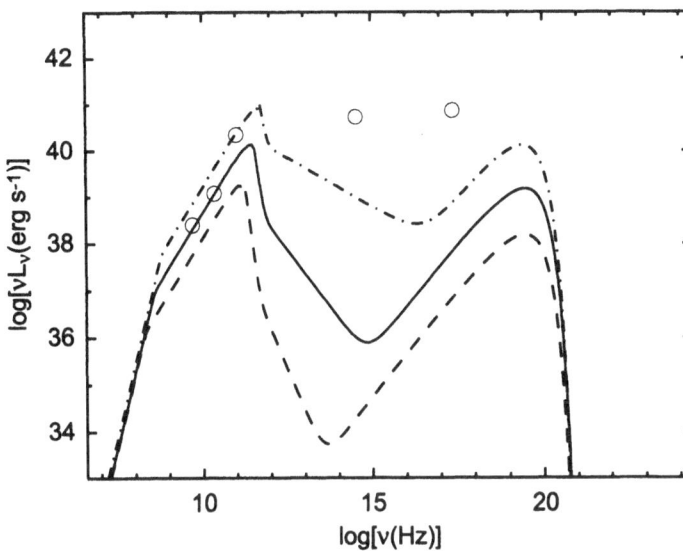

Fig. 1. Comparison of observed core fluxes at selected frequencies (open circles) with an ADAF model based upon that of Narayan & Yi (1995). The models are for accretion rates of $0.33\dot{M}_B$ (dashed line), $\dot{M}_B$ (solid line) and $3\dot{M}_B$ (dot-dashed line), where $\dot{M}_B = 0.2\,M_\odot\,\mathrm{yr}^{-1}$ is the Bondi accretion rate.

displays a comparison between observed core fluxes at selected frequencies (open circles) and ADAF models for mass accretion rates comparable to the Bondi rate (Reynolds *et al.* 1996; hereafter R96). The ADAF model, based upon that of Narayan & Yi (1995), displays three principal spectral components: a cyclo-synchrotron peak at radio frequencies (from the mildly relativistic thermal electrons), Comptonization of this synchrotron emission at IR through to UV frequencies, and a thermal bremsstrahlung peak at X-ray energies. Since we would like to compare this model to the actual emission from the central accretion flow, the observed core fluxes have been carefully drawn from the literature in order to minimize contamination by the underlying galaxy (see R96 for references to these core fluxes). However, contamination by the jet is unavoidable. Thus, these flux measurements must be treated as upper limits to the spectrum of the accretion flow.

As clear from Fig. 1, the ADAF model with the Bondi accretion rate is compatible with the observed upper limits to the spectrum of the accretion flow. Furthermore, most of the emission in the observed self-absorbed radio core may, indeed, originate from the ADAF and not the jet. However, any of the viable ADAF models fail to produce the observed optical to X-ray core flux. Emission from the core of the jet most likely dominates these bands.

We note here that the very existence of the ADAF solution in systems such as M 87 is vulnerable to several physical unknowns. Such solutions only exist below a critical accretion rate which is proportional to α^2, where α is the usual viscosity parameter of accretion disk theory. For Bondi accretion in M 87 to proceed via the ADAF mode requires that $\alpha > 0.1$. MHD simulations of the Balbus-Hawley instability suggest that α may be rather smaller, $\alpha \sim 0.01$, although firm conclusions are not yet possible due to the finite numerical resolution of these codes (Stone et al. 1996). A second issue is the survivability of the two-temperature plasma. Current ADAF models assume that the viscous dissipation heats mostly the ions rather than the electrons, and that the electrons are only coupled weakly to the ions via Coulomb collisions. If either of these conditions are violated, the electrons will copiously radiate and the ADAF solution will be modified or may even collapse back to the thin-disk case. These issues are currently the subject of active research by several workers (e.g. see Bisnovatyi-Kogan & Lovelace 1997).

3   Predictions and speculations

In the near future, direct evidence for or against the ADAF model of M 87 should be forthcoming. As mentioned above, the self-absorbed radio core of M 87 may well be entirely due to ADAF emission — at sufficiently high resolution (i.e., sub-milliarcsec) one should observe a self-absorbed structure with position angle perpendicular to the jet axis. Ultra-high resolution radio imaging with space-based extensions of VLBI (such as is possible with the recently launched VSOP) may facilitate this test. Secondly, one can search for the relativistically broad X-ray iron line which, in Seyfert galaxies, is thought to originate from the radiatively efficient accretion disk (Fabian et al. 1989; Tanaka et al. 1995). The detection of such a line with would strongly argue against the ADAF model (ADAFs are far too hot to allow the iron fluorescence line to form). AXAF will be the first instrument capable of spatially separating the AGN emission from the emission of the Virgo cluster thereby allowing the existence of such a line to be probed.

A fundamental mystery in the study of radio galaxies is the dichotomy between low power, edge-darkened sources (FR-I type; Fanaroff & Riley 1974) and high power, edge-brightened sources (FR-II type). In addition to the long-known dichotomy in radio properties, Zirbel & Baum (1995) and Baum, Zirbel & O'Dea (1995) found a striking difference in optical line properties between the FR-I class and FR-IIs. In short, for a given radio core power FR-II sources are much stronger optical line emitters than FR-I sources. Indeed, the line emission from FR-Is only slightly exceeds that expected from a quiescent galaxy of the same magnitude, whereas FR-IIs clearly possess powerful emission line regions associated with the AGN activity. Assuming the optical line emission to be driven by photoionization, the inference is that FR-IIs possess a powerful UV source whereas FR-Is of a similar radio core power do not. Among other possibilities, these authors speculate that the FR-I/FR-II dichotomy reflects the ADAF/thin-disk dichotomy. Our current picture of M 87 fits in with this speculation. X-ray spectral studies also lend circumstantial support to this idea. Several FR-II radio galaxies are known to possess relativistically broad X-ray iron lines which maybe a signature of radiatively-efficient thin disks (*e.g.* 3C 109: Allen *et al.* 1997; 3C 111: Reynolds *et al.* 1998). On the other hand, no FR-I sources are known with broad iron lines (although the observational problems associated with detecting any such line in these often faint sources are significant). Developments in space based VLBI and X-ray spectroscopy, as described in the previous paragraph, will help address this interesting hypothesis.

4   Summary

In this contribution, we have examined the ADAF model for M 87 by comparing the modern measurements of core flux to the spectral predictions of a simple ADAF model. We find that the very small radiative output of the M 87 core is indeed compatible with accretion at the Bondi rate if the inner regions of the accretion disk are operating in the ADAF mode. Most of the flux from the observed self-absorbed radio core may originate from the ADAF. However, all viable ADAF models under-produce the observed optical and X-ray core fluxes — these bands are most likely dominated by non-thermal emission from the jet. Motivated by this result, we reiterate the observationally-motivated speculation of Baum *et al.* that the FR-I/FR-II dichotomy is, physically, the ADAF/thin-disk dichotomy.

CSR acknowledges support from a PPARC studentship, the National Science Foundation under grant AST9529175, and NASA under grant NASA-NAG-6337.

References

Abramowicz M., Chen X., Kato S., Lasota J. P., Regev O., 1995, ApJ, 438, L37
Allen S. W., Fabian A. C., Idesawa E., Inoue H., Kii T., Otani C., 1997, MNRAS, 286, 765
Baum S. A., Zirbel E. L., O'Dea C. P., 1995, ApJ, 451, 88
Begelman M. C., 1986, Nat, 322, 614
Bicknell G. V., Begelman M. C., 1996, ApJ, 467, 597
Biretta J. A., Stern C. P., Harris D. E., 1991, AJ, 101, 1632
Bisnovatyi-Kogan G. S., Lovelace R. V. E., 1997, ApJ, 486, L43
Fabian A. C., Canizares C. R., 1988, Nat, 333, 829
Fabian A. C., Rees M. J., 1995, MNRAS, 277, L55
Fanaroff B. L., Riley J. M., 1974, MNRAS, 167, L31
Ichimaru S., 1977, ApJ, 214, 840
Narayan R., Yi I., 1995, ApJ, 452, 710
Rees M. J., Begelman M. C., Blandford R. D., Phinney E. S., 1982, Nat., 295, 17
Reynolds C. S., Di Matteo T., Fabian A. C., Hwang U., Canizares C. R., 1996, MNRAS, 283, L111

Reynolds C. S., Fabian A. C., Celotti A., Rees M. J., 1996, MNRAS, 283, 873

Reynolds C. S., Iwasawa K., Crawford C. S., Fabian A. C., 1998, MNRAS, submitted

Shakura N. I., Sunyaev R. A., 1973, A&A, 24, 337

Stone J. M., Hawley J. F., Gammie C. F., Balbus S. A., 1996, ApJ, 463, 656

Zirbel E. L., Baum S. A., 1995, ApJ, 448, 521

X-ray Variability in M 87: 1992 – 1998

D. E. Harris[1], J. A. Biretta[2], and W. Junor[3]

[1] Smithsonian Astrophysical Observatory
60 Garden St.
Cambridge, MA 02138, USA
[2] Space Telescope Science Institute
3700 San Martin Dr.
Baltimore, MD 21218, USA
[3] University of New Mexico
800 Yale Blvd. NE
Albuquerque, NM 87131, USA

Abstract. Beginning in 1995 June, we have obtained an observation of M 87 with the ROSAT High Resolution Imager (HRI) every 6 months. We present the measurements of X-ray intensity for the core and knot A through 1998 January. We find significant changes in both components. For the core, intensities measured in June 1995, December 1996, and December 1997 are roughly 30% higher than values obtained at three intervening times. For knot A, a secular decrease of approximately 15% is interrupted only by an intensity jump (3 σ) in December 1997. Because the background used for subtraction is probably underestimated, we suspect the actual variation is somewhat greater than these values indicate.

1 Introduction

The initial results of our X-ray monitoring of M 87 (Harris, Biretta, and Junor (1997), HBJ hereafter) were based on a re-analysis of two Einstein Observatory HRI observations, a ROSAT archival HRI pointing in June 1992, and the first two of our ongoing series of an observation every 6 months (June and December 1995). There we demonstrated that the core was variable at the 15% level and that knot A showed a gradual decline in intensity. In this paper we include data from 4 additional observations, with the most recent consisting of two segments: December 14, 1997 (8 ksec) and January 5, 1998 (20.6 ksec). We have also re-measured some of the earlier data to ensure that all data were treated uniformly.

2 Review of Problems and Measuring Techniques

There are three known problems affecting accurate photometry of ROSAT HRI data. The first is that the core and knot A (the two brightest features which we can measure with a 30 ksec observation) are separated by only 12″, thus precluding the use of the standard HRI aperture with r=10″. The second problem is that the central region of M 87 is X-ray bright with a complex emission distribution (Harris, Biretta, and Junor (1999)), thereby making it difficult to estimate the surface brightness that would have been observed at the locations of the core and knot A if they were absent.

The most serious problem however, is the image degradation which means that the effective point response function (PRF) is widened and distorted from its quasi-Gaussian shape (FWHM $\approx 5.5''$) to an irregular distribution with FWHM of $7''$ to $10''$. The degradation is caused by bad aspect solutions; sometimes associated with the spacecraft wobble and occasionally to time segments for which the aspect solution can be up to $10''$ (or more) away from the primary solution. This situation means that we have no assurance that our standardized measuring areas contain the same fraction of total source counts for each observation.

To mitigate the severity of these problems, we made two separate measurements. For the first, we take the ratio of net counts in circular apertures ($r=6''$) centered on the core and knot A. To first order, we expect that the effective PRF will be the same for these close sources. Consequently the major error of this approach is that we will measure different fractions of the total source counts in different observations. Since the background correction uses a region with a slightly lower surface brightness than that found adjacent to the core and knot A, the non-variable background component will increasingly dilute any variable feature as the effective PRF gets larger; *i.e.* fewer photons from the unresolved component will remain within the small measuring aperture.

The second measurement utilizes two adjacent rectangles rotated by $20°$. Each rectangle is $16''$ by $26''$ with their common border ($26''$) perpendicular to the line joining the core and knot A, and centered on the 'reference point'. The reference point is halfway between the core and knot A. Surrounding these two rectangles is the background 'frame' which is $10''$ wide. This background frame is used both for the circular and rectangular measurements. The geometry is depicted in Fig. 2 of HBJ. The larger area of the rectangle compared to the small circle is designed to include most of the source counts even when there is substantial degradation of the PRF. However, since it will include more of the non-variable background, the actual variability should be somewhat greater than that found by this technique.

As a control, we also measure the net counts in a circular aperture ($r=12''$) centered $45''$ to the SE of the reference point (PA=$110°$). This location is on a plateau in the X-ray brightness distribution. Since we use the same background frame as for the other measurements, the resulting value is always negative.

Finally, to accommodate any changes in quantum efficiency we employed a 'self-calibration' by measuring the count rate of the bright part of the cluster gas within a large circle with radius $276''$, but excluding the adjacent rectangles containing the core and knot A. For this measurement, we used a background annulus with radii of $280''$ and $300''$. All measured net count rates were then multiplied by a correction factor (always 5% or less) so that the cluster gas net count rate would be the same as that for our 'fiducial' observation of June 1995.

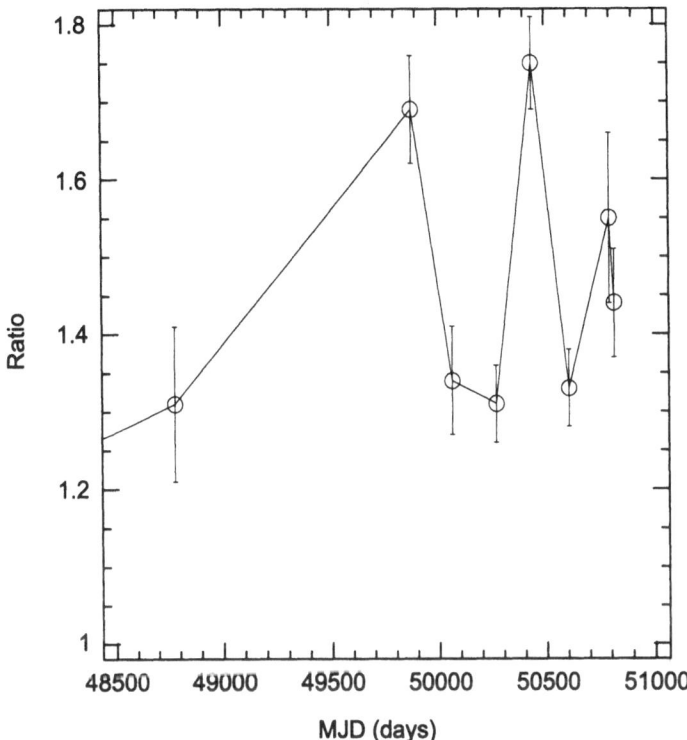

Fig. 1. The ratio of the core to knot A. Values are the ratio of net counts as measured in circular apertures with radius of 6″.

3   Results

The measured values are shown in Fig. 1 (the ratios) and 2 (the count rates). It can be seen that the major features of the ratio plot mimic the count rate variability for the core. This is consistent with expectations that the core will be more variable than knot A. The decrease in the intensity of the core following the observation of June 1995 (MJD = 49876) coincides with the decrease observed with the HST between MJD 49840 and MJD 49921 (see Fig. 3 of Tsvetanov et al. (1999)).

In Biretta, Stern, and Harris (1991), we analyzed the Einstein HRI data and argued that the X-ray emission from knot A was most likely synchrotron

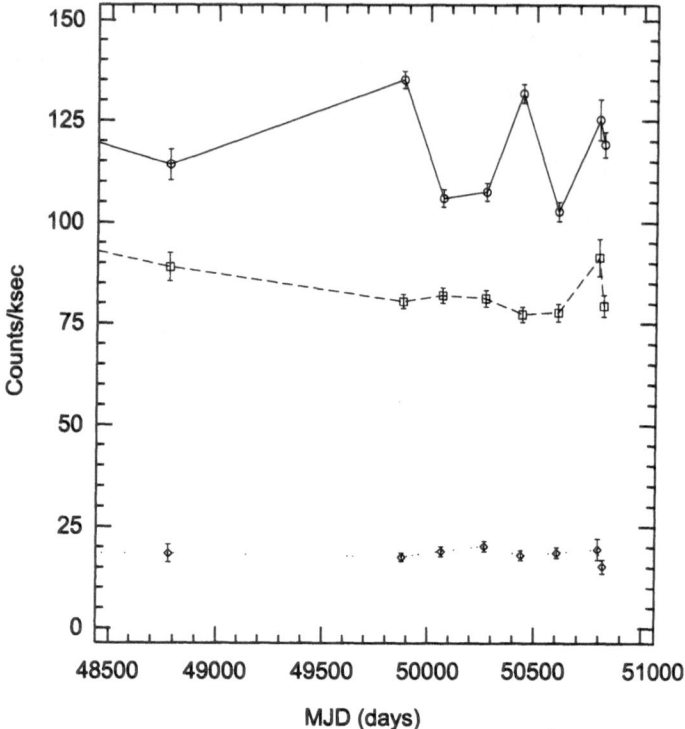

Fig. 2. The net count rates for the core, knot A, and a control region. The solid line is for the core, the dashed line is for knot A, and the control circle values are connected with a dotted line. Note that the control region is the net counts in a circle 45″ SE of the reference point, with the background from the frame. Therefore all control values are negative, but are plotted here as positive numbers. Thus the small apparent drop for the last measurement is actually in the opposite sense from those of the core and knot A. The lines running off the left side of the figure connect to the Einstein values (HBJ).

emission rather than thermal bremsstrahlung or inverse Compton emission. For a magnetic field strength of $\approx 200\ \mu$G (the minimum pressure field), we found that electrons with Lorentz energy factors, γ, of 2×10^7 were required and typical half lives were of order ten years. The observed decrease of knot A (Fig. 2) is consistent with these estimates, but we do not yet understand why only a small fraction of shocks produce enough $\gamma = 10^7$ electrons to generate an X-ray intensity detectable with current technology. We suspect that even the knot A shock (≈ 70 pc across as measured at radio and optical wavelengths) is not a uniform single entity, but may well display a complex brightness distribution at the highest energies where synchrotron losses are severe. Thus we expect that higher resolution X-ray observations will show occasional bright, compact components which will not persist many years (*e.g.* Biretta et al. (1999) find optical features with these characteristics). If the December 1997 increase in knot A is real, it could represent emission from such a feature. AXAF observations have been proposed to obtain 8 monthly exposures with the High Resolution Camera.

References

Biretta, J.A., Stern, C.P., and Harris, D.E. (1991): The Radio to X-ray Spectrum of the M 87 Jet and Nucleus, AJ **101**, 1632–1646

Biretta, J.A. et al. (1999) (this volume): *Ringberg Workshop on M 87* (Springer, Berlin, Heidelberg)

Harris, D.E., Biretta, J.A., and Junor, W. (1997): X-ray Variability in M 87 MN-RAS **284**, L21–L27

Harris, D.E., Biretta, J.A., and Junor, W. (1999) (this volume): *Ringberg Workshop on M 87* (Springer, Berlin, Heidelberg)

Tsvetanov, Z.I., Hartig, G.F., Ford, H.C., Kriss, G.A., Dopita, M.A., Dressel, L.L., and Harms, R.J. (1999) (this volume): The Nuclear Spectrum of M 87 *Ringberg Workshop on M 87* (Springer, Berlin, Heidelberg)

Closing Comments

Jean Eilek

New Mexico Tech
Socorro, NM 87801
USA

Abstract. As Peter Scheuer noted in the summary talk of a previous Ringberg meeting, I cannot possibly do justice to all speakers, and I probably cannot avoid doing injustice to some. Instead of a formal summary, then, what follows is my personal impression of this delightful meeting, and of the state of our knowledge of its topic, M 87.

1 Introduction

M 87 has been many things to many people over the years. It is a nearby laboratory for active galactic nuclei and black holes, for radio jets and radio sources, and for stellar dynamics. Its location in the center of the Virgo cluster shows us a nearby example of X-ray bright, cooling gas. The cluster itself is a nearby example of cluster dynamics and evolution. In the past few years, new data has been obtained on all aspects of this system. The data has come from radio, optical and X-ray telescopes, both new and old. This was therefore a very appropriate time to hold this meeting.

M 87 is also a galaxy which has fascinated me over the years. I have been involved with some problems about the galaxy at the research level, and I have followed other areas only as an interested outsider. In this article, I have taken advantage of my role as closing speaker, to share my reflections on the meeting, and also to share my thoughts and prejudices on the galaxy. I have arranged the topics by size, starting on the outside and working my way in.

2 The Virgo Cluster

Our picture of clusters is changing. Up to a few years ago, the clusters envisioned by most astronomers (well, most theorists at least) were static and relaxed. The picture has changed dramatically over the past few years. There is now good observational evidence for recent or ongoing mergers in several nearby clusters (*e.g.* A 754, Henriksen & Markevitch 1996, or A 2316, Feretti, Giovannini & Böhringer 1997). Norman reminded us that many cosmological simulations, especially at high Ω, also predict that mergers of subclusters into big clusters should be common today. Bingelli told us that the Virgo cluster, in particular, shows evidence for infall of two large clumps: the M 86 and M 49 regions. (The large-scale X-ray image, from Bohringer *et al.* 1994, agrees with this picture, although without the supporting kinematic data). This tells me that we cannot envision the Virgo cluster, even around M 87, as a static system. The motion of these outer clumps may well impact conditions in the inner part of the cluster.

How far away is the Virgo cluster? It seems to be jumping around. At this meeting, I noticed different speakers using distances of 15, 16, 17 and 20 Mpc. On a personal note, I often find myself trying to reconcile radio and X-ray data on M 87, and the diversity of assumed distances makes comparison of results from different authors rather frustrating. The true distance to Virgo may be difficult to pin down, as different indicators seem to give different answers (Neilsen presented one interesting method). However, this author humbly requests that we all agree, by fiat if necessary, on a usable distance, so that we could easily compare our results. Perhaps this can be the topic for another Ringberg meeting?

3   M 87 as a Galaxy

M 87 is an old friend to students of stellar dynamics and elliptical galaxy structure. I noticed at this meeting that some old questions about M 87 are still alive and still unanswered.

One of the ongoing discussions about M 87 has been, is there a stellar signature of a black hole? This question has been around, without resolution, since the work of Young *et al.* (1978) and Sargent *et al.* (1978) and later work of Dressler & Richstone (1990). While the focus for M 87 has passed on to the — apparently — smoking gun of gas kinematics in the nucleus (*cf.* §6), the question of a stellar signature of a central black hole remains important. However, as Dehnen reviewed, this question does not seem to have a unique answer. There seems (in my non-specialist view) to be enough flexibility in picking stellar orbits and populations, that one can still find a model without a central black hole if one wants, or to hide a central black hole if that is what one wants (*e.g.* Merritt & Oh 1997), although the price may be an unlikely stellar distribution.

Another old question was revisited at this meeting: how does one grow a black hole? Now that many nearby galaxies are being found to contain likely black holes (as Ford told us, also see Kormendy & Richstone 1995), this old question has again become timely. In particular, which came first, the black hole or the galaxy? There is some theoretical evidence (reviewed by Dehnen) that slow ("adiabatic") black hole growth in a pre-existing galaxy is inconsistent with the orbital anisotropies necessary to reconcile the stellar data in M 87 with the black hole mass suggested by the gas kinematics in the core. If this is the case, one needs to ask, was the black hole there before the galaxy? Did it form from a merger of two comparable-sized compact objects?

Another question might be, has M 87 itself undergone a recent merger? This is a fashionable topic in elliptical galaxy physics these days, and it seems directly relevant to M 87, although I have not heard it particularly applied to this galaxy. If one wanted to hunt for signs of a merger in M 87's past, one could invoke several indicators: the outer fan (Weil *et al.* 1997); the possible kinematic subsystem in the core (van der Marel 1994); the rising globular cluster velocity dispersion (Cohen & Ryzhov 1997); the two different globular cluster populations (Eslton & Santiago 1996); and the possibility that the central light profile is more consistent with recent merger of two black holes (Dehnen).

4   The Virgo Core

This is the transition region, on a scale $\sim 50\,\mathrm{kpc}$, between the active nucleus and jet of M 87, and the outer region where the stellar galaxy and its gaseous halo merge with the overall Virgo cluster. This transition region has hitherto gotten little observational or theoretical attention, and so I was thrilled by the new data which we saw at this meeting. We now have evidence that this inner core is an active and complex region. Our older picture which assumed a static core, or a slow, symmetric cooling inflow, must be modified.

On larger scales, the distribution of the X-ray loud gas matches that of the stars in M 87, suggesting a reasonably quiescent, settled system. However, the exciting new X-ray data (presented by Böhringer and by Harris) — show striking deviations on the scale of this transition region. The X-ray excess that roughly coincides with the radio "lobes" (Nulsen & Böhringer 1995) now appears in the ROSAT HRI data (from Harris) to be a sheet or bow shock which wraps around the galaxy. Could this is the ISM/ICM interface in a flowing cluster wind? In addition the inner X-rays are asymmetric, with some evidence of an X-ray "hole" around the inner radio lobes (Böhringer). The X-ray gas clearly knows about the radio plasma in the region of the core.

In addition, new radio data — high resolution total intensity from Owen, polarization and spectral data from Klein and Rottmann, and Faraday rotation data which I presented — reveal both the inner and outer lobes in great detail. It is now clear that the outer radio halo is inhomogeneous, and there is (to my eye) evidence of ongoing outflow from the galactic nucleus into the radio halo. New spectral data show that the particle aging question, with its connection to local re-acceleration of the relativistic particle population, is more complex than we realized. The Faraday rotation data show that there is a dynamically significant, ordered magnetic field in the very inner region. The fact that the X-ray excess coincides well with the bright knot complex in the radio jet suggests to me that the dynamics of the X-ray loud plasma are directly affecting the propagation of the radio jet. All of these results, put together, point to a turbulent, disordered plasma core.

For me, this meeting has raised more questions than it answered about the Virgo plasma core. In this new area, we have barely had time to digest these new results, let alone understand their implications for the physics.

We do have some starting points, however. Binney considered the dynamic and thermodynamic impact of the cooling catastrophe in the outer gas (noting the cooling radius ∼ 100 kpc, Peres & Fabian 1997, is well outside the core where disorder is observed). Several people discussed the energy budget in the core (Bicknell, Binney, Blandford, Owen and myself.) It seems clear that there are several possible power sources: the radio jet itself, gravitational energy released in an outer cooling inflow, the effects of ongoing mergers in the Virgo cluster. There are several radiative loss mechanisms: X-rays, emission line clouds, the radio source. In addition, energy is being transferred internally, from kinetic energy (both turbulence and directed flows) and probably from magnetic field dissipation, to plasma heating. Buoyancy and expansion of the radio lobes into the X-ray plasma also impact the energy budget. The thermodynamics of the system are far from simple. One is tempted to think of this as a weather problem and, following Binney's suggestion, possibly to look to the ISM in our own galaxy (multiphase, magnetized, and turbulent) for analogs.

5   The Jet

I have long been fascinated by the beautiful jet of M 87. The new data presented at this meeting convinced me that we theorists still have a long way to go in understanding it all.

5.1   The State of the Jet

Just when you think everything possible has been measured for this well-studied source, more data turns up. Dramatic new radio and optical data were presented at the meeting. Biretta showed us his latest work on the proper motions of the features in the radio jet. His careful work shows us that the features move independently of each other, with velocities not necessarily parallel to the jet axis. He has upped his previous apparent speed limit of ∼ $2c$, by finding speeds as high as ∼ $5c$ in knot D. Owen finds that the inner edge of knot A is still unresolved at 43 GHz, making it less than 3 pc in projected width. Biretta and Meisenheimer showed us the striking comparisons of radio (VLA) and optical (HST) images of the jet. I am most impressed by the slight deviations from uniformity that this comparison shows. While the radio-optical spectrum is generally uniform throughout the jet, there are some deviations, either on the edges of the jet (where the radio appears relatively fainter), or in bright spots (where the radio can be either fainter or stronger). Finally, Harris showed us that the inner X-ray feature, be it a bent sheet or a bow shock, seems to overlap the jet just at the knot A/C complex. Is this why the jet begins to disrupt there?

The quality of these data demand interpretation: what is the physical state of the jet? I was impressed by the level of specific discussion on the physical parameters of the jet (much more detailed than is usually the case in extragalactic astrophysics). The power, speed and angle of the jet are being pinned down. The total power, while still a model dependent quantity, seems to be $\sim 10^{43} - 10^{44}$ erg/s; a variety of physical arguments presented by Reynolds and by Bicknell agree on this range.

I was especially impressed by the convergence of a variety of arguments to determining the orientation of the jet. Radio-based arguments from Biretta (aspect angle of knot A, sidedness, feature speeds) agree with optical-based arguments from Macchetto and Axon (on the kinematics of the nuclear accretion disk). The result is, the angle of the jet seems to be forty-something degrees from the line of sight.[1]

Reynolds presented an intriguing argument on the matter content of the jet. He showed that the *pc-scale* jet must be a pair jet if $P_b < 10^{43}$. (Unfortunately, it is not clear if this can be continued to the kpc-scale jet; a nuclear pair jet may very well pick up ions in its passage through the local galactic ISM.) Another quantity which is not yet determined is the physical speed of the jet. The bright features are very likely to be signals (waves, shocks) rather than tied to the fluid. Several speakers noted that $\beta_{app} \sim 2$ requires $\gamma_{pat} \sim 2$ for the *pattern* speeds, and the higher β_{app} of knot D requires higher γ_{pat}. This is not the fluid speed, however. Simulations (*e.g.* those of Hardee) show that the fluid speed is likely to be higher than the pattern speed, if these patterns are normal modes or shocks. Robert Laing also reminded us that jet models with slower outer shear layers and faster cores are good matches to other extragalactic jets, and may well apply here too (one recalls the "shadow" down the center of the radio jet, first pointed out by Owen, Hardee & Cornwell 1989). Finally, the other possible estimate of flow speed is the jet sidedness. This old question was revisited but not resolved: is the lack of a detected counterjet (at $\sim 150/1$ level) due to beaming, to its intrinsic faintness, or to its nonexistence?

5.2   The State of the Theory

Once again, the observers are well ahead of the theorists. The high quality of the new radio and X-ray data justifies new work on the models. In particular, this data should enable we theorists to define and carry out specific tests of various models (such as models of jet structure, origin of the bright features, and particle acceleration and aging). I hope my fellow theorists will rise to the challenge.

Some new work has been done on large-scale models of the jet. Bicknell reported on analytic work developing the model in which the bright knots are internal shocks in a hydrodynamic jet. Hardee is continuing his thorough and careful study of numerical simulation of instability growth and normal

---

[1] This may run afoul of the is the higher $\beta_{app} \sim 5 - 6c$ seen in knot D, which requires a locally smaller angle, $\sim 20°$

modes of jets, and told us about work on relativistic and passive-field MHD flows. Such work is developing this particular picture (hydrodynamic, weak magnetic fields) in good detail and teaching us what such jets really can and will do. I am concerned, however, that we are forgetting the possibility — in my view probability — of magnetized plasmas in astrophysics. Indeed, our only truly local example of an astrophysical plasma — the earth's magnetosphere and the solar wind — is strongly magnetized. Some work has been done on the M 87 jet as an MHD jet (Villata & Ferrari 1995). The apparent helical structure of the bright features makes an MHD picture attractive to me personally; I hope research in this area will continue, so that in the future we can test both hydrodynamic and magnetohydrodynamic models of this jet.

One related question might be tractable now: what is the origin of the bright synchrotron filaments in M 87 (in the jet and also in the lobes)? Several speakers pointed out that such filaments arise in current numerical simulations only when the magnetic field is passive, and able to be dragged around by the fluid motions (*e.g.* Hardee, Clarke & Rosen 1997). One might conclude from this, that the magnetic field is unimportant in the M 87 radio jet (and lobes), and that fluid motions (including overpressure halos) account for the plasma structure and jet confinement. On the other hand, magnetic filaments are also seen in magnetically dominated situations, where they are likely due to resistive instabilities. Planetary magnetosheaths (*e.g.* Lui 1985 are one nearby example. Perhaps in the near future we can search for discriminants between magnetic-weak and magnetic-strong models.

I was surprised that another old issue was not given much attention at this meeting: *in situ* particle acceleration. Is it taking place in the M 87 jet, and if so, how? The extension of the nonthermal continuum into the optical has long been used to argue that *in situ* acceleration must be taking place in this jet. Heinz pointed out that "adiabatic" models (in which the relativistic electrons are transported out from the core without subsequent re-acceleration) can account for the jet surface brightness if the magnetic field is weak and if Doppler boosting is included in lifetime estimates. It remains to be seen, however, if this type of model can account for all of the new data. In particular, it will be interesting to see if this type of model can describe the new two-dimensional radio-optical spectral comparisons made possible by the combined HST/VLA data. In addition, there is synchrotron emission throughout the entire lobe, including the detached optical hot spot in the eastern lobe, reported by Meisenheimer. Can this be explained without *in situ* re-acceleration?

I personally expect *in situ* re-acceleration is taking place in the jet, and very likely elsewhere in the radio source (such as the eastern optical bright spot). In this context, we were reminded that other synchrotron sources have optical counterparts: Harris showed us 3C390.3 and Pictor A, and Röser showed us the jet in 3C 273 (see also Röser 1988). Such sources as these —

and M 87 is by far our best case to study, given the high quality of the new data — almost certainly require ongoing particle acceleration. The possibilities for acceleration are of course complex. Currently viable options include Fermi acceleration at quasi-parallel shocks, electrodynamic acceleration at quasi-perpendicular shocks, in reconnection layers or in double layers, and stochastic acceleration in strong turbulence. While none of these are simple mechanisms, the quality of the data on M 87 may make it once again an important laboratory for studying these options.

6   The Engine

In addition to its other roles, M 87 is the nearest galaxy with an active nucleus. People have been hunting for evidence of a central black hole in M 87 for at least twenty years (*cf.* §3). Finally, what appears to be a smoking gun turned up in emission line profiles from the central gas disk (Ford *et al.* 1994, Harms *et al.* 1991). Several speakers noted that the mass of the central black hole, inferred from this new data, is essentially the same as the first estimates of twenty years ago. Rather than being discouraged at what may appear a lack of progress, we should see a new challenge. Our job now is to understand the structure and kinematics of this region, and if possible to use the data to improve (dare I say test?) black-hole models of active nuclei.

6.1   Stalking the Black Hole

Several speakers presented HST data and careful analysis to determine the kinematics of the gas in the inner few pc of M 87. The initial HST results led people to think in terms of a simple Keplerian accretion disk, and the Keplerian model has been developed in detail to determine the kinematics of that disk (as presented by Axon, Macchetto, Ford). The picture has grown more complex, however. Dopita showed us that shock ionization is likely for this nuclear gas. In addition to going in circles (in the pc-scale disk), we now hear that the gas also flows in (perhaps along the larger-scale emission line filaments, according to Sparks), and out (as a wind from the disk, according to Ford). Adding more to this confusion, Biretta reminded us that VLB at 22 GHz has traced the jet down to 2000 AU from the core; there is also a relativistic plasma outflow in this region.

There was some interesting discussion on the question of how strong is the evidence for a black hole? While the HST evidence for a black hole is dramatic, it is still circumstantial. We have established there is a large, dark, gravitating mass, $\sim 3 \times 10^9\,M_\odot$ within a very small region (the emission line disk can be traced to $\sim 3\,\mathrm{pc}$). Tsvetanov presented optical variability data which constrain the nuclear flux to come from a region of $\lesssim 0.02\,\mathrm{pc}$. This is still much larger than any characteristic radius of a $3 \times 10^9\,M_\odot$ black hole. Thus, despite the high quality of the data and despite lots of hard work by lots of people, we still do not have an inarguable signature of a black hole. A skeptic could still argue for a cluster unusual and *dark* objects in the inner few thousand AU, although this seems astrophysically unlikely. Personally, however, I am content with the smoking gun that we have.

6.2   What Next for the Models?

Just how does a central black hole produce a jet? This remains one of the fundamental questions in the field; its complexity has made it both compelling and daunting. We must keep in mind that the plasma around the hole is almost certainly magnetized, the black hole is almost certainly rotating, and much of the initial jet energy and momentum may be carried by Poynting flux as well as by matter. None of these facts simplify the problem. Camenzind continues to do careful work on this problem, and I enjoyed hearing of his latest results. However, we are still far from having a unique model of the engine, and we are also still far from having robust observational signatures of these models.

On a more positive note, I was intrigued by the discussion at this meeting of why the engine does not behave as it should. That is, many of us have argued, and taught our students over the years, that "reasonable" efficiencies for converting $\dot{M}$ to radiation are $\sim 1 - 10\%$; and that black holes might "typically" accrete at a Bondi or Eddington rate. M 87 is not obeying these assumptions; it is much fainter for its hole mass than "it should be".[2] It may be faint because its accretion rate is less than we expect, or it may be faint because the accretion flow is a very inefficient radiator. Whichever turns out to be the case (and there was much diversity among the speakers and audience as to which might be the answer), it will be important to learn just how naive our standard assumptions are.

What excites me most in this area is that we now have enough data to explore the galaxy-engine connection. The gas disk we see is an outer disk, and seems to be slightly misaligned with the inner disk (assuming the jet is truly orthogonal to the inner disk). We cannot resolve the inner region of the (putative) accretion flow — that requires another factor of 100 in resolution, unlikely any time soon. What we can study, however, is the way in which the gas makes the transition from the galactic core region to the accretion region. Is the mass inflow steady? How does the motion of the gas connect to the local stellar orbits? How important is the magnetic field in the accreting gas?

---

[2] For that matter, evidence is accumulating that massive but dark black holes are common in the nuclei of many or most big galaxies. It is not just a problem with M 87.

Acknowledgements

This was a most enjoyable and productive meeting. I am grateful to all of the meeting participants who made the scientific atmosphere so positive. I am especially glad to thank Hermann-Josef Röser and Klaus Meisenheimer, who (once again) put in the work that allowed the rest of us to play.

References

Bohringer, H., Briel, U. G., Schwarz, R. A., Voges, W, Hartnet, G. & Trümper, J. (1994): Nat, **368**, 828.

Cohen, J. G. & Ryzhov, A. (1997): ApJ, **486**, 230-241

Dressler, A. & Richstone, D. O. (1990): ApJ, **348**, 120-126

Elston, R. A. W. & Santiago, B. X. (1996): MNRAS, **280**, 971-976

Feretti, L., Giovannini, G. & Böhringer, H. (1997): New Astron., **2**, 501-515

Ford, H. C., Harms, R. J., Tsvetanov, Z. I., Hartig, G. F., Dressel, L. L., Kriss, G. A., Bohlin, R. C., Davidsen, A. F., Margon, B. & Kochar, A. K. (1994): ApJ, **435**, L27-L30

Harms, R. J., Ford, H. C., Tsvetanov, Z. I., Hartig, G. F., Dressel, L. L., Kriss, G. A., Bohlin, R. C., Davidsen, A. F., Margon, B. & Kochar, A. K. 91994): ApJ, **435**, L35-L38

Hardee, P. E., Clarke, D. A. & Rosen, A. (1997): ApJ, **485**, 533-551

Henriksen, M. J. & Markevitch, M. L. (1996): ApJ: **466**, L79-L82

Merritt, D. & Oh, S. P. (1997): AJ, **113**, 1279-1285

Kormendy, J. & Richstone, D. (1995): ARAA, **33**, 581-624

Lui, A. T. Y. (1985): *Magnetotail Physics* (Baltimore: Johns Hopkins University Press)

Nulsen, P. E. J. & Böhringer, H. (1995): MNRAS, **274**, 1093-1106

Owen, F. N., Hardee, P. E. & Cornwwell, T. C. (1989): ApJ, **340**, 698-707

Peres, C. & Fabian, A. C. (1997): preprint

Röser, H.-J. (1988): in K. Meisenheimer & H.-J. Röser, eds, *Hot Spots in Extragalactic Radio Sources* (Berlin: Springer-Verlag), 91-114

Sargent, W. L. W., Young, P. J., Boksenberg, A., Shortridge, K., Lynds, C. R. & Hartwick, F. D. A. (1978): ApJ, **221**, 731-744

van der Marel, R. P. (1994): MNRAS, **270**, 271-297

Weil, M. L., Bland-Hawthorn, J. & Malin, D. F. (1997): preprint

Vilatta, M. & Ferrari, A. (1995): A&A, **293**, 626-639

Young, P. Westphal, J. A., Kristian, J., Wilson, C. P. & Landauer, F. P. (1978): ApJ, **221**, 721-730

List of Participants

Name		Address
Arp	Halton	Max-Planck-Institut für Astrophysik Karl-Schwarzschild-Straße D85748 Garching Germany arp@mpa-garching.mpg.de
Axon	David J.	Space Telescope Science Institute 3700 San Martin Drive Baltimore, MD 21218 USA axon@stsci.edu
Bicknell	Geoffrey V.	Mt. Stromlo & Siding Spring Observatories ANU / Private Bag Weston Creek, A.C.T. 2611 Australia gvb@maths.anu.edu.au
Binggeli	Bruno	Astronomisches Institut Universität Basel Venusstraße 7 CH-4102 Binningen Switzerland binggeli@astro.unibas.ch
Binney	James	Department of Physics Keble Rd. Oxford, OX1 3RH United Kingdom j.binney1@physics.oxford.ac.uk
Biretta	John	Space Telescope Science Institute 3700 San Martin Drive Baltimore, MD 21218 USA biretta@stsci.edu
Blandford Roger		California Institute of Technology 130-33 Pasadena, CA 91125 USA rdb@tapir.caltech.edu
Böhringer Hans		Max-Planck-Institut für extraterrestrische Physik Giessenbachstrasse -85748 Garching Germany hxb@mpe-garching.mpg.de

Name		Address
Camenzind Max		Landessternwarte
		Königstuhl
		D69117 Heidelberg Germany
		M.Camenzind@lsw.uni-heidelberg.de
Cramphorn Conrad		Max-Planck-Institut für Astrophysik
		Karl-Schwarzschild-Straße
		D85748 Garching Germany
		conrad@mpa-garching.mpg.de
Crane	Phil	European Southern Observatory
		Karl-Schwarzschild-Straße
		D85748 Garching Germany
		PCrane@eso.org
Dehnen	Walter	Theoretical Physics
		1 Keble Road
		Oxford, OX1 3NP United Kingdom
		w.dehnen@physics.ox.ac.uk
Dopita	Michael	Mt. Stromlo & Siding Spring Observatories
		Private Bag
		Weston A.C.T. 2611 Australia
		michael.dopita@anu.edu.au
Eilek	Jean	New Mexico Institute for Mining and Technology
		Socorro, NM 87801 USA
		jeilek@nrao.edu
Ford	Holland	Space Telescope Science Institute
		3700 San Martin Drive
		Baltimore, MD 21218 USA
		ford@stsci.edu
Hardee	Philip E.	University of Alabama
		Department of Physics & Astronomy
		Tuscaloosa, AL 35487 USA
		hardee@venus.astr.ua.edu
Harris	Daniel	Center for Astrophysics
		60 Garden Street
		Cambridge, MA 02138 USA
		harris@cfa.harvard.edu
Heinz	Sebastian	University of Colorado
		JILA
		BOX 440
		Boulder, CO 80309-0440 USA
		heinzs@bogart.colorado.edu

Name		Address
Klein	Uli	Radioastronomisches Institut der Universität Auf dem Hügel 71 D53121 Bonn Germany uklein@astro.uni-bonn.de
Laing	Robert	Royal Greenwich Observatory Madingley Road Cambridge CB3 0EZ U.K. rl@ast.cam.ac.uk
Macchetto	Duccio	Space Telescope Science Institute 3700 San Martin Drive Baltimore, MD 21218 USA macchetto@stsci.edu
Massaglia	Silvano	Osservatorio Astronomico di Torino Strada Osservatorio 20 I-10025 Pino Torinese Italy massaglia@ph.unito.it
Matsumoto	Hironori	Cosmic Ray Group Dept. of Physics Kyoto Kitashirakawa-Oiwake-Cho Sakyo-ku Kyoto 606-01 Japan matumoto@cr.scphys.kyoto-u.ac.jp
Meisenheimer	Klaus	Max-Planck-Institut für Astronomie Königstuhl 17 D69117 Heidelberg Germany meise@mpia-hd.mpg.de
Neilsen	Eric	Johns Hopkins University Department of Physics & Astronomy Baltimore, MD USA neilsen@pha.jhu.edu
Norman	Michael	Max-Planck-Institut für Astrophysik Karl-Schwarzschildstraße D85740 Garching Germany norman@ncsa.uiuc.edu
Owen	Frazer	National Radio Astronomy Observatory P.O. Box 0 Socorro, NM 87801 USA fowen@nrao.edu
Reynolds	Chris	JILA University of Colorado Campus Box 440 Boulder, CO 80309-0440 USA chris@rocinante.colorado.edu

Name		Address
Röser	Hermann-Josef	Max-Planck-Institut für Astronomie Königstuhl 17 D69117 Heidelberg Germany roeser@mpia-hd.mpg.de
Rottmann Helge		Max-Planck-Institut für Radioastronomie Auf dem Hügel 69 D53121 Bonn Germany rottmann@astro.uni-bonn.de
Scheuer	Peter	Mullard Radio Astronomy Observatory Cavendish Laboratory Madingley Road Cambridge CB3 0HE United Kingdom pags@mrao.cam.ac.uk
Sparks	William	Space Telescope Science Institute 3700 San Martin Drive Baltimore, MD 21218 USA sparks@stsci.edu
Tsvetanov Zlatan		Johns Hopkins University Department of Physics & Astronomy Baltimore, MD 21218 USA zlatan@pha.jhu.edu

Subject Index

Springer
and the
environment

At Springer we firmly believe that an international science publisher has a special obligation to the environment, and our corporate policies consistently reflect this conviction.

We also expect our business partners – paper mills, printers, packaging manufacturers, etc. – to commit themselves to using materials and production processes that do not harm the environment. The paper in this book is made from low- or no-chlorine pulp and is acid free, in conformance with international standards for paper permanency.

 Springer

Lecture Notes in Physics

For information about Vols. 1–491
please contact your bookseller or Springer-Verlag

Monographs

For information about Vols. 1–14
please contact your bookseller or Springer-Verlag